Draw the Lightning Down

Draw the Lightning Down

*Benjamin Franklin and Electrical Technology
in the Age of Enlightenment*

MICHAEL BRIAN SCHIFFER

*With the assistance of
Kacy L. Hollenback and Carrie L. Bell*

University of California Press

BERKELEY   LOS ANGELES   LONDON

© University of California Press
Berkeley and Los Angeles, California

University of California Press, Ltd.
London, England

© 2003 by the Regents of the University of California

Library of Congress Cataloging-in-Publication Data

Schiffer, Michael B.
   Draw the lightning down : Benjamin Franklin and electrical
technology in the age of Enlightenment / Michael Brian Schiffer, with
the assistance of Kacy L. Hollenback and Carrie L. Bell.
       p.   cm.
   Includes bibliographical references and index.
   ISBN 0-520-23802-8 (alk. paper).
       1. Electrical engineering—History—18th century.   2. Electricity—
History—18th century.   3. Franklin, Benjamin, 1706–1790.
4. Enlightenment.   I. Hollenback, Kacy L.   II. Bell, Carrie L.
III. Title.
TK16.S35   2003
621.3'0973'09033—dc21   2002156533

Manufactured in the United States of America

11   10   09   08   07   06   05   04   03
10   9   8   7   6   5   4   3   2   1

For Annette, my loving wife and best friend

Contents

Figures

Preface

I began to envision a project on eighteenth-century electrical technology sometime in 1992. In search of the earliest motors for my 1994 book on electric automobiles, I came across a reference to Benjamin Franklin's "electrical jack," an electrostatic motor. I tracked down the original source and read it with rapt attention; I also did some desultory reading about eighteenth-century electricity. At the time, this was no more than a pleasant and brief diversion from electric automobiles. But one impression registered deeply: the eighteenth century was an electrical age, with fascinating technology that played many roles in diverse people's lives. Someday, I promised myself, I would tell the story of this technology. But after I finished the car book in 1993, another project—developing an artifact-based theory of human communication—consumed my attention for nearly five years.

At long last, in 1998 I plunged into eighteenth-century electricity. From the very beginning I planned to adopt an *archaeological* approach to this subject by exploring, and attempting to explain variations in, the entire gamut of electrical technology. I would not only look at early physics or medical electricity or lightning conductors but would also seek information on all major—and even some minor—electrical things that people created and used in the eighteenth century. On the basis of mostly secondary sources, I prepared one scholarly paper on competitions among electrostatic, electrochemical, and electromagnetic technologies. During the summer of 1999 I drafted the first outline of this book as well as several very rough chapters, but the book lacked a coherent framework that could easily accommodate the incredible variety of people, activities, and technologies that made up eighteenth-century electricity. One day, while waiting for my wife, Annette, in a shopping mall in Flagstaff, Arizona, it finally came to me.

I would recount the early history of electrical technology and then trace the changes it underwent as people adapted it to the performance requirements of different activities in various science and nonscience communities.

With the rudiments of this framework in hand, I began intensive research, employing mainly primary sources, during the 1999–2000 school year. In early spring I fleshed out the "technology transfer" framework in a paper using eighteenth-century electrical technology as a case study. This framework turned out to be more than adequate for handling the social, behavioral, and technological variation I encountered, but the book still lacked a "hook" for the nonspecialist. Thus I decided to highlight the activities of Benjamin Franklin, who was, after all, one of the most significant figures of eighteenth-century electrical science and technology: along with his friends and acquaintances, Franklin took part in the activities of every electrically based community. As a senior fellow at the Dibner Institute for the History of Science and Technology, Massachusetts Institute of Technology, I was able to finish this project in the fall of 2001.

As I did with my other books on electrical history, I targeted this work beyond specialized academic audiences. From the beginning I believed that the story of early electrical technology was too exciting to remain the well-kept secret of a tiny intellectual elite. Although my goal as a writer is to reach people who are curious about how our electrical world began, I also expect some historians and philosophers of technology and science to find the book of interest, as will the growing number of scholars in American studies, cultural studies, and science-technology-society studies who look seriously at technology—and the materiality of human life. Finally, colleagues in my home disciplines of anthropology and archaeology, including historical and industrial archaeologists, might find something useful between these two covers. A note to specialists: I emphasize that this work is an *archaeological* study of a particularly fascinating technology that just happens to be accessible through materials in the historical record.

This book has been a pleasure to research and write, for it taxed all of my scholarly skills and never left me bored. I hope the reader will likewise find it of consuming interest.

—|ı|—

The electrical literature of the eighteenth century is rather fertile ground for historians of science and technology. In recent decades many eminent scholars, including I. Bernard Cohen, John Heilbron, Willem Hackmann, and Roderick Home, have cultivated it. Their projects differed from mine, but these men made my work much easier by furnishing bibliographies as well as insightful analyses. Indeed, the reader of the notes will appreciate that in places I rely on discussions by Cohen, Heilbron, Hackmann, and others; the judicious use of secondary sources is normal practice in archaeology, for we are reluctant to repeat work done well.

The vast majority of research for this book was conducted, from 1999 to 2001, at three libraries: the Bakken Library and Museum of Electricity in Life, Minneapolis, Minnesota; the Dibner Library in the National Museum of American History, Smithsonian Institution, Washington, D.C.; and the Burndy Library of the Dibner Institute, MIT, Cambridge, Massachusetts. These libraries exist through the foresight and generosity of two men, Bern Dibner and Earl Bakken, whose interests in electrical history compelled them to collect early materials long before it became the fashion. I thank the people who staff these libraries, especially Elizabeth Ihrig (the Bakken), Judith Nelson, Larry Leier, and Anne Battis (Burndy), and Ronald Brashear (Dibner), for they are kind, helpful, and efficient. Elizabeth Ihrig also patiently copied countless drawings, many of which I have incorporated into the figures.

Supplemental research was carried out on primary sources at various libraries of Harvard University, Cambridge, Massachusetts; and the National Library of Medicine, Bethesda, Maryland. I also made use of the University of Arizona libraries.

In partial support of this project, I received a visiting scholar fellowship at the Bakken Library, October 2000; was awarded a Riecker Fellowship by the Department of Anthropology, University of Arizona, in the fall of 2000; and (as noted) was a senior fellow at the Dibner Institute during the fall of 2001.

For sundry assistance, I thank Lucia C. Stanton, who kindly supplied information on Thomas Jefferson's electrical artifacts; Londa Schiebinger and Charles Bazerman, who furnished useful references; and Annette Schiffer, who tracked down a number of factoids on the Internet.

I greatly appreciate the contributions of the following people, who commented perceptively on various drafts or partial drafts of this book: Carrie Bell, Elizabeth Cavicchi, Janet Griffitts, Kacy Hollenback, Alberto Martínez, William Pencak, Elizabeth Perry, Robert Post, John Riddle, Annette Schiffer, Frances-Fera Schiffer, Louie Schiffer, Mary Stephenson, and William Walker. I am especially indebted to Robert Post, whose intellectual generosity knows no bounds. I also thank Davis Baird, Elizabeth Cavicchi, Ann Johnson, Alberto Martínez, and Martin Reuss for stimulating discussions about the materiality of scientific activities. Michael Tite, a physicist and archaeologist, coached me on electrostatics. And my research assistants Kacy

Hollenback and Carrie Bell contributed immeasurably to this project's success through their labors in the libraries at the University of Arizona.

Stanley Holwitz, an assistant director of the University of California Press, had an unwavering faith in this project and helped bring it to a successful conclusion. I am also grateful for the fine work of the production staff at UC Press, especially Dore Brown, Laura Pasquale, and Jessica Grunwald. In copyediting the manuscript, Edith Gladstone enhanced the gracefulness and clarity of countless sentences.

Once again my wife, Annette, has furnished encouragement, moral support, and feedback throughout the course of this project. Above all, she has made my life a joy.

1. The Franklin Phenomenon

The name of Benjamin Franklin resonates with most Americans. Not only is Franklin prominently mentioned in high-school and college history texts, but many cities and towns also have a Franklin Avenue, Franklin Insurance Company, or Ben Franklin Crafts store. His face is familiar—on postage stamps, busts in museums, and, of course, $100 bills. And Franklin's seemingly timeless sayings still grace calendars, magazine articles, and advice columns. In the pantheon of great Americans, Franklin looms large as a patriot and tireless diplomat, the founding father who, had he not died during Washington's first term, might have succeeded him as president.

Behind today's memorials was a man who led a long and complicated life. Like a brilliant-cut diamond, Franklin had many facets: as printer and publisher, founder of countless public institutions from library to fire department, and even scientist. But did he really snatch lightning from the clouds? Was the famous kite experiment (Fig. 1) merely a myth that padded the already impressive résumé of America's "renaissance man"? In the history of electrical science and technology, did Franklin actually earn any more than a tiny footnote?

Franklin (1706–1790) lived during much of the first age of electricity, which lasted from around 1740 to 1800. Not only did important electrical principles receive their first formulation then, but many people outside science acquired some familiarity with electrical technology, from lightning conductors to medical devices. Franklin, working on electricity in the late 1740s and early 1750s, in fact contributed crucial theoretical insights, ingenious experiments, and new technologies; and, yes, he did fly the kite. By the mid-1750s, Franklin had become a famous scientist whose ideas propelled new experiments and set the agenda for theoretical discussions throughout the West.

Figure 1.   Benjamin Franklin
snatches lightning from the
clouds. (From the author's
collection.)

But Franklin did not labor alone. Not only did he do his experiments in
collaboration with close friends in Philadelphia, but he also participated
in a far-flung network of European investigators linked to one another by
visits, letters, publications, and the exchange of technology. That Franklin spent many mature years in England and in France enabled him to participate intimately in European high culture and natural philosophy. He
was a towering figure whose encouragement and advice were highly val-
ued. Franklin and his friends, in the Old and New Worlds, helped usher in
the first electrical age.

My foray into the first age of electricity was guided by two striking
facts. First, people in the eighteenth century invented thousands of electrical things, from lights to motors to musical instruments. Second, the
groups creating and using electrical technologies came from diverse
nations, social classes, occupations, and religions. In view of these facts, I
aimed to learn more about the varieties of eighteenth-century electrical
technology—what these artifacts looked like, what materials they were
made of, and how they worked—and to discern their roles, utilitarian and
symbolic, in the activities of different groups. This book reports my findings, which I have woven around the significant contributions of Benjamin
Franklin and his friends. To underscore the diversity of the people who cre-

ated the first electrical technologies, I furnish throughout a number of brief biographies.

⊣॥⊢

Before proceeding, I want to consider a few key ideas and terms—beginning with technology. What is technology anyway? The high priests of our supposed "information age" would have us believe that not until the advent of computers did humanity boast anything worthy of the name technology. And that view is echoed in the mantra of educational administrators, "let's put technology in the classroom." In 1999, at a large state university in the American Southwest, the provost preached these very words to the faculty senate. While most of the congregated professoriate nodded in agreement (or nodded off), an irreverent archaeologist stood up and retorted, "But I already use technology in the classroom: blackboard, erasers, and chalk." Needless to say, the provost was not amused.

Like others in my discipline—whether or not they voice irreverent thoughts, as I did—I study how technologies, as artifacts, related to the lives of the people who made and used them, seeking meaningful *behavioral* patterns in the human-made world. Archaeologists, especially, frown on narrow conceptions of technology. Having in our purview the entire span of humanity's existence, we see technology in all of its material and behavioral diversity, whether it be the million-year-old chipped-stone tools of the Lower Paleolithic or today's multibillion-dollar Hubble Space Telescope. Technology includes not just the "high tech" of a particular era or place, but all technologies—high, medium, and low. Thus, blackboard, erasers, and chalk are technologies as surely as anything disgorged by Silicon Valley. For archaeologists and many other scholars, technology—as hardware—is any kind of artifact.[1] In reporting on the first electrical technologies, then, I do not confine my discussions to high-tech devices, for some of the most consequential electrical things were rather simple.

An archaeological perspective also emphasizes that technologies pervade virtually every human activity, from rebuilding a diesel engine to praying in church to deconstructing Derrida.[2] The chapters below demonstrate that electrical technologies in the eighteenth century participated in many activities—scientific, political, educational, medical, and recreational. What is more, *any* activity can impel the development of a new technology or spur the invention of new varieties of an old technology. Although electrical technology originated in scientific activities, it became highly differentiated in many other activity contexts.

Because activities involve people as well as technologies, I focus on the groups—let us call them "communities"—whose activities employed electrical artifacts.[3] Depending on the nature of its activities, each community acquired a particular constellation of electrical technologies. For example, *electrophysicists* used the kinds of equipment that made possible cutting-edge experiments in physics. Likewise, *electrotherapists* acquired appliances for administering electricity to patients. For ease of analysis and communication, I have taken the liberty, historically speaking, of naming the electrical communities of the eighteenth century (see below).[4]

By defining communities in terms of activities carried out and artifacts employed, archaeologists create groups that often have overlapping memberships. Indeed, different electrical communities often shared many of the same members. Franklin, for example, was an important participant in the electrophysics and property-protector communities. The activity-artifact orientation also creates socially diverse groups. Electrical communities contained Protestants and Catholics, reactionaries and revolutionaries, physicians and clerics, and people of royal and peasant heritage. Moreover, some community members were citizens of nation-states unified under one crown, such as England and France, whereas others lived in the many kingdoms within Germany and Italy; some of these polities participated in early industrialization, yet others were almost entirely agrarian. By focusing on these cross-cutting—often overlapping—communities, and their members' activities, we can come to appreciate the variety and functions of eighteenth-century electrical artifacts.

As indicated already, technologies seldom remain confined to their community of origin. Electrical technology was at first quite localized, the specialized artifacts of a handful of natural philosophers, the electrophysicists.[5] From these modest beginnings, electrical technology was rapidly adopted by other communities, including those of disseminators (science lecturers) and electrotherapists. The members of each recipient community discovered the strengths and weaknesses of electrical technology in their own activities. Not surprisingly, when shortcomings in performance characteristics—that is, behavioral capabilities—such as ease of use, turned up, community members modified the technology, inventing new functional varieties.[6] Electrotherapists, for example, discovered that the devices used in physics experiments were too clumsy for use with human patients. Thus, they designed new artifacts for applying electricity to ailing limbs and organs. Likewise, although the major scientific effects of electricity could be illustrated with a few very simple and unremarkable devices, disseminators developed hundreds of complex and fascinating objects that could enthrall an audience. The members of electrical communities proliferated an amazing assortment of things tailored to perform varied—often specialized—functions in their own activities.

Although enormously varied and important, the electrical technology of the eighteenth century—known today as "static" or "electrostatic"—gets short shrift in many histories, regarded as a dead end that did not, and could not, produce today's electrical wonders.[7] However, a close reading of eighteenth-century works and an examination of surviving artifacts disclose a technology of far richer roles in eighteenth-century societies than it is usually assigned. As one observer put it in 1766, "does it not appear . . . that we are living in an electrical century?"[8] Not only did the electrical technology of the Enlightenment touch a surprising number of human lives and bodies, but it also laid a foundation for technologies important today. For example, experiments with telegraphy raised the possibility of a communications revolution. Moreover, modern electrical devices for producing heat, light, and rotary motion all have eighteenth-century antecedents. In addition, electrical experimenters invented the spark plug and furnished the fundamental principles of xerography. Also, the first people engaged in electrical engineering were at work in the late Enlightenment, designing lightning conductors—one of Franklin's inventions. Nonetheless, even at the end of the eighteenth century, few homes of ordinary people contained any electrical technology.

This book describes how, in the hands of many inventors in many communities, electrical technology became differentiated. I aim to bring to light and make intelligible the entire compass of electrical artifacts in the Enlightenment.[9] In presenting the story of eighteenth-century electrical technology, structured by the community-activity-artifact orientation, the chapters below touch on many intriguing topics. Among the subjects we encounter are the roles of the human body (dead and alive) in scientific experiments and the significant involvement of religion in electrical technology, as well as the use of models for formulating and evaluating generalizations in science and engineering and the constraints on developing and commercializing large-scale electrical systems. Also, in fleshing out the activities of community members, I digress often because there are so many interesting stories to tell about the people who helped bring about the first electrical age.

The main period covered by this book, the late eighteenth century, coincides—not coincidentally—with the *High Enlightenment*. A time of widespread belief in the benefits of education and in the possibilities of human progress through rational thought and experiment, the flowering of the late

Enlightenment was both cause and consequence of the first electrical age.[10] And no one better embodied and gave voice to Enlightenment ideals than Benjamin Franklin.

—|ı|ⱶ—

Benjamin Franklin and his friends, the latter both in Europe and the American colonies, were central figures in most electrical communities. But there was little in Franklin's first decades to suggest that he would one day achieve great success in science.[11]

Like many citizens of England chafing under the state religion, the Church of England, Franklin's father, Josiah, and his wife, Anne, immigrated to the American Colonies in hopes of practicing their dissenting faith without fear or fetters. They arrived in New England with their three children in 1683 and in due course begat four more offspring. Anne, however, died in her thirty-fourth year giving birth to a seventh child. Not many months later, Josiah took a second wife, Abiah Folger, of Nantucket; together they would raise ten more children. The couple's last son, Benjamin, was born in Boston on January 17, 1706; he was the youngest son of the youngest son, going back at least five generations. To support his ample family, Josiah made and sold soap and candles; frugal and successful at his trade, Josiah even managed to buy a house and accumulate modest savings. Benjamin remembered his father—almost fifty years his senior—as someone with great mechanical gifts, musical ability, and sound judgment.

Benjamin from the first showed an aptitude, not only for mechanical tinkering, but also for reading and learning, and so his father enrolled him in grammar school, anticipating that he would come to serve the church. However, after two years of schooling, Benjamin was summoned home to help in his father's business. And so ended his formal education; all else Benjamin learned on his own, spending the coins that occasionally came his way on books. But the young Benjamin, working with his father, did not take to the tallow trade.

Casting about for a more suitable trade and knowing that Benjamin, still only twelve, loved books, Josiah saw an opportunity when an elder son, James, set up a print shop in Boston. Benjamin, who had "a Hankering for the Sea," was indentured to his brother until age twenty-one.[12] With the memory fresh of one son already lost at sea, Josiah feared that young Benjamin, unless bound to James by contract, might also heed Neptune's call.

Toiling in his brother's shop, Benjamin soon became a proficient printer. In his spare time, sometimes late into the night, he would read borrowed books and write verse. After his father once pointed out shortcomings in his writing style, Benjamin set about polishing his prose. He worked hard and, in time, his writing achieved an enviable elegance and clarity. As a teen Benjamin also became a vegetarian, but he did not always shun meat in later years, appreciating that it was an inconvenience for his hosts. With the money he saved on food—he was now responsible for his own board—the printer's apprentice bought more books. When his brother James founded a newspaper, *The New England Courant*, Benjamin wrote anonymously a number of pieces, at night sliding them under the door of the print shop; these articles found favor among James's learned friends.

But not all was bliss between the brothers. James treated Benjamin like an employee, not a family member, and sometimes beat him. Even decades later, Franklin recalled his brother as a "harsh & tyrannical" taskmaster; this experience, he emphasized, "impress[ed] me with that Aversion to arbitrary Power that has stuck to me thro' my whole Life." [13] After printing a piece critical of the Massachusetts Assembly, James was jailed for a month and ordered to cease publishing his paper. Seeking a way to circumvent this decree, he decided to put out the paper under his brother's name. To make this stratagem appear legitimate, James officially freed Benjamin from indenture but then indentured him anew in a secret agreement.

Appreciating that James could never make public the revised indenture, Benjamin took advantage of his brother's predicament and announced that he was leaving the print shop. A furious James visited other Boston printers and urged them not to hire the young runaway. So at age seventeen Benjamin departed Boston; he sold some of his books to raise a little money and arranged for passage by boat to New York. He slipped away in secret, becoming estranged not only from James but also from his father.

There was no job in New York either, so Franklin traveled to Philadelphia after hearing about a possible opening for a printer's assistant. He found employment in an ill-run print shop owned by Samuel Keimer and also secured reasonable lodging. Keimer was no intellectual, and the witty and sarcastic Franklin enjoyed baiting him. Frugal like his father, Franklin managed to save some money and even had time for a social life. He met and courted Deborah Read, whom he later took as his wife, but without benefit of formal marriage.

Around 1727, Franklin and Hugh Meredith, another of Keimer's employees, set up their own print shop with the help of Meredith's father. Two years later, drawing on loans offered by kind acquaintances, Franklin bought out his partner; at age twenty-three he was at last his own master. Not long afterward, perhaps atoning for past sins, including costly and inconvenient "Intrigues with low Women," Franklin embarked on a project to achieve moral perfection.[14] To furnish a framework for guiding his own behavior along a righteous path, Franklin enumerated a list of virtues and accompanying precepts, which I present in its entirety:

1. Temperance.
   Eat not to Dullness.
   Drink not to Elevation.

2. Silence.
   Speak not but what may benefit others or yourself.
   Avoid trifling Conversation.

3. Order.
   Let all your Things have their Places.

4. Resolution.
   Resolve to perform what you ought.
   Perform without fail what you resolve.

5. Frugality.
   Make no Expense but to do good to others or yourself: i.e. Waste nothing.

6. Industry.
   Lose no Time. Be always employ'd in something useful. Cut off all unnecessary Actions.

7. Sincerity.
   Use no hurtful Deceit.
   Think innocently and justly; and, if you speak, speak accordingly.

8. Justice.
   Wrong none, by doing Injuries or omitting the Benefits that are your Duty.

9. Moderation.
   Avoid Extremes. Forbear resenting Injuries so much as you think they deserve.

10. Cleanliness.
    Tolerate no Uncleanness in Body, Clothes or Habitation.

11. Tranquility.
    Be not disturbed at Trifles, or at Accidents common or unavoidable.

12. Chastity.
    Rarely use Venery but for Health or Offspring; Never to Dullness, Weakness, or the Injury of your own or another's Peace or Reputation.

13. Humility.
    Imitate Jesus and Socrates.[15]

Propounded in his many writings and repeated by others, these virtues reached a large audience. In its essentials, the Franklinian code of conduct remains familiar to us today, the backbone of American common sense.

In his new printing business, Franklin worked hard, earning a stellar reputation well above Keimer and Bradford, Philadelphia's other printers. Clearly outclassed, Keimer withdrew from the business and sold his newspaper to Franklin, who promptly renamed it in 1729 the *Pennsylvania Gazette*. To supplement earnings from newspaper sales, Franklin solicited advertisements. According to one biographer, Carl Van Doren, Franklin was then America's best writer, and he put his craft to use often.[16] He not only wrote much of the newspaper but also penned letters to the editor, which in turn he answered. But, having learned a lesson from his brother's experience, Franklin did not then challenge civil authority.

Even as the printing business grew and thrived, Franklin made time for other activities. In one well-known move, he founded in 1731 the Library Company, the first subscription library in America, the model for all that followed. The Library Company still exists in Philadelphia, and among its holdings are many books that belonged to Franklin. In 1732 he inaugurated *Poor Richard: An Almanack*, which furnished another outlet, in addition to books and newspapers, for his prolific prose. He wrote a pamphlet on the causes of house fires and means to avoid them; this led to the formation of a volunteer fire department whose members were required "to keep always in good Order and fit for Use, a certain Number of Leather Buckets . . . which were to be brought to every Fire."[17]

Franklin, continuing his program of self-improvement, studied languages; he quickly mastered French and also achieved some proficiency in Italian, Spanish, and Latin. By this time he had also taught himself arithmetic. Later in the decade, Franklin was appointed postmaster of Philadelphia. Around 1740 he invented and began to market a metal stove that, in comparison to a fireplace, more efficiently extracted heat from burning wood. The Franklin stove is still manufactured today. By his late thirties, Franklin had become a prosperous businessman, a founder of important civic institutions, and an influential citizen of the Pennsylvania colony. In a few more years, he would also become a scientist and diplomat.

Surprisingly, the Philadelphia printer of humble background had the basic skills needed in the eighteenth century to work at the frontiers of physics: a mechanical knack (for nearly all physicists then were experimenters), the financial means to acquire necessary equipment and supplies, the ability to read and write in at least one language proficiently, and a boundless enthusiasm for learning and discovery.[18] Most scientists then were merely literate people with an overwhelming curiosity about the natural world who had the wherewithal to indulge their interest. The scientific literature was as yet small, and so a person could master a subject in not many months or at most a few years. As we see in later chapters, Franklin, adhering to his various virtues, did not allow a latent scientific passion to take command of his life until he reached middle age. By then—the late 1740s—other workers had already begun to cultivate the fertile field of electrophysics.

In the next chapter, I turn to the men who laid the earliest foundations of electrical science and technology. In addition, chapter 2 affords me the opportunity to present some basic principles of electrostatics that are indispensable for understanding the remainder of the book. Chapter 3 discusses the florescence of the *electrophysics* community in the 1740s, showing the state of electrical science and technology as Franklin found it; also mentioned are a few important technologies that members of this community invented in subsequent decades.

The bulk of the book, in chapters 4–11, presents overviews of each electrical community through discussions of major activities, prominent community members, and new technologies: in writings and public lectures, *disseminators* brought electrical technology to the attention of people outside electrophysics (4); *collectors and hobbyists* acquired electrical technology for display and use at home (5); *electrobiologists* adopted electrical technology for research on plants and animals (6); in clinical medicine, *electrotherapists* treated patients with electrical devices (7); *earth scientists* used electrical technology for investigating terrestrial and atmospheric processes (8); *property protectors* designed, installed, and used Franklin's invention, the lightning conductor, to safeguard structures (9); *new alchemists* incorporated electrical technology into chemical experiments (10); and *visionary inventors* came up with product ideas and prototypes that conceivably could have been used in activities beyond science, such as electric lighting, but significant societal barriers prevented the development and commercialization of an electrical infrastructure to power them (11).

In chapter 12, I generalize and discuss the technology-transfer framework that helped me investigate the differentiation of electrical technology in the Enlightenment. This chapter may be of interest mainly to other scholars who study technological change.

Let us now turn to the very beginnings of our modern electrical world.

2. In the Beginning

The first technologies we recognize as "electrical" arose in the activities of natural philosophers who, today, might be called physicists. What distinguishes their works from scholastic philosophy, with its unresolvable debates, and from other systems of knowledge, like religion, is a reliance on experiment. Although not without faith in authorities, scientists use experiments to create and evaluate purported knowledge about empirical phenomena. Many people—including this author—would claim that experimentation is a necessary part of scientific research.[1]

Of course it all depends on how experiment is defined. My definition is simple: an experiment is an activity in which the investigator manipulates a technology, in a laboratory or other setting, to make evident some empirical phenomenon or "effect."[2] Defined this broadly, an experiment (not unlike the French term *expérience*) can include activities such as observing through a microscope or telescope as well as electrifying a substance or flying a kite. In preference to the term "instrument" and in conformity with much eighteenth-century usage, I label the technology of scientific activities "apparatus." Regrettably, instrument implies to some people a technology that is elaborate, specialized, or expensive.[3] But electrical science prior to 1800 was mainly about producing effects using any *things* that were suitable. As we shall soon see, electrical apparatus included a surprising array of mundane objects and materials, everything from glass tubes to beeswax to sirloin steak.

In addition to the scientist's active participation in an experiment, assistants—usually anonymous—also had close interactions with the technology. In many eighteenth-century experiments, the human body sometimes served as a reservoir or conductor of charge and also as a sensor for the sounds, sights, and feelings that electricity elicited. Whether we include the human participant as part of an apparatus (as I sometimes do), an experiment consists of a sequence of choreographed interactions among people—the investigator and assistants—and artifacts.[4]

The effects emerging in an experiment are, in several nontrivial senses, a human creation. Because people invent and manipulate apparatus, an experiment is simply a complex and transitory cultural production. Moreover, an effect created experimentally might exist nowhere else on earth, perhaps nowhere else in the universe.[5] We could even question whether it is primarily "natural" phenomena that physical scientists study in their laboratories.[6] In any event, one of the most daunting tasks facing the scientist is to relate the often unnatural effects that arise in experiments to natural phenomena in the external world. The generalizations that scientists fashion to make these connections—experimental laws and theories— impose meaning on the effects created by an experiment's people-artifact interactions.

In the present chapter, I introduce some basic principles of electrostatics along with the technologies that contributed to their creation. And to illustrate the points made above, I situate both principles and technologies in the experimental activities of the earliest electrical investigators.

⊣ı⊢

Although the precise beginnings of electrical science are contestable, no one doubts that William Gilbert (1540–1603), an Englishman, carried out the first sustained and influential research on electrical phenomena.[7] Born into a family of comfortable means, the eldest of eleven children, Gilbert graduated from Cambridge University as a medical doctor in 1569 and was elected in 1573 to fellowship in the Royal College of Physicians. A healer of some repute, Gilbert received a permanent appointment to minister to Queen Elizabeth.

Her Royal Highness apparently looked kindly upon Gilbert's philosophical investigations, which included chemistry, magnetism, and electricity, for she awarded him an annual pension. After eighteen years of work and a vast number of original experiments, Gilbert in 1600 published his findings on magnetism and electricity. Written in Latin with a typically ponderous title, the book is known today simply as *De magnete*; surprisingly, an English translation did not appear until 1893.

De magnete is a remarkable product of scientific activity. Although more than 90 percent of the book treats magnetism, as its title proclaims, in the twoscore pages on electricity Gilbert reported new experimental effects, presented cogent critiques of earlier electrical theories, and offered an ingenious theory of his own.[8] He also set the research agenda and a style of theorizing that some investigators would follow in the early eighteenth century. Finally, Gilbert invented electrical technologies that influenced later generations of researchers.

When Gilbert began his experiments, there was almost nothing deserving the name of electrical science; only the effects seen in everyday life had been reported. Fortunately, one of these effects is easy to devise from simple apparatus at home and will allow readers to appreciate, firsthand, what was known when Gilbert began his research.

First, assemble the appropriate artifacts. Rummage through your home, looking for some amber, perhaps in an old pendant or earrings. If there is no amber lying around, then you may cheat, historically speaking, by using a plastic comb or the plastic lid of a yogurt container. Then cut a piece of paper into twenty or thirty tiny squares, no more than one-sixteenth of an inch on a side. Now, rub a knuckle on the amber or plastic gently but swiftly a dozen times or so. Finally, bring it close to, but not touching, the paper specks. If all goes well, you should witness a most remarkable effect: electrical attraction. As if by magic, pieces of paper are drawn to the rubbed—that is, "electrified"—amber or plastic. In today's electrical jargon, we say that its surface has been "charged," meaning that it has been given—in this case through friction—an excess of positive or negative electricity relative to its stable, neutral state.[9]

Prior to Gilbert, natural philosophers knew that amber and a few other materials, when rubbed, would attract straw or chaff. This effect joined a catalog of miscellany that all seemed to involve "attractions," including the falling of bodies to earth and the pull of magnets on iron and on each other. Such diverse attractions had themselves, in earlier centuries, attracted the curiosity of natural philosophers, who had not sorted out these phenomena.

One of the most important operations of science is the crafting of categories that distinguish a phenomenon of interest from others that it resembles.[10] In formulating such categories, the scientist exercises creative judgment informed by cultural traditions, intuition or an aesthetic sense, previous research, and—not insignificantly—the experimental apparatus. Gilbert showed in experiments that inserting an object between a magnet and a piece of iron did not appreciably affect the magnet's pull (unless the object was another piece of iron); in contrast, interposed objects did disrupt electrical attractions. On this basis he distinguished magnetic attractions from electrical ones; he also differentiated electrical attractions from gravity, vacua, and chemical processes. By these simple moves, Gilbert isolated a set of reasonably coherent phenomena that could be managed experimentally and that apparently corresponded to phenomena outside his laboratory.

Above all, Gilbert was an astute generalizer who constructed experimental laws on the basis of effects he produced with his apparatus.[11] Acknowledging that previous workers had observed that amber and jet, well rubbed, both attract small pieces of straw and chaff, Gilbert generalized that such attraction extends to all "light corpuscles"—that is, lightweight particles.[12] Thus, nothing in the nature of straw or chaff makes them uniquely susceptible to electrical attraction. Rather, a much larger class of substances, *all lightweight things* "save flame and objects aflame, and thinnest air," are subject to its power.[13]

But the generalization process did not stop there, for Gilbert had tested countless other materials, seeking those he could electrify by rubbing.[14] These latter substances, which included "diamond, sapphire, carbuncle [ruby], iris stone, opal, amethyst . . . beryl, rock crystal" and also "clear, brilliant glass," he termed "electrics."[15] In contrast, no amount of rubbing could cause nonelectrics—for example, the hardest woods, bone, and metals—to acquire the attractive power.

Gilbert's generalizations, constrained by the effects he could produce with his apparatus, turned out to be incomplete. Later investigators using different apparatus observed different effects and generalized accordingly.[16] Those who employed insulators to electrically isolate the rubbed substance were able to conclude that all solid substances, including metals, can be charged by friction. Thus, the distinction between electrics and nonelectrics is no longer used today. Nonetheless, it helped Gilbert generalize the observations based on his experimental technologies.

Gilbert also fashioned a simple compasslike device, which he called a *versorium,* for indicating whether a rubbed object was an electric. The versorium was simply a tiny metal arrow, free to pivot on the sharp point of a spindle. If a substance had acquired a charge after rubbing, it would cause the needle to "revolve."[17] Although seldom adopted by later investigators, the versorium is regarded as the first electrical measuring instrument.[18]

On the basis of his new experimental laws dealing with electrics and nonelectrics, Gilbert pondered theoretical questions. What gives electrics alone the ability to attract nearby lightweight substances? And, through what mechanism do electrics exert a force on other objects? Earlier thinkers had suggested that an electrified object rarefies the air around it, causing a small wind that sweeps in nearby substances. Gilbert, making good use of additional experimental observations, decisively refuted these theories.

Gilbert sought a mechanical explanation for attraction, for he believed, in advance of Descartes, that "no action can be performed by matter save by contact."[19] But what kind of fine matter could intervene between an electric and something attracted to it, and where did that matter come from? He suggested that all electrics share an origin in watery substances and are in fact "concreted humor[s]."[20] When electrics are rubbed, they release a subtle and rarefied form of matter, an effluvium. Extending to a nearby particle, the electric effluvium creates a unity with it. If the particle is sufficiently lightweight, then the result is attraction, which he likened to the well-known affinity between two wet objects brought together. Gilbert's ingenious theory explains electrical attraction and why it is confined to electrics: they alone are concreted humors. But it appears to rule out electrical repulsion, an effect reported later in the seventeenth century, which we encounter shortly. Significantly, the effects Gilbert produced with his apparatus tightly constrained his theory. Had he succeeded in electrifying metals, his theory could not have assumed the shape it did, unless he were willing to regard metals as congealed humors.

Gilbert's interest in attraction was shared by many natural philosophers who followed him, including Newton and Descartes. Isaac Newton (1642–1727), in his magisterial synthesis on mechanics (*Principia mathematica*), concluded that offering hypotheses to explain the attractive forces—his focus was gravity—was futile. He argued that a natural philosopher could at best describe, mathematically, how these forces operated. In his later book on optics, however, Newton ended up positing an "aether" that permeates all substances, even empty space, and serves as a medium for the transmission of forces to distant objects.[21] René Descartes (1596–1650) and his followers went much further, believing, like Gilbert, that mechanical explanations were always necessary. Thus, Cartesians posited elaborate invisible entities—whirling vortices—to convey the equally invisible forces.

Natural philosophers of the early eighteenth century who studied electricity, even those who pledged allegiance to the Newtonian program, often proposed Cartesian mechanisms reminiscent of Gilbert's fine effluvium. Later investigators suggested that an electrified object was surrounded by an "atmosphere," which exerted force on nearby objects. Not until the late nineteenth century, after increasing acceptance of Michael Faraday's "field" concept, would scientists fully adopt the Newtonian program and cease worrying about the ultimate causes of electrical (and magnetic) attraction.

Ironically, modern physics with its zoo of hypothetical force-carrying particles has a distinct Cartesian flavor.

—|‖—

Another early contributor to electrical science and technology was Otto von Guericke (1602–1686), the *Burghermeister* (mayor) of Magdeburg, in Prussia. Using technologies of his own design, Guericke put on scientific demonstrations that both surprised and mystified his audience. One of the most colorful and creative figures of seventeenth-century science, Guericke is of special interest here because his limited electrical experiments, done mostly with unexceptional objects, brought to light previously unobserved or neglected effects. Guericke is also credited, erroneously I believe, with building the first machine to generate electricity.[22]

Trained as an engineer, Guericke is best known for inventing an air (vacuum) pump. His interest in doing so was to create on earth a microcosm of the heavenly void. And he did. Nowhere else on earth at the time did vacua exist save in barometers and in the vessels evacuated by air pumps.[23] Hence Guericke's experiments disclosed many new effects, such as the demonstration that sound cannot traverse a vacuum whereas light can.

Guericke also used vacua in outlandish displays before groups of dignitaries. Indeed, a request from the Holy Roman emperor called forth his most famous and dramatic demonstration, put on at Regensburg in 1654 before members of Ferdinand III's Imperial Diet. Wishing to impress the assembled officials with the weight of air, he joined, by means of a specially treated leather gasket, the edges of two copper hemispheres, making a sphere nearly 2 feet in diameter.[24] After evacuating the sphere with his pump, Guericke harnessed to each hemisphere a team of eight horses. Such is the force of atmospheric pressure that the horses, struggling and straining, were only sometimes able to pull the hemispheres apart. Yet Guericke could easily part the hemispheres by opening a stopcock.

In putting on public displays—in contrast to the electrical lecturers who would emerge in the eighteenth century—Guericke made little effort to explain the effects. As Charles Bazerman recently noted, "the nonphilosophic citizens of the town . . . were left to wonder at the amazing powers of their burghermeister."[25] He carried out some demonstrations at his home, using apparatus hidden below the floor, operated by unseen assistants or by nature. In one instance, Guericke constructed a water barometer in a pipe 30 feet tall. A tiny wooden manikin rose and fell with the water level, its finger pointing to the current barometric pressure. Not surprisingly, some Magdeburgians believed that their mayor was a sorcerer. Guericke's wizardry, in the tradition of Renaissance natural magic, was an asset in the political arena.[26] Surely someone who could command forces unseen was fit to lead a community. In later centuries, even today, it is the magical quality of electrical displays that makes them so appealing to audiences; even after hearing "scientific" explanations for the observed effects, some people go away believing they had witnessed the exercise of occult powers.

Guericke's electrical investigations, like those of Gilbert, were only a tiny and seemingly insignificant part of his experimental activities. Although he was aware of Gilbert and other earlier workers in electricity, Guericke did not frame his experiments or results in electrical terms. Rather, his interest was in illustrating, through electrical demonstrations, a grand theoretical system of the world.[27] An essentially animistic interpretation of everything, Guericke's theory cobbled together elements from Aristotle, Protestantism, and natural magic. Effects, in his system, resulted from the exercise of an object's (or a person's) innate powers or "virtues." The latter term would also be used by later investigators, as in "electrical virtue," to denote the ability of an object that possessed it to exert force on another.

In an effort to demonstrate various virtues, Guericke constructed a model of earth. He took a spherical glass vessel, about the size of a child's head, and filled it with powdered sulfur. Then he heated the globe, melting the sulfur, which cooled to a hard spherical mass. After he broke the glass, liberating the sulfur globe, he inserted a rod through its diameter. In this configuration, the globe could be set in a wooden frame for conducting experiments (Fig. 2). Directly under the globe, and close to it, he placed "all kinds of shreds or bits of leaves, gold, silver, paper, hop plants and other tiny particles."[28] When Guericke rubbed the globe with a dry hand, the sulfur acquired a charge and attracted the small particles. When he rotated the globe 180 degrees, the charged particles remained in contact with the globe, demonstrating, according to Guericke, the "conserving" virtue he equated with gravity.[29] He also removed the globe from its frame, parading it about using the rod as a handle.

This curious device has been regarded—almost since Guericke's time—as the world's first electrical machine (or electrostatic generator, as it would be termed today). Indeed, later authors have looked at Guericke's illustration and assumed that the wooden cradle was used to hold the sphere while he rotated it for charging.[30] But rotating the globe was not part of the charging process; Guericke turned the globe only when he wanted to display the conserving virtue. Had he wished routinely to spin the globe, he would have placed a crank on one end, not a handle; he was far too clever an engineer and experimenter to have come up with a design so patently

Figure 2. Otto von Guericke's sulfur globe and its cradle. (Adapted from Dibner 1957, 15. Courtesy of the Burndy Library, Dibner Institute, MIT.)

clunky. On the basis of the design of Guericke's apparatus as well as his description of its actual operation, I conclude that it was not in fact an electrical machine.[31] Yet it is entirely possible that later investigators, perhaps inspired by Guericke's illustration, created true electrical machines that generated charge on a continuously rotating object.

While rubbing his globe, Guericke also heard sounds and, in the dark, observed flickers of light. These effects did not detain him—they merely revealed the "sound-" and "light-producing" virtues—for other effects held more interest. In experimenting with various objects, he found that a feather would float above the charged sulfur globe. Somehow, an electrified object could repel as well as attract other substances. Earlier observers had seen but not appreciated electrical repulsion as such. And if you now repeat the amber- or plastic-rubbing experiment, you may notice something new: occasionally a speck of paper is first attracted to, then repelled by, the charged object. Previously fixated on attraction, electrical investigators found repulsion unfathomable, but in the eighteenth century they fashioned theories to explain it.

Further experiments brought to light another effect. Guericke took a linen thread over 20 inches long. On one end he attached a pointed piece of wood that had been affixed to a tabletop; the other end of the thread dangled below. When he brought the charged globe near the point of the wood, the free end of the thread moved toward an object placed nearly an inch away. The unmistakable conclusion was that an electric charge can be communicated through a series of objects.[32] In this case, the charge was communicated by *conduction*. Metals, of course, are the best conductors, but moisture in wood and linen thread also enables them to conduct somewhat. (Later we will encounter *induction*, another mechanism for communicating charge.)

The publication of Otto von Guericke's book in 1672, in Latin, brought him into contact with scientists elsewhere, although his air pump generated the most interest. The English scientist Robert Boyle (1627–1691), sometimes called the father of chemistry, used an air pump to show that electrical attraction operates in a vacuum.[33] Today in the town of Magdeburg, Guericke is a celebrated ancestor, whose name adorns a university. In electrical circles he remains renowned for the endless research possibilities opened up by his supposed invention of the electrical machine.

-|i|⊢

William Gilbert, Otto von Guericke, and other founders of electrical science were amateurs who for the most part pursued their projects in isolation. By "amateurs" I mean only that they earned a living through occupations other than scientist. In the seventeenth century, there were few paid positions for people who wanted to spend their days, and sometimes nights, manipulating apparatus in search of new effects. Unlike scientists today, early natural philosophers did not have a peer group, in a recognizable scientific community, for exchanging and winnowing information. Critical feedback on work came mainly after reports were published. As the seventeenth century drew to a close, however, the dilettantish and individualistic basis of natural philosophy began gradually to give way to new social roles and institutions that would figure greatly in the transfer of electrical technologies among communities. No one illustrates these important changes in science better than the Englishman Francis Hauksbee.[34]

In the early years of the eighteenth century, Francis Hauksbee (1666–1713) was performing a series of electrical experiments. These investigations were remarkable, not only for their results, but also because Hauksbee was being paid to do them by the Royal Society of London. Founded in 1662, the Royal Society for Improving Natural Knowledge was a learned society sanctioned by the Crown. Its counterpart in France, established in 1666 by King Louis XIV and his minister, Jean Colbert, was the Royal Academy of Sciences.[35] During the late seventeenth and eighteenth centuries, dozens of official or quasi-official scientific societies came into existence, particularly in Europe, modeled after the elite institutions of London and Paris. Given their legal standing and royal imprimaturs, the societies in principle exercised some control over science in their territories. But their main power was in granting recognition to the "true" natural philosophers by admitting them to membership and publishing their reports.

That European states played a part in establishing or sanctioning scientific societies is understandable in view of the perpetual strife on that continent. Perhaps, some monarchs believed, these strange natural philosophers—part scholar, part magician—might conjure up something useful. Beginning with Francis Bacon, a contemporary of William Gilbert, the ideology of science had promised practical dividends from research activities, but the results in the seventeenth century had been meager, little more than improved lenses and better clocks. And yet these objects had military applications—lenses in naval telescopes and clocks for calculating longitude at sea—raising the specter of rivalry over inventions that could be useful in war. A monarch with his or her own scientific society could easily keep track of significant discoveries elsewhere and, if necessary, match them. Thus, perhaps in fear of falling behind militarily, many a sovereign fostered the establishment of a scientific society, adding it to the arsenal of defensive weapons.

After the English and French societies came into being, monarchs elsewhere had an additional motivation to establish national scientific societies. European kings and queens participated in a pan-European courtly culture, which stipulated the adoption of the latest fashions in dress and dance, music and literature, and the decorative arts. As the quest for new knowledge and learning became fashionable in the early Enlightenment, a nation without a scientific society was a nation that progress might leave behind.

Clearly, a strong scientific society advertised a monarch's commitment to the progressive ideology of the Enlightenment, a potent symbol of modernity. A case in point: as part of his larger enterprise to turn Russia into a modern European nation, Czar Peter the Great established in 1725 the St. Petersburg Academy of Sciences.[36] However, because Russia had few home-grown scientists at that time, the czar was obliged to import some from the West and pay them handsomely. Among the foreigners who joined the St. Petersburg academy in the eighteenth century were Leonhard Euler, one of the world's most distinguished mathematicians, and Franz Ulrich Theodosius Aepinus, who researched electricity and magnetism. With such luminaries, the St. Petersburg academy achieved credible standing in the emerging international community of science, reflecting favorably upon the Russian monarchy.

In addition to national societies, a great many regional and local scientific societies were also established, especially after 1750, but most were far less prestigious than the national societies of the great powers. In the American colonies, Benjamin Franklin fomented the formation of a club called the Junto in 1727, which brought together literate Philadelphians to ponder matters literary, civic, and scientific. It was in fact the Junto's members in 1731 who organized the Library Company of Philadelphia.[37] Acting on another proposal by Franklin, the Junto also brought into existence, in 1744, an early incarnation of the American Philosophic Society; the latter was officially founded in 1768.[38] The Philadelphia-based institution, however, never attained the prestige of the elite European societies.

Throughout the Western world, scientific societies took their place alongside other learned societies of the Enlightenment, bringing together people who shared certain intellectual or practical interests. There were, for example, societies that catered to students of literature and languages and others that indulged interests in painting, architecture, and ancient history. Among the practical societies were those dedicated to agriculture, economics, medicine, and the mechanical arts.

Even the scientific societies dabbled in practical affairs. The Paris academy, for one, was entrusted with reviewing patent applications, making maps, conducting economic surveys, and developing standardized weights and measures. Not uncommonly, societies offered prizes for inventions of obvious utility, such as improved chronometers. Perhaps because magnetic compasses were also important in navigation, in 1742 the Paris academy offered a prize for "the explanation of the attraction of the magnet and iron, the pointing of the magnetized needle toward the north, its declination and inclination."[39] Believing that the entries submitted were not sufficiently meritorious, the society withheld the prize, reopening the competition in 1744 with the same result. Finally, on the third try in 1746, the prize was split among the best of the previous years' entries. Needless to say, prizes both stimulated and channeled research while serving the state's agenda.[40]

Most scientific societies, centered in urban places, were part of "high culture"; indeed, membership might facilitate access to royal and noble persons. Membership was by election, and the applicant usually had to obtain sponsorship from an esteemed member of the club. For the elite among scientists, such as Isaac Newton and Christian Huygens, belonging to one society was not enough to support their exalted status. They belonged to many societies, especially the most prestigious.[41] Ambitious scientists jockeyed to accumulate foreign memberships in the quest for recognition, prestige, and social power. And they did not hesitate to flaunt their affiliations.

For example, the title page of Antoine Lavoisier's 1793 opus, *Elements of Chemistry,* boasted that he was a "Member of the Academies and Societies of Paris, London, Orléans, Bologna, Basel, Philadelphia, Haarlem, Manchester," etc. Lavoisier's illustrious status in science, well earned by his studies in chemistry, regrettably did not spare him the wrath of the revolutionaries (Lavoisier is discussed in chapter 10).

As a member of an academy or society, a person occupied a new social role, that of natural philosopher—the term "scientist" was not coined until 1840.[42] Although many people who pursued science in the eighteenth century remained amateurs, in this new role they achieved recognition for their experimental activities. Societies created social groups and gave their members a sense of belonging, an identity beyond that conferred by landholdings, family lineage, and occupation.

—|||—

Despite occasional contributions to their sponsoring polities, most scientific societies enjoyed less than lavish support. In fact, the finances of many were precarious; only the wealthiest could afford spacious buildings, laboratories, libraries, shops, and a permanent staff. Hauksbee was one of the fortunate ones, appointed to the post of chief experimentalist by Newton himself, then president of the Royal Society in London, which at that time was relatively prosperous. As part of his duties, Francis Hauksbee was required to present public demonstrations. This is a more challenging job than it might seem, for Hauksbee had to come up with a constant stream of ideas for new experiments each week and build the necessary apparatus. As John Heilbron, a historian of early electricity, noted, this unceasing demand for novelty could exhaust even the most creative experimenter.[43] Fortunately, Hauksbee had help: Newton, the most renowned living scientist, was more than a figurehead president of the Royal Society and offered suggestions for new experiments relating to his then-current interest in light and optics.

Little is known about Hauksbee's early years, but he claimed to lack a "Learned Education."[44] This was not, however, a bar to his success as an electrical experimenter, for he was a quick study and a skilled maker of apparatus. Although electrical experimenters in the pre-1740 period used relatively simple apparatus, Hauksbee—the professional—was the main exception. In the course of fashioning impressive displays, he documented several important effects and built the first real electrical machines.[45]

Hauksbee's earliest experiments, carried out in the first decade of the eighteenth century, dealt with a number of apparently unrelated effects involving the production of light. When a mercury barometer was shaken, as had long been known, specks of light appeared where the mercury, sloshing around in the vacuum at the top, came into contact with the glass wall. Urged by Newton to investigate this phenomenon further, Hauksbee used a vacuum pump to recreate the "mercurial phosphorus" and devised other technologies to produce light.[46] He found, for example, that a glow would appear when a woolen pad rubbed against amber in an evacuated glass globe.

To create this effect, Hauksbee built an ingenious and sophisticated device that combined both vacuum pump and electrical machine.[47] Inside the glass receiver of a vacuum pump he mounted a spindle whose upper end, emerging from the glass, was connected by means of a small pulley and rope to a large pulley. By turning a crank on the large pulley, Hauksbee could make the spindle rotate rapidly. On the spindle he could mount various substances for testing in a vacuum. Thus, he fastened a string of amber beads on the outer circumference of a wooden ring and placed the latter on the spindle. A piece of wool was wrapped on a brass plate that pressed, with "a moderate force," against the amber beads.[48] When the crank was turned, the woolen pad rubbing against the spinning amber beads gave off light.

Employing this apparatus, Hauksbee substituted many materials for both the amber and wool to learn if they, too, could produce light. In one experiment, he replaced the amber beads with a small glass vessel.[49] Cranking up the machine, he found that a "vivid" purple light emerged from the area of contact between the woolen pad and glass, and sometimes the light spread around the glass like a halo.

The discovery that rubbed glass could also produce light was important because it contributed to the design of the first dedicated electrical machine, the prototype for most made in the eighteenth century.[50] Hauksbee built a sturdy wooden framework to hold two large, vertically mounted pulleys with cranks.[51] The large pulleys were connected by belts to tiny pulleys on short shafts that rested on mounts fastened to the top of the wooden frame. Using this general configuration, Hauksbee could place a variety of glass vessels between the shafts. In one case, he inserted an evacuated glass globe, about 9 inches in diameter. When the pulleys were cranked, the globe rotated rapidly and could be rubbed with ease.[52] With his hand pressed against the spinning glass, Hauksbee encountered a most dramatic effect: the inside of the globe glowed purple. So much light was produced, in fact, that it illuminated a wall at least 10 feet away and allowed him to read large print.[53]

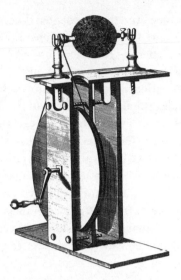

Figure 3. Francis Hauksbee's
one-crank electrical machine.
(Adapted from Hauksbee
and Whiston 1714, pl. 6. Courtesy of the Bakken Library,
Minneapolis.)

From this auspicious beginning, Hauksbee embarked on many more experiments, from which he concluded that frictional electricity produced not only attraction and repulsion but also "sparks" and light.[54] He also once substituted a glass cylinder for the globe that, when evacuated and rubbed by hand, produced additional effects.[55] For some experiments, he used a machine with only one large pulley and crank, but even this version required two hands: one to turn the crank, the other to rub the glass globe (Fig. 3). An assistant most likely turned the crank, but Hauksbee placed his own hand on the spinning glass vessel, setting a precedent that many would follow. The creation of new electrical effects clearly required not only the experimenter's mind but also his body.

In addition to inventing the electrical machine, Hauksbee came up with another important glass-based technology for generating charge. He procured a tube of fine flint glass 30 inches long and about 1 inch in diameter. After rubbing the tube briskly with a piece of paper, he found that it strongly attracted small pieces of brass leaf (brass pounded very thin), sometimes from as far as a foot away.[56] Hauksbee also discovered that he could feel the electrical effluvia when he brought the charged tube near his face; this sensation resembled fine hairs being pushed against the skin. Later investigators would refer to this effect as the "electric wind."

Needless to say, after each new discovery in the laboratory, Hauksbee put together a demonstration and presented it to assembled members of the Royal Society. His findings, rapidly published in the Royal Society's *Philosophical Transactions* (from 1704 through 1709), brought that institution to the leading edge of electrical science. Hauksbee also published in 1709 a book summarizing his electrical experiments; translated into several languages, it made his work accessible to many would-be experimenters on the continent.[57] A second edition was published posthumously in 1719, perhaps revised by his nephew—also named Francis Hauksbee—who was an instrument maker (see chapter 3). Glass-based electrical machines, modeled on Hauksbee's, would be widely adopted in the decades ahead as the principal generator of charge.

$$-|{\mid}|{\vdash}$$

The many variations of "electric light" produced by Hauksbee were surprising and impressive, but another Englishman, Stephen Gray (1666–1736), created effects so wondrous that they eventually captured the imagination of electrical experimenters and a large lay public alike.[58] Perhaps more important, Gray invented a host of technologies that would become commonplace in the apparatus of many electrical experimenters.

Gray was born in Canterbury and took up the family trade as a dyer. In his middle years, he devoted much time to science, making original observations with microscopes and telescopes. As a senior citizen, Gray resided at the Charterhouse, a London home for respectable pensioners of meager means. Unlike most of his peers, who frittered away their last years in decent idleness, Gray was a busy man, nurturing a passion for research on electricity and carrying out original experiments. His findings, published from 1720 to 1736 in the *Philosophical Transactions*, gained him international renown. At long last the Royal Society extended Gray membership in 1732, only four years before he died.

Above all, Gray's work shows that, in the earliest stages of a science, an investigator can produce noteworthy effects without complicated or expensive apparatus. Typical were Gray's experiments that tested various substances in order to expand the list of electrics. No apparatus could be more rudimentary: to test a material, he merely rubbed it between his thumb and forefinger and observed whether it attracted small particles or clung to his finger. On the basis of these manipulations, the list of electrics grew to include, for example, linen and wool, wood shavings, silk ribbon, leather, a hair from Gray's wig, "the fine Hair of a Dog's Ear," and ox gut.[59]

Gray also showed that a rubbed glass tube, like that first used by Hauksbee, was a versatile technology for supplying charge. Gray's glass tube was slightly over 3 feet long and about 1.2 inches in diameter.[60] Having noticed that the rubbed tube gave off light (that is, a spark) when another object approached, he was eager to see if the tube could also communicate electricity to nearby things without contact. Indeed it could. From a piece of packthread he suspended various objects above the tube, including a shilling, copper teakettle, sandstone, and chalk—and all acquired electricity.[61] An even more surprising effect appeared after he placed a cork stopper in the end of the glass tube. When Gray brought a down feather—very light and frizzy—near the cork, the cork alternately attracted and repelled it many times.[62] How bizarre! In these experiments, electricity had passed—without benefit of actual contact—by induction.

To put it mildly, induction is a strange process that depends on the electric field that surrounds any charged object. Surprisingly, that field can *induce* a charge—often the opposite charge—on anything nearby. Thus, induction can communicate charge not only to or from conductors, such as coins, but also to insulators—objects that ordinarily resist the flow of electricity by conduction. Although conduction and induction are seemingly simple effects, in the operation of an apparatus they can combine in myriad ways, making it difficult for the investigator to explain exactly what is going on. And in fact many investigators working on electricity in the eighteenth century never understood the role of induction in their own experiments.[63]

—|||—

Curious to see how far he could send electricity, Gray in subsequent experiments strung together various objects, including fishing pole and copper kettle, and learned that the practical limit was set by the size of his room at the Charterhouse. Continuing the experiments in larger venues—sometimes exciting charge on a solid glass cane, at other times on his glass tube—Gray succeeded in communicating electricity many dozens of feet.[64] He was able to claim success when the end of his line attracted bits of brass leaf.

Gray's research piqued the interest of Granville Wheler, a wealthy fellow of the Royal Society, who encouraged him to extend his studies and probe the limits of electrical communication. Working at Wheler's home, this pair must have been an odd sight, stringing up, throughout the estate, packthread lines suspended by silk cords. Finding that the silk suspenders gave way under the weight of the long pieces of packthread, the two gentlemen tried hanging the packthread by loops of brass wire. The latter, it turned out, sapped the charge by conducting it into the wooden supports, making communication over the packthread impossible. The two men learned a profound practical lesson: some materials, especially metals, communicate electricity better than others. In subsequent experiments they found a way to use silk suspenders after all. Although beginning to formulate a distinction between conductors and insulators, Gray and Wheler did not go that far. Their experiments in electrical communication, however, did reach many hundreds of feet. That electricity could be conducted over long distances, perhaps without limit, was the effect that later planted the seeds of electric telegraphy (see chapter 11).

In assembling apparatus, the pensioner Gray only occasionally required technologies specially made by others; usually he enlisted objects—and people—close at hand. The Charterhouse was apparently home not only to old men and curious objects, but also to young boys, perhaps charity cases, who might be coaxed into volunteering for scientific research. In one set of experiments, Gray incorporated a Charterhouse boy into his apparatus. The eight-year-old lad, who weighed 48 lbs., was suspended from the ceiling of Gray's room by strong insulating cords that formed two slings. Chest and thighs resting on the slings, hovering horizontally facedown about 2 feet above the floor, the subject was deemed ready to participate. When Gray brought the charged glass tube near the bare soles of the boy's feet, his face—its expression not recorded—attracted bits of brass leaf at a distance of more than 10 inches; the boy could also attract brass leaf from two fishing rods held simultaneously in his outstretched arms. These intriguing experiments demonstrated that an insulated person could be charged to a considerable degree and transmit that charge to other objects. But what about animals?

He soon had an answer. In experiments with a white cock—first alive and then dead—as well as "a large Sirloin of Beef," suspended on silk cords and charged by an excited glass tube, Gray found that animals could also serve as a reservoir of charge.[65] As an alternative reservoir in some experiments on sparks, Gray employed an iron bar, 4 feet long and around .5-inch in diameter, which was suspended from silk cords.[66] Having received charge from a rubbed glass tube, the iron bar could serve as an experiment's immediate source of electricity. In the late 1740s, insulated metal bars and tubes, known as "prime conductors," would become a mainstay of electrical apparatus; some would reach gigantic proportions. In another precedent, Gray showed how the investigator himself could serve as an electrometer. For example, when he used a hand or cheek to discharge an iron rod, the strength of the "Pain of pricking or burning" indicated the charge's magnitude.[67] Surprisingly, many investigators throughout the eighteenth century used their own sensory organs to measure charges.

Gray also discovered the advantage of placing an object to be electrified on a cake of resin, which insulated it from ground.[68] In one experiment he made two resin cakes, about 8 inches in diameter, on which a standing person could be electrified. The resin cake became a fixture in apparatus used by many subsequent investigators, especially those who wanted to incorporate into their experiments a charged person.

Later experiments conducted separately on resin, sulfur, beeswax, and other classic "electrics" revealed another surprising effect.[69] Gray began by melting the substance in an iron ladle, allowing it to cool in place. Then he reheated the ladle slightly, which yielded a hemispherical solid. Even without rubbing, these objects after cooling exhibited some "Attractive Vertue," which he assessed with a thread on a stick—a serviceable electrometer more sensitive than brass leaf.[70] Gray found that if he placed the hemispheres in "black Worsted Stockings," the objects retained the charge for at least a month. He referred to this effect, so simply produced with domestic objects, as "perpetual attractive Power."[71]

The legacy of Gray's research was more than new electrical effects, for he employed technologies that allowed others easily to repeat and extend his experiments at no great expense. With merely a glass tube an investigator could excite a considerable "Force" of electricity and communicate it to nearby objects of every kind—even people—sometimes over long distances. Gray also showed the necessity of insulating the entire experimental system—with, for example, suspenders of silk and supports of resin cakes—to prevent charge from leaking to other objects and to ground.[72] And with an insulated iron rod, an investigator could easily accumulate a sizable charge for use in sundry experiments. Finally, by communicating electricity through a series of objects, strung together, Gray established a pattern for building what would become known in the following decade as electrical "circuits." Gray was not a gifted generalizer, much less a theorist; he merely asked insightful questions about electrical effects and, in the course of answering them, created simple but significant technologies.

—|||—

The publication by the Royal Society of Gray's experiments, with their host of innovative technologies and new effects, catalyzed an interest in electricity both in England and on the continent.[73] One Frenchman, Charles-François de Cisternai du Fay (1698–1739), picked up where Hauksbee and Gray had left off.[74] Wealthy and well educated, Charles du Fay came from a long line of military men; not surprisingly, he accepted the family calling and even fought as an infantryman in Spain. But in 1723, after more than a decade of soldiering (he had joined up when only fourteen), du Fay, exploiting family connections, secured a position as chemist on the staff of the French Academy of Sciences. There he published on countless subjects, including phosphorescence, and even took an interest in botany. Presumably the latter led to his appointment in 1732 as director of the king's botanical garden.

Shortly after taking over the royal garden, du Fay began electrical experiments, some in collaboration with Jean-Antoine Nollet (whom we encounter in chapter 3 as Benjamin Franklin's most severe and persistent critic).[75] Although du Fay's research is marked by explicit hypothesis testing and punctuated by frequent generalizations, his immediate contributions to electrical technology itself were limited. His main source of charge was the rubbed glass tube popularized by Gray; in experiments that repeated work by Hauksbee on the production of light, du Fay did use an electrical machine with a hand-rubbed glass globe.[76] Once he also employed an insulated iron bar, which he charged with an excited glass tube.

In early experiments, du Fay showed that an excited tube could impart a charge to any material—including metals. He found that the best way to test materials was to place them on a small glass stand.[77] This simple insulating stand, alone among du Fay's inventions, would become standard equipment for many electrical experimenters. In later years this technology would be termed a "universal discharger."[78]

Du Fay's most original and significant finding arose in experiments with the attraction and repulsion of gold leaf. He first found a familiar pattern: when exposed to a charged tube, the gold leaf is first attracted to the tube, then repelled. But, intrigued by the effect, du Fay played around with charged specks of gold leaf, presenting them to other electrified substances, including rock crystal, glass, and gum copal. These manipulations revealed a surprising effect: a piece of gold leaf repelled by charged glass is nonetheless attracted by gum copal and other resinous materials.[79] On the basis of such three-body exercises, many carried out on a glass stand or on a versoriumlike device made from a wooden ruler, du Fay in 1733 proposed that electricity comes in two varieties. He named them *vitreous*—which arises on rubbed glass—and *resinous*, which resins such as gum copal or amber acquire.[80] He also noted that nonelectrics like gold can acquire either kind of electricity.[81] Not many years later, Benjamin Franklin would relabel du Fay's two *kinds* of electricity as two *states* of a single electricity, positive and negative (or plus and minus), corresponding to a deficiency or excess of electrical fluid.

Having identified two kinds of charge, du Fay proceeded to formulate a brilliant generalization: objects possessing the same charge repel each other, whereas those having opposite charges attract.[82] This experimental law was framed in the most general terms possible, its reach extending far beyond the specific objects that du Fay had manipulated. Although derived from experiments on utterly simple apparatus, this law enabled du Fay and others to explain many puzzling interactions among charged objects. And it remains to this day a cornerstone of electrical science.

Not all of du Fay's generalizations fared as well.[83] One experiment in particular gave rise to a conclusion that confused some students of electricity for decades. Inside a glass vessel du Fay placed some bits of gold leaf. When he brought a charged tube near the glass, the pieces of gold leaf moved about. The obvious conclusion to du Fay was that electricity can pass through glass.[84] However, we know that glass does not conduct electricity; induction caused the effect observed by du Fay. Regrettably, du Fay—and especially his influential collaborator Nollet—did not understand the role of induction in many effects.

$$-|\,|\,|-$$

The electrical research carried out before 1740 was remarkably fruitful, yielding a plethora of new phenomena. With the notable exception of the salaried instrument maker, Hauksbee, early experimenters were able to coax important effects from uncomplicated apparatus cobbled together from familiar materials and objects. These appropriate technologies can scarcely be regarded as scientific instruments, yet their use created an entirely new and significant branch of physics.

The technologies invented by Hauksbee, Gray, and du Fay would soon be taken up, modified, and augmented by a sizable community of investigators, the electrophysicists, who in the 1740s sought to create new effects and new principles. They invented increasingly complex and expensive apparatus, but the yield of "new physics"—effects not previously observed—was surprisingly meager.

People in many walks of life, fascinated by the possibilities discussed widely in the media inside and outside science, began experimenting with the new technologies in diverse activities, from public lecturing to medicine. These activities, incorporating often redesigned technologies, became the basis of many new electrical communities. As a foundation for the remainder of the book, let us consider the activities of the electrophysicists and their new technologies.

3. A Coming of Age

Prior to the mid-1740s, the electrophysicist's involvement in the process of generating charge was rather intimate. An investigator like Gray or du Fay began most experiments by vigorously rubbing a glass tube with his own hand. Those who used electrical machines required an assistant to turn the crank; in addition, someone—often the experimenter himself—had to place a hand on the revolving glass vessel. Electrophysics was, literally, hands-on science.

Soon, however, investigators fashioned new kinds of electrical machines, many of which dispensed with the hand-rubbing routine. Electrophysicists of the 1740s also perfected the prime conductor and invented the Leyden jar (the first true capacitor). Together, these three technologies—electrical machine, prime conductor, and Leyden jar—comprised the core components of an electrical power system that would serve electrophysicists as well as people in many other communities. In the following decades, electrophysicists continued to invent generator designs along with a host of accessories. Some of these new technologies would be transferred to other communities, where their use contributed to the creation of new effects, new principles, and still more new electrical things.[1] This chapter explores the major technologies developed by electrophysicists from 1740 to about 1800.[2]

‑|I|⊢

Beyond bestowing prestige and ratifying a new social role, scientific societies did much for their members. An important activity was the holding of periodic meetings for reporting and discussing new findings. In addition to reading their own papers before the group, members sometimes sponsored the work of neophytes and lesser luminaries. Benjamin Franklin, for example, often received scientific reports as letters, which he then forwarded to the Royal Society for presentation. Letters read at meetings were frequently published in proceedings, transactions, and memoirs, and in a few general scientific journals.[3] These publication outlets gave scientists new ways to disseminate their findings, by which they could flaunt their intellectual prowess as well as demonstrate priority in discoveries.[4]

By the mid eighteenth century, most societies had worked out agreements for the exchange of publications with other societies. This was momentous because it established for scientists a formal mode of international communication. A telling case in point: du Fay learned the details of Gray's work, the major inspiration for his own research, by reading the *Philosophical Transactions* of the Royal Society. Likewise, electrical experiments carried out in England and France became quickly known to investigators in Germany and Italy, and vice versa. Because of this relatively good international communication, at times the Germans were the innovators in electrical science and technology, at other times the French, English, or Italians were in the vanguard.

In perusing the contents of society publications from the eighteenth century, we cannot help marveling at their variety. A typical issue of the French academy's annual volume held papers that reached far and wide over scientific and mathematical subjects, including anatomy, chemistry, botany, arithmetic, algebra, geometry, astronomy, geography, hydrography, and mechanics.[5] Because publications of other national societies were equally eclectic in content, a dedicated reader could keep abreast of the latest findings in every physical and biological science.

More surprisingly, one person often published on several subjects. Because science had not yet fragmented into *professional* territories, each occupied by specialists who had to undergo lengthy and formal training and who jealously guarded their borders, people with scientific curiosity could tackle any—and all—problems that struck their fancy, whether they concerned solar eclipses, salamanders, the origin of minerals, or human deformities. Commonly, people wrote papers in totally different fields, and many also published moral and philosophical tracts. The record holder for diversity of interests was perhaps Charles du Fay, who published on every scientific subject recognized by the French academy.[6]

Research on a given subject such as chemistry and botany was carried out, usually, by a widely dispersed community of investigators. Membership in a given community depended on a person's familiarity with previous research—indicated in citations to earlier investigators.[7] But, more important, the scientist had to possess and use skillfully a core of apparatus.

Thus, the chemist had his glassware, distillation equipment, and furnaces, a listing of which can be found in chemistry textbooks of the late eighteenth and early nineteenth centuries.[8] For the botanist, it was the plant specimens in a herbarium and the tools, such as microscopes, for studying them. Likewise, the electrophysicist had an electrical machine and its accessories. Membership in more than one scientific community required, of course, access to the core apparatus of each. Later in this chapter we encounter the versatile Martinus van Marum (1750–1837), botanist, geologist, electrophysicist, and an inventor of "big science"—eighteenth-century style. Fortunately, he had the resources of a wealthy museum at his disposal and so was able to acquire the technologies for studying many subjects.

Within the overarching community of electrical investigators, smaller and narrower research communities rapidly arose, built around apparatus that enabled particular studies in chemistry, biology, even earth science. Although some electrical apparatus served essentially unmodified in several communities, the larger pattern was the proliferation of specialized electrical devices, especially accessories, whose performance characteristics were tailored to the members' activities.

Sometimes an individual with interests in many subjects transferred electrical technologies from one community to another. The Englishman Joseph Priestley (1733–1804), who had carried out experiments in electrophysics and published in 1767 a massive work on electricity, introduced the use of electrical technology into pneumatic chemistry. Likewise, Benjamin Franklin, printer, publisher, prolific inventor, and writer of witty prose, not only took up electrophysics but adapted electrical technology for earth-science experiments, the most memorable involving his celebrated kite (chapter 8). Like Priestley and Franklin, many investigators were members of several electrical communities and contributed to the transfer of technology among them.

That scientific societies encompassed the entire gamut of physical and biological sciences also had important implications for technology transfer. Obviously, investigators could read about a new technology in a society publication. In addition, they could often witness, in society lectures, the demonstration of new apparatus—a highly salient process for transferring information about technologies. In either case, readers or observers might attempt to acquire or make the new apparatus. Given the myriad interests of their members, scientific societies were important places for transferring knowledge—and technologies—within and among communities.[9]

People from the industrial and commercial worlds, usually not scientists themselves, sometimes attended a society's lectures and demonstrations.

Thus, the public presentation of science—electricity in particular—furnished opportunities to transfer scientific knowledge and technologies of science into industrial and commercial communities.

-||⊢

Scientists were not the only people who closely followed society publications and demonstrations. Since the middle of the seventeenth century, instrument makers—some of them also scientists—had begun to manufacture and sell apparatus such as telescopes, microscopes, vacuum pumps, surveying tools, mathematical instruments, products for magic shows, medical devices, even eyeglasses.[10] As soon as a new technology or accessory was reported, instrument makers assessed its commercial potential and sometimes brought it quickly to market. Thus, instrument makers made the technologies of science highly visible and accessible to a diverse public.

Because most instrument makers sold many kinds of articles to people of varying occupations and interests, instrument shops and catalogs were also an important mechanism of technology transfer. Like many a modern shopper, the visitor to an eighteenth-century instrument shop might have entered in search of one device but left with others. Instrument makers could be found in most large cities on the continent, even in the American colonies, but Great Britain was the center of instrument making in the eighteenth century; indeed, London alone hosted, on average, more than sixty firms.[11] The larger London workshops were beehives of craft activity, employing dozens of skilled artisans who turned out both standard products and customized designs.[12]

Several important instrument makers had an abiding interest in electricity, which enhanced their ability to offer, throughout the eighteenth century, electrical technologies invented in many communities. And in a few cases, instrument makers such as Edward Nairne in London and Johann Gütle in Germany invented and commercialized their own apparatus.[13] Instrument makers also published experiment books, sometimes even lengthy treatises on electricity; not surprisingly, such works often ended with a detailed catalog and price list.[14]

Probably the first instrument maker to bring electrical technology to market was Francis Hauksbee (the younger), who commercialized his uncle's electrical machines along with many other apparatus for physics experiments.[15] Sold in a shop on Fleet Street, in London, these technologies were advertised in a book—published in 1714—containing physics lectures by William Whiston and lavish illustrations of Hauksbee's wares. The electrical machine, which came with accessories for repeating some of the senior Hauksbee's experiments, was the one-crank variety; regrettably, no prices were supplied.[16]

Several other instrument makers marketed electrical things before 1740. Among the most significant were Petrus van Musschenbroek in the Netherlands and Jean-Antoine Nollet in France; both were also electro-physicists.[17]

—|ı|—

Petrus van Musschenbroek (1692–1761), along with his brother Jan, came from a long line of artisans and instrument makers.[18] Although Jan mainly carried on the daily operation of the family business, Petrus had many other interests. He was educated as a physician at the University of Leiden and even practiced medicine for a time; he also earned a philosophy doctorate. In 1719 Petrus accepted a professorship at Duisburg, where he lectured on medicine, mathematics, and philosophy. Later he was awarded professorships at the Universities of Utrecht and Leiden; the latter was the first Dutch university, founded in 1575—a prestigious post indeed.

At Leiden University, where he remained from 1740 until his death, Musschenbroek propounded Newtonian physics in very popular lectures, illustrating the basic principles with apparatus made in the family shop. He also compiled his lectures into several books; translated from Latin into Dutch, French, English, and German, these texts helped spread knowledge of the Newtonian system beyond its early strongholds in England and the Netherlands.

In the 1739 edition of Musschenbroek's *Essai de physique*, there is a sizable instrument catalog containing hundreds of items. Near the end of a long string of apparatus used in pneumatic experiments were a few electrical things, including several kinds of Hauksbee-type globe machines and a glass tube which, when rubbed, "revealed in a surprising manner its electrical virtue."[19] Sales of these electrical devices were doubtless encouraged by Musschenbroek's popular lectures and books.

Jean-Antoine Nollet (1700–1770) was one of the most celebrated figures of eighteenth-century French science.[20] Born into the French peasantry, he was singled out for a career in the clergy because of his considerable intelligence. Although he earned a theology degree in Paris along with the title of *abbé*, his true calling was more earthly. Fascinated by the possibilities of applying natural philosophy to the practical arts, Nollet in 1728 joined a group of kindred spirits called the Société des Arts. His association with that group brought him into contact with important natural philosophers, including Charles du Fay. From 1731 to 1733 the senior scientist took Nollet under his wing, permitting him to assist in experiments, especially electrical ones.[21] Nollet also accompanied du Fay on trips to England and the Netherlands, where he was introduced to prominent Newtonian physicists. Nollet never adopted a Newtonian perspective, for like du Fay he maintained a Cartesian concern with mechanical causes for action at a distance. However, Nollet did accept and apply the experimental method with gusto.

The death in 1734 of Pierre Polinière, France's most famous physics lecturer, created a vacuum that Nollet eagerly filled. Unable to afford scientific instruments, Nollet began to manufacture them himself, usually making duplicates to sell at a profit. In this way he amassed an enormous instrument collection that he employed in lectures covering all of physics. Through these lectures Nollet eventually attracted the notice of the French monarchy and put on dazzling displays for the royal family and their guests. Of these electrical doings in the French court, the Spanish electrophysicist Joseph Vazquez y Morales wrote that "no other phenomenon of physics had achieved so much applause, so much admiration."[22]

Not surprisingly, after du Fay's death in 1739 Nollet acquired the mantle of France's leading electrical investigator and was inducted into the French academy. A few years later, he was awarded the new chair of physics at the University of Paris along with assorted lectureships at other schools. Nollet's path from peasant to college professor and academician signaled that, even in the heyday of the French monarchy, occasionally someone without land-based wealth could achieve high social standing.

A prolific writer, Nollet not only published a massive work on physics—essentially the text of his lectures—and several books on electricity, but near the end of his life he wrote *L'art des expériences*, which included details on constructing and using scientific apparatus.[23] *L'art des expériences* solidified his reputation for being, according to Aimé-Henri Paulian's 1781 physics dictionary, "the greatest man that France has produced in the art of conducting experiments."[24]

Nollet's passion was electricity. In addition to developing an extensive program of experiments and displays, he also fashioned an electrical theory. It was based on the belief that electricity is a very subtle fluid, probably related to fire or light, which surrounds an electrified object.[25] This fluid (or atmosphere) has a complex motion: one part (the *effluence*) moves away from the object, while the other part (the *affluence*) simultaneously moves toward it.[26] This dual motion, which mainly accounted for attraction and repulsion, was interpreted by many as a two-fluid theory of electricity.[27]

In piecing together the elements of his theory, experiment by experiment, Nollet—like most of his contemporaries—failed to distinguish be-

Figure 4. Nollet's electrical machine. (Adapted from Dibner 1957, 57. Courtesy of the Burndy Library, Dibner Institute, MIT.)

tween different kinds of electrical communication. Impressed that a charged object could cause another object to move, even when both were separated by glass, Nollet, following du Fay, drew the obvious conclusion: glass and other solid objects were permeable to the electric fluid.[28] With the benefit of hindsight, we appreciate that this electrical communication results from induction, not conduction—a distinction that Nollet never appreciated.[29] As we shall see shortly, Nollet's theory was soon called into question by electrophysicists familiar with the Leyden jar and the one-electricity theory of Franklin.

Nollet published his earliest instrument catalog in 1738; appearing at the end of the first, and very brief, edition of his physics lectures, it presented no prices.[30] But the catalog contained a formidable array of physics apparatus, 345 entries in all. There were 31 electrical items, including machines, accessories, and materials for performing experiments devised by Hauksbee, Gray, and du Fay; among the offerings were silk suspenders for electrifying living things.

In his experiments and demonstrations, Nollet employed a large electrical machine of pleasing aspect and generous proportions. A sizable glass globe was rotated rapidly by a wooden wheel, more than 3 feet in diameter, which could be cranked from either side (Fig. 4).[31] To generate charge, the investigator or an assistant placed his hands upon the spinning globe. This electrical machine apparently sold well, if we can judge by its common illustration in the works of other French investigators. Nollet's instrument sales, along with his lectures and books, stimulated interest in electricity and also helped promote the experimental method in French physics.

Musschenbroek and Nollet hardly held a monopoly on sales of electrical apparatus. By the mid-1740s, many other instrument makers, especially in Germany and England, began to manufacture electrical things. The availability of off-the-shelf electrical technologies—that investigators sometimes purchased, sometimes copied—contributed to the rapid expansion of the electrophysics community throughout the Western world and was a fillip to the formation of other electrical communities.

—|ı|⊢

During the late 1730s and early 1740s, a handful of German electrophysicists, mostly university professors, took up the development of new electrical technology, especially machines and accessories. These men had many motives for inventing electrical devices. Some doubtless hoped that more powerful electrical machines, like more powerful microscopes and telescopes, would create new effects or render old effects easier to study. Others, however, sought to make machines whose operation required less direct human involvement.

Another impetus for invention derived from a performance deficiency of Hauksbee-type electrical machines: they could not be used conveniently for a wide range of experiments. Indeed, Hauksbee had designed rather specialized machines for studying a narrow range of electrical phenomena. Charge was created on the surface of a glass globe, on an immovable machine, and that was precisely where the researcher had to work. There was no easy way to draw off the charge, accumulate it, and apply it to nearby accessories. Perhaps that is why Gray (and du Fay) used a glass tube as the source of charge, for at least it could be easily taken where needed—to the distant end of a long packthread line or the feet of a suspended boy. As the German professors strove to repeat and extend the studies of Gray and du Fay, they fashioned, piece by piece, a general-purpose electrical power system that could be used instead of rubbed glass tubes.

Georg Bose of Wittenberg University, probably the first person appointed to a professorship on the basis of electrical research, sometimes used a metal tube in place of a person as an accumulator of charge.[32] In one experiment of 1743, Bose suspended from silk cords a telescope tube 21 feet long and 4 inches in diameter. When he brought one end very close to the spinning glass globe, the tube accumulated a large charge. A metal tube, sometimes merely a gun barrel, suspended in this manner or supported by glass pedestals would become the familiar prime conductor of virtually all later electrical machines.[33] Beginning in the 1750s, the ends of the tube were fixed with hemispheres or knobs in order to eliminate the loss of charge from the tube's edges.

Early prime conductors had a problem: on occasion the swinging metal tube would strike the globe, smashing it to smithereens. To solve this problem, investigators at first affixed to one end of the tube a bundle of metallized threads, which could even touch the glass without causing harm. Although efficient collectors of charge, thread bundles were replaced by metal combs having a small number of widely spaced, pointed teeth. In some machines, the comb was an integral part of the prime conductor, but in others the comb and prime conductor were joined by a bracket or wire. However, not all machines employed combs. In another common design, a chain hung from the prime conductor, collecting charge from the globe's surface. With prime conductor and collecting comb or chain, electrical machines gave investigators a versatile technology that could supply charge for almost any experiment they might envision.

Another noteworthy invention of the German professors was the mechanical rubber. Seeking a replacement for the human hand pressed against spinning glass, Johann Winkler, a professor of Latin and Greek at Leipzig, apparently took a cue from Hauksbee's earliest electrical machine. Hauksbee had used a small mechanical rubber that pressed against various test materials *inside* a glass globe. Winkler enlarged the rubber and mounted it *outside* the globe. The rubber was a small cushion, made of leather or linen stuffed with wool or other soft material, which was held against the glass by an adjustable wooden fixture.[34] In later designs, Winkler spring-loaded the cushion to compensate for unevenness in the glass's surface. It was soon discovered that the application of a mercury amalgam to the rubber could increase the charge's strength; not surprisingly, instrument makers also sold amalgam.[35] Although mechanical rubbers with amalgam were widely adopted, especially in Germany and England, in a surprising number of cases users of electrical machines persisted in pressing the glass with their own hands.

The German professors created many one-of-a-kind machines, their designs almost whimsical. In search of higher power, Bose once made a machine with three glass globes—the largest 18 inches in diameter—all rotating at once. This machine yielded a charge so great that its shocks could cause bruising.[36] But three globes were trifling compared to the amount of glass in a behemoth machine fashioned by Johann Winkler: it incorporated eight glass cylinders, and each had a cushion.[37] With their collectors connected together, these compound electrical machines were perhaps the earliest instance of parallel circuitry.

Winkler also sought ways to generate electricity rapidly with as little human effort as possible.[38] In hopes of eliminating at least one assistant, he built machines operated not by crank but by treadle. For one design, a rather tall machine, he claimed that the pedal could be depressed 106 times a minute. Another of Winkler's treadle-operated machines moved a long glass tube up and down against a stationary rubber. To achieve this motion, which replaced human hands, the machine required a complex mechanism—worked by human feet. Treadle-operated machines, which might not have seemed like a labor-saving invention to other investigators, were little more than curiosities.[39] True labor-saving designs would have required a shift from human operators to other prime movers. Not until the late nineteenth or early twentieth century, however, would small steam engines, internal combustion engines, and electromagnetic motors mount a serious challenge to people power. For motive power at an intermediate scale between springs and falling weights on the one hand, and beasts of burden and water wheels on the other, inventors in the Enlightenment recognized no substitute for human arms and legs.

Some of the professors' new technologies were commercialized quickly by German instrument makers, and soon descriptions of these machines and accessories appeared in the *Philosophical Transactions* of the Royal Society.[40] Not surprisingly, British instrument makers also seized the opportunity to bring some of these inventions to market. By 1747, for example, Francis Watkins, who was the official optician for the prince and princess of Wales, sold in London a compact globe machine with rubber, prime conductor, and chain collector; it was even advertised to be "portable."[41]

‐|| ‐

When they were not tinkering with electrical machines, the German professors were busily perfecting accessories that displayed many delightful effects. These inventions were mostly adapted from the apparatus of Hauksbee, Gray, and du Fay. One new effect, easily produced by the higher power of the new machines, caused a sensation: a spark, often from a finger, could ignite flammable substances such as alcohol vapor. That electricity had affinities with fire entered both scholarly and public discourse.

Accounts of the German professors' experiments appeared in newspapers and magazines, in England and on the continent, and contributed to a flurry of interest in electrical things. Among discussions of foreign affairs, public policy, history, and religion, *The Gentleman's Magazine* covered electricity, beginning in 1745 with an historical article that highlighted the recent research in Germany. In its generous use of terms like "found" and "discovered," this article emphasized that human agency, in the persona of an ingenious professor, could bring to light new phenomena "as surprising as a miracle." [42] The formerly arcane goings-on of electrical experimenters had become not just newsworthy but significant. After all, electricity— more than any other science in the eighteenth century—gave substance to Enlightenment beliefs in the possibility of human progress. In the decade from 1745 to 1755, *The Gentleman's Magazine* alone published on average about ten pieces a year on electricity, ranging from short letters to long articles, which reported new technologies and effects, scientific principles, and medical applications.

Alerted by the printed word, people in many corners of the Western world gravitated to public lectures to witness the marvelous electrical phenomena firsthand. In addition to physicians, ministers, and merchants, even members of the nobility turned up, expecting to be enlightened. As *The Gentleman's Magazine* noted in 1745, the new phenomena were "so surprising as to awaken the indolent curiosity of the public, the ladies and people of quality, who never regard natural philosophy but when it works miracles"; indeed, electricity became "the subject in vogue." [43] Impressed by stunning demonstrations, some spectators even visited instrument shops and bought electrical machines, accessories, and experiment books.

The engrossing electrical spectacles were sometimes complex, their explanations requiring mastery of subtle interactions among electricity's many effects. Eventually, however, electrophysicists found that many of the new phenomena could actually be explained by a small number of experimental laws established prior to 1740.

―||―

One invention of the 1740s, the Leyden jar, did cause an uproar in the electrophysics community because its operation at first seemed incomprehensible. Unlike the mechanical rubber, prime conductor, and comb, which derived from deliberate efforts by electrophysicists to redesign apparatus, often by trial and error, the Leyden jar arose by accident—not once but several times.

Accident, chance, serendipity are all words we use to describe effects that turn up unexpectedly during experiments, but these labels are misleading. Obviously, some might point out, a mind prepared by prior experience and knowledge is needed to appreciate the importance of novelty, to distinguish it from mere noise or annoyance. Yet all other factors constant, certain accidental discoveries seem to be more probable than others. Something else is going on.

To understand the sort of "chance" discoveries so common in eighteenth-century electrical science, let us consider experimental activities—or at least some of them—as a form of play. A uniquely mammalian behavior pattern, play is especially well developed among the higher primates, including humans. Monkeys and apes during play manipulate the objects of their surroundings, juxtaposing them in countless novel combinations, observing the outcomes. In this way are revealed new phenomena, new effects. Play opportunities are especially rich for humans because we fill our immediate environments with artifacts of all sorts, which furnish endless manipulative possibilities. Some behavioral sequences arrived at through play, though at first executed randomly or haphazardly, have strongly reinforcing outcomes and so may be repeated, adopted by other individuals, and even transferred to different activity contexts and communities.

A scientific laboratory is a place crammed with artifacts—some ordinary, others extraordinary—where play activities can reveal new effects and reinforce the behavioral sequences that produced them. And "play" does seem an apt description of Stephen Gray's manipulations of glass tube, fishing pole, or live chicken. That play produces new technologies in a variety of activity contexts is widely acknowledged, but play's part in scientific discovery and invention has not been as thoroughly explored.[44] After all, when discoveries and inventions get written up, they are made to appear far less playful and much more purposeful than in fact they were.

When many scientists in a community adopt a new technology, regardless of its original context of invention, there often follows a period of vigorous play and rapid discoveries. The more people who play with the new technology, the more likely the discovery of certain effects becomes; sometimes the appearance of an effect is highly probable. Chance is at work, and so is the scientist's fund of knowledge and creativity, but both are channeled by the manipulative possibilities of the laboratory apparatus. Had Galileo not discovered the moons of Jupiter, surely they would have been found by the second or third astronomer to point a telescope at the giant planet.[45] And so, in accounting for the scientific discoveries that led—by "chance" or "accident"—to the creation of new electrical technologies, we should consider not just theory-laden "prepared minds" but also the laboratory apparatus that channels the investigator's play activities.[46] No invention il-

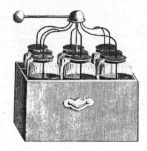

Figure 5. Leyden jar with *left*, discharger, and *right*, battery. (Adapted from Adams 1799, pl. 4, figs. 65 [right] and 66 [left]. Courtesy of the Bakken Library, Minneapolis.)

lustrates the importance of playful manipulations in science better than the Leyden jar.

—|ı|⊢

The Leyden jar was the first true capacitor, capable of storing a sizable charge. It was made from a glass jar or bottle—"the thinner the better." [47] The jar's exterior was covered with metal foil, such as lead or tin; the interior contained a brass chain resting on the bottom. The chain's upper end was brought through a cork stopper and terminated in a metal hook or ball (Fig. 5, left). Instead of a chain, some Leyden jars contained metal filings or foil, lead shot, mercury, or water with a wire protruding through the stopper.

The essential design feature of the Leyden jar, like all capacitors, was an insulator—today called a "dielectric"—sandwiched between two conducting materials. When one conductor was connected to an electrical machine's prime conductor and the other was grounded, the Leyden jar could accumulate its charge. Because the charge was retained for hours, even days, Leyden jars became the immediate source of electricity for many experiments.

The invention of the Leyden jar may have been accidental, but it was an accident waiting to happen. In repeating certain experiments of Gray, du Fay, and Bose, which showed that water could be electrified, many electrophysicists had in their laboratories glass vessels filled with water. Ordinary water, containing dissolved minerals, conducts electricity. Another conductor—often a grounded one at that—is the human hand holding the container. When an investigator, glass vessel in hand, connected the water to a prime conductor, perhaps with a chain, the container acquired a charge. Subsequent discovery of the stored charge, as a smart shock, merely required the person to touch the chain—still in contact with the water—with the free hand. This surprising and sometimes painful effect, arising from simple manipulations of objects common in electrophysicists' laboratories, was discovered "by accident" several times.

The Leyden jar takes its name from the most well publicized of these discoveries, which is associated with the laboratory of the famed physicist and instrument maker Petrus van Musschenbroek at Leyden University.[48] A frequent visitor to Musschenbroek's laboratory was Andreas Cunaeus, a lawyer entranced with electrical experiments. One day in 1746, seeking to electrify water at home, Cunaeus held a jar of the liquid in his hand and brought it to the prime conductor. Testing the charge on the water, Cunaeus found that it was horrifically greater than he expected.

Informed by Cunaeus of this surprising outcome, Musschenbroek tried the experiment several days later, employing a glass globe in place of the jar. The result was the same: the great professor received a jolt so severe that he was still shaking several hours later; he vowed never again to take a shock in this way. Always the academic scientist, Musschenbroek quickly published a report of the experiment, and credit for the discovery flowed to him. A German, Ewald Jürgen von Kleist, was actually first to encounter the effect, but his muddled description of the experiment rendered its comprehension and repetition by others difficult.[49] And, compared to the great Musschenbroek, von Kleist was an unknown.

British investigators were alerted to the Leyden jar and its terrible shocks by several letters published in the *Philosophical Transactions*. Johann Winkler's account was perhaps the most dramatic. Repeating Musschenbroek's experiment had caused in his body "great Convulsions." And it "put my Blood into great Agitation; so that I was afraid of an ardent Fever; and was obliged to use refrigerating Medicines. I felt a Heaviness in my Head, as if I had a Stone lying upon it. It gave me twice a Bleeding at my Nose."[50] Winkler also delivered two shocks to his wife, who became almost too weak to walk. Ironically, the kindly classics professor chastised other investigators for using charged Leyden jars to torture birds; apparently, however, it was acceptable to torture one's wife. The British public also learned, in early 1746, about the Leyden jar's awesome power in a two-paragraph account appearing in *The Gentleman's Magazine*.[51] The Leyden jar, commercialized by many instrument makers, could be readily acquired by electrophysicists as well as members of other communities. Also, the jars were simple enough to be easily made at home.

Playing with their new toys, electrophysicists discovered that Leyden jars brimmed over with new effects. When a jar was discharged by a metal conductor, it gave up its charge in a peculiar, oscillating manner.[52] In a fraction of a second, the flow of electricity reversed direction countless times. Unbeknownst to workers in the eighteenth century, the alternating current produced by a capacitor's discharge generated electromagnetic waves; a century and a half later, these would become the basis of radio communication. In the eighteenth century, however, no one had a receiver.

—|||—

Among the many people who pondered the action of the Leyden jar was Benjamin Franklin. As his scientific biographer, I. Bernard Cohen, points out, Franklin's interest in science preceded his involvement in establishing the Junto. On his first trip to England, at age nineteen, Franklin met many important scientific figures, including Hans Sloane, Newton's successor as president of the Royal Society. The young man from the colonies had also hoped to encounter the aged Newton himself, but it was not to be. On the return voyage, Franklin recorded in a journal many observations on meteorological conditions, the ocean, and astronomical phenomena. Franklin later took an interest in botany, participating in plant exchanges between the Old and New Worlds; it was he who introduced Americans to the rhubarb.[53]

The Library Company's British agent was Peter Collinson (1694–1768), like Franklin an enthusiast of science. Collinson, who had been a member of the Royal Society since 1728, was in the habit of bestowing gifts on the fledgling American enterprise and included in his first shipment from London a book by Newton and a gardening dictionary.[54] But it was Collinson's present in 1746 of a glass tube that would alter Franklin's life and the course of electrical research. Although Franklin had already been exposed to a few rudimentary electrical experiments in a Philadelphia lecture by Dr. Archibald Spencer, the Collinson gift catalyzed his decision to retire from direct involvement in business and pursue his scientific interests full-time.[55]

The wealth generated by his businesses would, Franklin expected, free him from distractions to pursue "Philosophical Studies and Amusements."[56] Others, however, came to view Franklin as a man of leisure who could be enlisted in various worthy causes. Believing that service to one's community should have the highest priority, Franklin reluctantly became more deeply involved in the affairs of Philadelphia, the Pennsylvania colony, and eventually the United States. In August 1751, four months after his first major work on electricity was published in London, he became a member of the Pennsylvania Assembly, serving until 1764.[57] Franklin's period of relatively uninterrupted electrical research, which began in 1747, lasted less than five years. After the early 1750s, he conducted relatively few experiments and seldom published on electricity. Nonetheless, he kept up on new discoveries and continued to offer his friends theoretical interpretations and suggestions for new experiments. Moreover, his seminal ideas were promoted by a vigorous group of followers, the Franklinians.

Decades later, Franklin recalled that Collinson had sent him, along with the glass tube, a description of the recent German electrical research. This account almost certainly was none other than the first article on electricity published in *The Gentleman's Magazine* in 1745.[58] Some of Franklin's earliest experiments were variations on German themes. In tackling the puzzle of the Leyden jar, however, Franklin's theoretical insights and inventions were stunningly original.

While experimenting, Franklin counted on the assistance of several close friends, particularly Thomas Hopkinson, Philip Syng, and Ebenezer Kinnersley (we encounter Kinnersley again in chapter 4). A group this large was in fact needed for Franklin's first reported experiment, in which he pondered the passage of charge among three people: two standing on resin cakes and the third on the ground.[59] On the basis of this experiment and others, Franklin formulated the distinction between positive and negative charges but did not cite the prior—and seemingly identical—distinction by du Fay between vitreous and resinous electricity.[60]

Throughout the eighteenth century, both Franklin's and du Fay's terminologies coexisted but not always peacefully. These terms came to have obvious nationalistic connotations in priority disputes, but they also reflected different electrical theories. As we saw, du Fay and his followers, especially Nollet, maintained that there were two *kinds* of electricity. For Franklin, however, there was but one kind of electricity, which could be manifest in two *states*—positive or negative—depending on whether a charged object had a surplus (+) or deficiency (−) of electrical fire.[61] Franklin's terms embodied a mathematical way of thinking: the charge on an object was simply a departure (plus or minus) from its natural state of neutrality.[62] An object could be brought to one of these altered states by friction or by communication from something already electrified.

In the eighteenth century, Franklin's theoretical conceptions and terms would prove to be the most fertile by far; they remain with us today. (Now, however, "negative" is regarded as a surplus of electrons—the carrier of negative charge—and "positive" an electron deficiency). In pursuit of "amusements," Franklin and friends also invented a host of intriguing accessories for displaying electrical effects; these were adopted by many people throughout Europe for use in public lectures and demonstrations.

Franklin's one-electricity theory and clever experiments enabled him to solve the puzzle of the Leyden jar. First he observed that the charges on the jar's inside and outside conductors were always opposite and equal.[63] Next he noted that only external connection of the two conductors would restore their natural "equilibrium"; the charge did not pass through the glass.[64] After numerous and varied discharges of Leyden jars using simple apparatus that included people, wires, corks, books, and wax insulators, Franklin began to entertain the hypothesis that the charge resided not in the jar's conductors—water on the inside, metal foil on the outside—but on the glass itself.[65]

Franklin tested this conjecture in several striking experiments. After charging one jar, he poured its water into a second but uncharged jar. The latter jar, even after its infusion of water, exhibited no signs of electricity. However, when the original jar was refilled with fresh water from a teakettle, it delivered a shock, indicating that the charge could not have resided in the water.[66] Franklin next sandwiched a pane of glass between two sheets of lead, creating the equivalent of a flat Leyden jar—in modern parlance, a plate capacitor. He charged it by connecting one lead sheet to the prime conductor and grounding the other, then removed both from the glass. Putting his finger on the glass, Franklin was able to detect "only very small pricking sparks" but felt them in many places on the pane.[67] When the lead sheets were again placed in contact with the glass, and another conductor brought near, they yielded a large discharge. The capacitor's conductors—the larger their area, the better—merely served to communicate charge efficiently to and from the glass. In constructing the first plate capacitor (examples of which were still used in radios of the early twentieth century), Franklin demonstrated beyond a reasonable doubt that the Leyden jar's charge is actually stored on the glass. No longer was the operation of the Leyden jar—and all other capacitors—a mystery.

In experiments with plate capacitors and Leyden jars, Franklin also invented "an electrical battery." He found that interesting effects were produced when capacitors were connected together, as batteries, in different configurations. On the one hand, when capacitors were wired in series, the overall charge that could be stored drastically decreased.[68] On the other hand, the stored charge increased when a battery's capacitors were wired in parallel (Fig. 5, right). Franklin appreciated that connecting large numbers of Leyden jars in parallel could furnish a limitless source of charge: "There are no bounds (but what expence and labour give) to the force man may raise and use in the electrical way: For bottle may be added to bottle *in in-finitum* and all united and discharged together as one, the force and effect proportioned to their number and size." [69] Needless to say, electrical bat-teries—some of immense capacity—became a common (and sometimes dangerous) appliance in many laboratories. Investigators could make their own batteries or buy them from instrument makers; small ones sold for a few pounds, but larger ones with many jars fetched a small landholder's in-come at £100.[70]

In playing with different combinations of Leyden jars, electrophysicists could not help recognizing that a spark's length and thickness varied inde-pendently. For example, many Leyden jars connected in parallel dramati-cally increased the thickness of the spark without affecting its length. Grad-ually, over the decades, spark length would be labeled electrical "intensity" or "tension," and thickness would come to be called the "quantity" of elec-tricity. Today tension is also termed voltage or electromotive force, and quantity has become current.

—|||—

Franklin's writings on electricity were first dribbled out in articles appear-ing in *The Gentleman's Magazine.* The magazine's editor, Sylvanus Urban, also arranged for the publication of several pamphlets; appearing from 1751 to 1754, they included Franklin's most important statements. In the following years the pamphlets metamorphosed into a real book that was published in several editions, including translations into French, German, and Italian.[71] Most electrophysicists appreciated the elegance of Franklin's one-electricity theory and his explanation of the Leyden jar. In a review of Franklin's book, William Watson, England's most knowledgeable electro-physicist, said that hardly anyone "is better acquainted with the subject of electricity" than Franklin.[72]

But not everyone was enamored with Franklin's ideas. Watson's coun-terpart in France, Jean-Antoine Nollet, was at first mystified by this sup-posed Franklin of Philadelphia and believed he was a hoax concocted by his enemies. Indeed, the comte de Buffon, the esteemed French naturalist and no friend of Nollet, had arranged for the translation of Franklin's work into French.[73] After conceding that the Philadelphian was no phantom, Nollet published a series of letters attacking Franklin's one-electricity theory (in-cluding the interpretation of the Leyden jar) and supporting his own. Al-though Franklin began composing a reply to Nollet, he eventually decided against entering the fray.[74] Obviously, he preferred to spend his limited time away from public service on new experiments. Franklin also believed that anyone could repeat his research and verify his findings, whereas some of Nollet's results were highly suspicious. Perhaps Franklin was reluctant to accuse Nollet in public of faking data—a conclusion to which he and others had come.[75] Doubtless, he regarded Nollet's theory to be so patently incorrect and so ill-supported by honest experiment that it would fall of its own dead weight.

Although Franklin did not respond to Nollet's criticisms, with his blessings the Franklinians did; sometimes they modified the one-electricity theory to account for effects, such as attraction and repulsion, which the original version did not explain well. After Nollet's death in 1770, there were few champions of his superannuated two-electricity theory. Uncharacteristically Franklin gloated that, with one exception, Nollet "liv'd to see himself the last of his Sect."[76] Nollet's ideas rapidly lost ground to Franklin's because the latter's easily explained most known electrical effects, accommodated novel effects as they were reported, and laid a foundation for new technologies. Above all, Franklin's ideas pointed the way to innumerable new experiments.

The self-made man from Philadelphia, modern poster boy for the Enlightenment, rapidly rose to prominence in scientific circles in England and on the continent. The Royal Society awarded him its highest honor for scientific achievement, the Copley Medal, in 1753—the same year he received honorary degrees from Harvard and Yale.[77] Three years later, Franklin became a fellow of the Royal Society, upon unanimous election, and was excused from making the "customary Payments"—around 25 guineas.[78] Curiously, and perhaps not by coincidence, in 1772—shortly after Nollet's death—Franklin was elected a foreign member of the French academy. And, as Cohen has argued at length, it was precisely Franklin's fame as a natural philosopher that ensured his easy entry into the highest echelons of European diplomacy and increased the clout of the colonies—and, later, the credibility of the fledgling American nation.[79]

—|||—

Unlike many electrophysicists, Benjamin Franklin failed to describe the major parts of his apparatus. We know that in addition to the glass tube furnished by Collinson, Franklin and his Philadelphia friends early on employed electrical machines. Franklin does mention that his collaborator Syng invented a very simple machine in which a crank was attached directly to an axle passing through a glass globe, which supplied enough electricity in just a few turns to fully charge a Leyden jar.[80] But, in contrast to European practice, he did not illustrate the new machine. Other electrical machines allegedly owned by Franklin reside today in several museums, but they are unremarkable; all employ glass cylinders of modest size.

Although cylinder and globe machines enjoyed considerable commercial success, electrophysicists continued to confect new designs.[81] The most important was the plate machine, which creates charge by rubbing a rotating glass disk. The historian Willem Hackmann has probed the plate machine's history and found it very confusing, because there were many likely inventors.[82] For people interested in pinpointing precisely who made the first example of any new technology, the uncertain or multiple authorship of an invention is a problem. But, in view of the earlier discussion about play, we should not be surprised that the plate machine was invented more than once in the 1750s and 1760s; after all, glass plates were not uncommon objects in kitchens and dining rooms and thus could be easily appropriated for the laboratory.

One of the plate machine's inventors was Jan (or John) Ingen-Housz (1730–1799).[83] Born in the Netherlands, he was educated as a physician but also studied some physics at Leiden under Musschenbroek. Despite having established a successful medical practice in his hometown of Breda, Ingen-Housz's far-ranging interests drew him in the 1760s to England. There he met many scientific luminaries, including Priestley and Franklin, the latter becoming a close friend. Ingen-Housz achieved great fame—and fortune—for his safe methods of smallpox inoculation. The Austrian royal family, having been successfully inoculated, granted Ingen-Housz a generous pension, which enabled him to devote more time to research in biology, chemistry, and physics. Although Ingen-Housz is best known today as the "discoverer" of photosynthesis, his many contributions to electrophysics were far from insignificant.

In the mid-1760s, Ingen-Housz built a prototype plate machine employing an ordinary glass stand—a dish supported on a pedestal.[84] He showed this machine to Franklin, who encouraged him to continue developing it. But before Ingen-Housz could perfect the design, he discovered that a plate machine was already being sold by Jesse Ramsden, an instrument maker in London.

Ramsden sold all sorts of scientific apparatus, from barometers and hygrometers to air pumps and electrical machines; he also offered a variety of spectacles. In 1766 Ramsden published a 12-page pamphlet describing his new electrical machine, which employed a circular glass plate instead of a globe or cylinder.[85] The glass plate stood upright in a wooden frame, fixed on an axle that could be rotated by a crank. Fastened in the wooden frame on both sides of the glass were two adjustable rubbers, top and bottom. Charge was drawn from the immediate vicinity of the glass plate by points attached to the two arms of a Y-shaped prime conductor. A Leyden jar also accompanied the machine.

It is not apparent how or why Ramsden came up with this novel design; nor was its intended market evident. Ramsden's pamphlet did illustrate and discuss a variety of accessories used in popular displays of electrical effects, but the machine also came with a Lane electrometer, used mainly in electromedical applications. Perhaps it was designed as a general-purpose machine. In any event, Ramsden made no special claims for his machine's performance.

In subsequent decades, other instrument makers throughout Europe copied and embellished Ramsden's elegant design, sometimes using two glass plates, and it enjoyed enormous success (for examples of plate machines, see Figs. 8 and 17). The plate machine, it would appear, had several major performance advantages over the older designs. First, the rubber's pressure occasionally caused the globes and cylinders, made of thin glass, to explode, spraying shards on participants and spectators. A *well-adjusted* plate machine was much less likely to suffer a similar fate.[86] Second, the plate machine could be scaled up greatly to generate high power. It is unlikely, however, that small plate machines were appreciably more powerful than globe and cylinder machines of comparable size; after all, usable power was also strongly influenced by the capacity of the battery.[87] Third, as a new design with a very distinctive visual performance, the plate machine carried the cachet of cutting-edge electrical science. It became the latest fashion, the must-have machine for members of several communities.

But the plate machine was more expensive to manufacture and so had a higher price tag. It could also be temperamental. Not only did a plate machine sometimes arc and lose charge to ground, but cushion adjustment was critical: too little pressure and insufficient charge was generated; too much pressure and the cushions became efficient disk brakes, preventing the plate from rotating easily; and uneven pressure could break the plate. The cushions, mounted in a complex assembly, also had to be removed and cleaned periodically. Many of these problems were solved by new designs in the 1790s, but the revised machines were even more expensive. Apparently, the plate machine was a mixed blessing. Only in some electrical communities did members need to have the most "modern" machine, and, in any event, few applications of electricity required extraordinary power. People seeking a reliable, inexpensive, and unassuming source of electricity continued to buy globe and cylinder machines—despite the danger of explosion—and augmented their charge with batteries. Some people, of course, owned several kinds of machines.[88]

The plate machine itself became highly differentiated as inventors in many communities played with its possibilities. One Italian, Salvator Dal Negro, constructed in 1799 a modular plate machine by attaching eight wedge-shaped pieces of glass to the circumference of a wooden wheel.[89] The final diameter of the disk was in excess of 3.5 feet, and Dal Negro claimed that such machines could grow without limit. The only construction problem the inventor acknowledged was that of securely fastening the glass plates to the wheel. The big machine worked well enough for Dal Negro to conduct some experiments, but none broke new ground.

Gütle also invented several unique plate machines. Having built insulated grips for demonstrating how metals could be easily electrified by friction, Gütle put together a small, well-insulated plate machine employing a copper disk. He also created a plate machine that used a woolen disk stitched into a wooden hoop; this he sold in one- and two-disk varieties.[90]

Franklin's friend Ingen-Housz also came up with a strange plate machine using pasteboard disks covered with varnish. His idea was to employ a material much less expensive than plate glass. Ingen-Housz actually built a prototype in which four pasteboard disks, each 4 feet in diameter, were mounted on a single axle. With rubbers of cat's fur, the machine produced a spark between 1 and 2 feet long.[91] The addition of a prime conductor thickened the spark so much that Ingen-Housz declined to receive many shocks, and visitors would take no more than one. Informed about this invention by Ingen-Housz, Franklin wrote back, "I like very much your Pasteboard Machine."[92] Other investigators copied Ingen-Housz's design, but apparently no instrument maker brought it to market.

Ingen-Housz was also responsible for inventing another kind of machine that was to have a far greater influence in the eighteenth century—and much later: the drum or band machine. Ingen-Housz's first effort used a continuous ribbon of silk, covered with velvet and stretched over the circumferences of two large wooden disks; the rubber was cat's fur.[93] He proposed that more powerful varieties of this machine could be made by increasing the area of the band. This suggestion was followed by several continental investigators whose machines used wide taffeta belts rolling over sizable wooden drums.[94] A collector could be placed inside or outside the taffeta belt, between the drums, for drawing off charge and communicating it to a prime conductor or Leyden jar. Drum machines were sold by Gütle and became somewhat popular in Germany.[95] Although use of the drum machine by electrophysicists disclosed no new effects, its design principles would surface again in the van de Graaff electrostatic generator of the twentieth century.[96]

The previous examples barely scratch the surface of the intriguing designs for electrostatic generators created in the eighteenth century. Most were one-of-a-kind machines, prototypes built to demonstrate the feasibility of a design concept. Even so, they enable us to appreciate the endless variety of electrostatic generators that can be built, each favoring a particular configuration of performance characteristics.

In fashioning accessories for experiments, electrophysicists were equally inventive. They created literally hundreds of electrical things; most were employed in highly specialized experiments and were neither copied nor commercialized. Almost none, in the end, produced any new physics. I focus here on just two interesting inventions, the electrophorus and the torsion balance.

-||-

Italians had long taken an interest in electrophysics, but not until the latter decades of the eighteenth century did the cutting edge of electrical science and technology move to those Mediterranean climes. The person most responsible for bringing Italy to the forefront of electrical research was Alessandro Volta (1745–1827). After Franklin, he was the Enlightenment's most esteemed electrophysicist.[97]

Volta was born in Como to a family steeped in service to the Catholic church. His father, Filippo, had been a Jesuit but withdrew in favor of marriage to Maddelena de' conti Inzaghi and the fathering of a large family. Although only Alessandro, among seven surviving children, ignored the church's call, he was educated at Catholic colleges. After beginning a fruitful collaboration with Giulio Cesare Gattoni, a rich friend who maintained a laboratory and museum of natural history, Volta decided at age eighteen to pursue electrophysics.

In 1769 Volta published his first dissertation on electricity, *De vi attractiva ignis electrici,* in which he proposed that most effects could be explained by just one attractive force. Repulsion, for example, was interpreted simply as attraction away from an object. Ingenious but flawed, this theory failed to attract followers, and even Volta himself eventually abandoned it. The most useful contribution of the early dissertation was an analysis of induction, whose implications would play a role in the design of the electrophorus.

Like so many Enlightenment scientists, Volta's explorations ranged over many subjects, including meteorology, pneumatics, and the electricity of living things (see chapter 6); he even wrote poetry. In 1776 Volta discovered methane, the simplest hydrocarbon and the main ingredient of our "natural gas." His reputation growing, he was appointed professor of experimental physics at the University of Pavia, a position he held for almost four decades. At Pavia, Volta assembled a large instrument collection and gave lectures so popular that they required construction of a larger hall. For his many contributions to scientific theory and technology, Volta received numerous honors, including memberships in the elite societies of London, Paris, and Berlin; in 1794 he was awarded the Royal Society's Copley Medal. Volta's meager university salary was supplemented, in later years, by a pension from Napoleon and a princely salary for serving in the senate.

Volta was in many ways a prototype of the modern experimental physicist, seeking in the laboratory evidence for or against specific theories. However, unlike the apparatus of modern physicists, his creations tended to be simple and inexpensive. In seeking to disprove a hypothesis offered by the rival Italian electrophysicist Giambatista Beccaria, an outspoken Franklinian, Volta constructed the electrophorus. Beccaria had been puzzled by the fact that the discharge of an insulator was incomplete. This effect was manifest, for example, in the discharge of the Leyden jar, which was not instantaneous; only many alternations in the flow of electricity would completely dissipate the jar's charge. Beccaria conjectured that the charge on the glass rebounded because the electricity had been, in his term, "revindicated." Using Franklinian arguments, Beccaria explained that the glass, after discharge, would still not be in electrical equilibrium and so could be discharged again, many times.[98] Unconvinced, Volta believed that if he could prolong the effects of the supposed vindicating electricity, Beccaria's hypothesis would lose all plausibility. The electrophorus, a novel combination of venerable electrical technologies, was designed to advance Volta's theoretical agenda.

In a letter to Joseph Priestley, in 1775, Volta reported his invention of the electrophorus, the *elettroforo perpetuo* (Fig. 6).[99] It consisted of an insulator in the shape of a large disk, made from turpentine, wax, and resin. The disk, when briskly rubbed, perhaps with cat's fur, acquired a large charge that it could hold for many days. But this was nothing new: decades earlier Stephen Gray had shown that a resin cake retained a charge for weeks. However, the electrophorus had a new wrinkle in the form of a wooden lid covered with tin foil. When the lid, which had an insulating handle, was placed on the resin disk and grounded, the tin foil acquired by

Figure 6. Volta's electrophorus. (Adapted from Volta 1926, figs. 1 and 2. Courtesy of the Burndy Library, Dibner Institute, MIT.)

induction a charge opposite to the disk's. After removing the ground connection, the operator could then pass this charge to a Leyden jar's knob and, by repeating this process, charge a Leyden jar without appreciably depleting the electrophorus's electricity. This effect seemed magical, a charge nearly inexhaustible; surely, many believed, this must be new physics.

The invention of the electrophorus, like the Leyden jar, caused a sensation because at first no one could figure out how it worked—except Benjamin Franklin. The Philadelphian was informed about the surprising electrophorus in a letter from his friend Jan Ingen-Housz.[100] In reply, Franklin simply stated that "as far as I understand it from your description, it is only another form of the Leyden Phial, and explicable by the same principles."[101] Indeed it was. In 1778, Ingen-Housz, employing Franklin's theory of the Leyden jar, showed that the electrophorus's operation resulted merely from a combination of well-known effects—the storage of charge by an insulator, the peculiar property of resinous materials to stubbornly retain charges, and the inductive communication of charge to a conductor.[102] And, although Volta believed that the electrophorus, with its perpetual charge, would vanquish Beccaria's vindicating electricity, the latter also had a surprising resilience; the theoretical controversies continued.[103]

The electrophorus, though not resolving theoretical disputes, did capture the imagination of many electrophysicists as well as members of other electrical communities. Perhaps the most unusual of the new kinds of electrophorus that appeared was Gütle's. Instead of resin, his small electrophorus employed a cat's pelt as the insulator.[104] Presumably this bizarre technology could have been scaled up if the researcher had access to the pelt of a lion or tiger. Another German, Joseph Webers, wrote an entire book showing how the electrophorus could substitute for an electrical machine in providing power for a large series of experiments.[105] The electrophorus was also the starting point for inventions that would have a profound effect on twentieth-century technologies (see chapter 11). Needless to say, instrument makers found the electrophorus to be a lucrative product and offered it in many sizes.

Some electrophorus fanciers built enormous examples. Believing that an instrument of generous dimensions could clarify previously obscure effects, the electrophysicist Georg Christoph Lichtenberg assembled an electrophorus whose disk reached a diameter of 80 inches.[106] Around 1781, an electrophorus was built in St. Petersburg, reportedly by order of the Russian empress. Made from 80 lbs. of Spanish wax and 180 lbs. of pitch, it measured 9 feet long and 4.5 feet wide. The empress's electrophorus was doubtless able to induce impressively large charges.[107]

Surprisingly, even in royalty-sized versions, the electrophorus yielded only familiar phenomena; this was symptomatic of the state of electrophysics in the last decades of the eighteenth century. Despite the efforts of many able investigators, electrophysics had apparently become a sucked orange, the invigorating juices of new effects long gone.[108] But, as we shall see shortly, there was no lack of attempts to build technologies that might produce new physics.

—|||—

In contrast to the surprising electrophorus, which rapidly found its way from electrophysics into other electrical communities, many specialized accessories of the late eighteenth century attracted little attention. The torsion balance, though not widely adopted, is of interest here because of its purported significance in the history of electrophysics. In the nineteenth century its invention would be enshrined, especially by French physicists, as the defining moment in the emergence of modern electrical science. However, in the eighteenth century the torsion balance was mostly ignored—even by electrophysicists—for its use merely confirmed a principle almost everyone had already accepted.

The torsion balance was invented by Charles Augustin Coulomb (1736–1806).[109] A Frenchman trained as an engineer, Coulomb served for twenty years in seven different posts on military projects, including a long stint in Martinique. In the context of engineering activities, Coulomb nurtured a scientific interest in friction and other mechanical topics; soon he was publishing ideas and experiments. Coulomb's prize-winning work on friction earned him in 1781 election to the Royal Academy of Sciences, and this enabled him to take up permanent residence in Paris. Although still actively consulting on engineering issues of interest to the French state, Coulomb's scientific studies intensified as he turned to topics in magnetism and electricity. These investigations bore fruit, and from 1785 to 1791 he read to the Royal Academy a series of seven memoirs. In these works he established himself as an uncompromising Newtonian: not concerned with the causes of action at a distance—either magnetic or electric—Coulomb devised the torsion balance merely to quantify the effects of these forces.

Electrophysicists had long suspected that, like gravity, electrical attraction and repulsion followed an inverse-square law.[110] In 1779, this belief was given quantitative support by Charles Stanhope (Lord Mahon). Remarkably, Stanhope was admitted as a fellow of the Royal Society at age nineteen, not because he was a precocious scientist but because his father, an earl and also a fellow, had used his influence. Many members of the European nobility owned electrical things, but Charles Stanhope was among the few who actually used them to carry out original research. Stanhope's experiments on quantifying electrical force involved no new technologies; he required only a pith-ball electrometer, a brass conductor 40 inches long, and a large prime conductor.[111] Although the apparatus was simple, Stanhope's geometrical arguments in support of the inverse-square law—based on changes in a charge's polarity with distance from the two conductors—were not.[112] Clearly, there was still room for someone to quantify electrical forces directly using a sensitive, special-purpose instrument, and that is precisely what Coulomb did.

The path to the balance began with Coulomb's invention of an ultrasensitive compass to measure daily changes in the earth's magnetic field. Instead of resting the needle on a pivot, which introduced frictional forces, he suspended the needle from a silk thread. As the needle turned, it imposed on the thread only the most negligible twisting or torsional force. In further experiments, he found that the needle could also be suspended by a fine piano wire. To protect the needle from air currents, Coulomb enclosed the compass in a glass cylinder.

When measuring electrical force, Coulomb modified the torsional compass by substituting for the needle a pith ball on a counterbalanced horizontal stalk (Fig. 7).[113] A knob attached to the top of the wire on the outside of the cylinder turned the stalk. Another pith ball, on a rod, received a charge outside the device and, inserted in the plane of the stalk's rotation, shared its charge with the first pith ball. Now having identical charges, both balls separated. Because Coulomb could change the distance between the pith balls by adjusting the stalk, he was able to measure variation in the re-

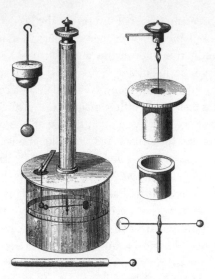

Figure 7.   Coulomb's torsion balance.
(Adapted from Coulomb 1785, pl. 13.
Courtesy of the Bakken Library,
Minneapolis.)

pulsive force. Controlling for the effects of torsion on the wire, he found
that the force varied in inverse proportion to the square of the distance between the balls. This ingenious technology had at last allowed electrophysicists to quantify a common belief.

Coulomb's technical triumph was a fitting conclusion to decades of conjecture and imperfect measurements. But Coulomb also stood, with others,
at the beginning of a thoroughly Newtonian electrophysics, for he invoked
no effluvia, no Cartesian vortices, not even electrical atmospheres to explain
electrical action at a distance. Electrical force was simply a force, like gravity, whose actions could be quantified even if its cause remained unknown.
In the nineteenth century, with the emergence of the new mathematical
electrophysics in France, Coulomb would come to occupy a special place in
the roster of distinguished ancestors.[114]

Although a few instrument makers included a torsion balance in their
catalogs, this expensive device was not a general-purpose electrometer. For
the latter, electrical experimenters had many choices, most of which were
simpler, cheaper, and more robust than Coulomb's invention.[115] These
ordinary electrometers also "measured" the force of repulsion. Usually, a
metal ball mounted on a pivot could swing away from a similarly charged,

fixed ball; the angle subtended by the moving ball, read from a built-in scale, indicated the charge's magnitude. As we shall see in later chapters, specialized electrometers were also invented by members of other communities and marketed by instrument makers.

⊣║⊢

The Netherlands in the seventeenth century had been a center of world trade, the arts, and tulip culture. Several of its scientists, including Christian Huygens and Anton van Leeuwenhoek, achieved world renown. And in the early eighteenth century, the Netherlands produced a number of distinguished natural philosophers, including the important Newtonians Willem Jacob 's Gravesande and Petrus van Musschenbroek. But, with the rise of other centers of trade, high culture, and the sciences in the late eighteenth century, the Netherlands risked becoming a backwater.[116] Although there was no shortage of local scientific societies or of natural philosophers, world-class discoveries were uncommon. One concerned citizen and natural philosopher, Martinus van Marum, hoped to reestablish the Netherlands as a hearth of scientific innovation; his vehicle would be an extraordinary electrical machine.[117]

Unlike Nollet and Franklin, Martinus van Marum (1750–1837) was born to well-off, middle-class parents in Delft, a town famous for its pottery.[118] In fact, van Marum's father, though educated as an engineer, was for a time master potter of his own workshop. In 1764, the family moved to Groningen, where Martinus, who had already shown himself to be a stellar student, entered Groningen University. There he studied diverse subjects including medicine and natural philosophy; he developed a special love for botany under the tutelage of Professor Petrus Camper. For his contributions to plant physiology, van Marum received a doctorate on August 7, 1773; just two weeks later, he was awarded a second doctor's degree in medicine.

Not surprisingly, van Marum's first two publications were on plants, but his third, in 1777, was on the design of a new electrical machine. What had caused this dramatic change of interests? Camper retired in 1773, and van Marum had been promised his professorship. As it turned out, however, the position went to someone less qualified but perhaps better connected. Embittered, van Marum decided to forge his reputation in a new field, electricity, which seemed to offer endless opportunities for someone so industrious and ambitious.

In the meantime, van Marum had to make a living, which he did by practicing medicine in Haarlem, a center of intellectual activity and home to the Hollandsche Maatschappij der Wetenschappen, a scientific society founded in 1752. A year after induction into the society, van Marum in 1777 was appointed director of the society's scientific collections. To house the ever-growing assemblages of specimens, the society bought a large house in which, incidentally, van Marum was able to live rent-free. Soon he married Joanna Bosch, whose ample inheritance freed the physician from all concerns with mundane financial matters.

The society's house had been bought with money loaned, at 2.5 percent interest, by the very wealthy Pieter Teyler van der Hulst. When the public-spirited Teyler died in 1778, his will directed the formation of a foundation, which in turn funded the establishment of several societies, including Teyler's Tweede Genootschap, dedicated to the study of natural philosophy, poetry, history, and other subjects. In 1784 the Teyler Foundation set up a museum and library, and van Marum was appointed director.

The Teyler Museum was housed in a new structure built to accommodate the display and demonstration of scientific instruments.[119] With the ample resources and facilities now at his disposal, van Marum decided that his first project would be to build apparatus that could resolve, finally, the interminable disputes about the nature of electricity. Surely success in this endeavor would bring fame to the Netherlands, to Haarlem and the Teyler Museum, and to van Marum—doubtless still smarting from his rebuff by Groningen University. No ordinary apparatus would do, of course, for clarifying the nature of electricity would require nothing less than the world's largest electrical machine. Like other electrophysicists, he believed that "if one could acquire a much greater electrical force than hitherto in use, it could lead to new discoveries."[120]

The Netherlands at that time hosted one of the ablest instrument makers in all of Europe, John Cuthbertson, an Englishman who in 1768 had set up shop in Amsterdam.[121] Already a manufacturer of plate electrical machines, Cuthbertson was hired in March 1783 to build the Teyler Museum's monster machine.[122] Under van Marum's close supervision, Cuthbertson assembled a generator that even today—it remains on display at the Teyler Museum—elicits gasps of wonder with its two glass plates 65 inches in diameter (Fig. 8). Before tackling the Teyler behemoth, Cuthbertson had constructed machines with 33-inch plates. In fact van Marum had purchased one of these generators for the Teyler Museum in 1781. But as Cuthbertson attempted to scale up this design, he encountered unexpected mechanical and electrical problems. After many experiments and much redesign

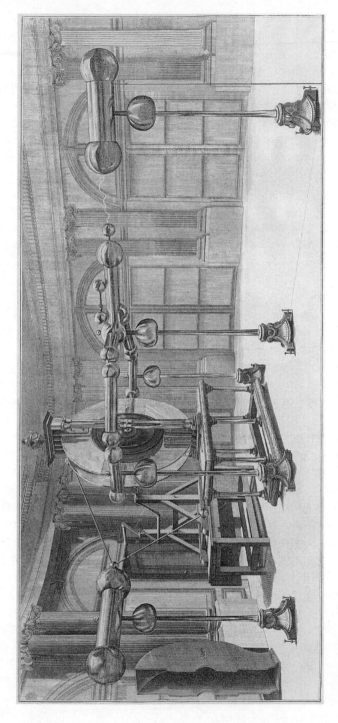

Figure 8. The Teyler Museum's giant electrical machine. (Adapted from Dibner 1957, 45. Courtesy of the Burndy Library, Dibner Institute, MIT.)

in his Amsterdam shop, Cuthbertson completed the machine, nearly ten months late; he delivered it to the Teyler Museum the day before Christmas 1784. At 3,250 guilders, this machine was probably the most expensive scientific apparatus built up to that time.

Not only were the glass plates the largest that could be cast at that time, but the five massive brass conductors—supported on glass pillars almost 5 feet tall and tipped by enormous knobs—dwarfed people standing in their shadows. To give motion to the machine, van Marum employed two assistants to turn the crank; during lengthy experiments, however, he put four men to work. A fitting counterpoint to the image of men wrestling machine was the ornamental carving on the wooden elements. In every respect, electrically and visually, the Teyler machine had been designed to impress.

Charge was generated by eight cushions, capable of precise adjustment, pressing against the plates. And the charge they produced was formidable, with sparks between the conductor knobs reaching a length of 2 feet and the thickness of a pen's quill. It is estimated that a spark this long is equivalent to between 330,000 and 500,000 volts. The charge was so strong that it could be registered on a sensitive electrometer dozens of feet away. To vastly augment the machine's usable power, van Marum could employ several batteries, one of which contained 135 large jars.

Although van Marum expected that the Teyler machine would help resolve theoretical disputes in electrophysics, it did not. The character of the machine's massive sparks supported the one- or two-electricity theories equally well.[123] And prevailing theories were safe because the experiments yielded no new effects to challenge them. Although failing to furnish important findings in electrophysics, van Marum's machine did see significant service in other electrical communities.

"Big science" implies big, or at least grotesquely expensive, apparatus; on this criterion alone, the Teyler machine qualifies. But big science also means collaborative science, as investigators—perhaps from different communities—pool their diverse talents and experimental designs to take advantage of an apparatus's unique performance characteristics. In this respect as well, the Teyler machine—that is, van Marum's management of it—was an early example of big science.[124] First of all, van Marum had local collaborators. Using the enormous power that the machine and its batteries provided, Adriaan Paets van Troostwyk, a merchant with scientific interests, and Johann R. Deimann, a medical doctor, carried out with van Marum—and his unacknowledged assistants—many experiments in biology and chemistry (see chapter 10). Second, as caretaker of the Teyler machine, van Marum issued an international call for suggestions about new experiments. Alessandro Volta, among others, responded with ideas, and these experiments were dutifully performed.

Sometimes foreign scientists actually visited Haarlem to participate in the research. One visiting scholar, Marsilio Landriani, arrived in Haarlem at a most inopportune time, when the cushions needed cleaning. Occupied with other business, Cuthbertson could not come from Amsterdam to do the job, and so Landriani, expectations of great new science unfulfilled, returned to Italy.[125] In order to prevent similar embarrassments, van Marum redesigned the rubbers so that the cushions would be easier to maintain. A somewhat sensitive Cuthbertson took offense at this tinkering, and the fruitful working relationship between him and van Marum came to an unpleasant end.[126]

Despite occasional disappointments, the Teyler machine under van Marum's management served several electrical communities, with members from many nations, for more than two decades.[127] Countless international visitors put Haarlem on the scientific map. The electrochemical experiments in particular brought the Netherlands, and the Teyler Museum, to the forefront of Lavoisier's "new chemistry." Although the monster machine hammered out little new physics, it did at last deliver the scientific acclaim that van Marum so desperately desired. Perhaps van Marum's most memorable accolade came when the aged Benjamin Franklin, whom he met in Paris in 1785, gave his electrical researches high praise.[128]

-|ı|ⵏ-

The lives in science of Franklin and Nollet, in particular, give us a glimpse into the profound social changes taking place during the middle of the eighteenth century. Both men began life in very modest circumstances yet managed to enter the scientific societies' elite. And members of the scientific elite were among the elite of the Enlightenment, joining other luminaries in literature, fine arts, and the political realm. Although the meteoric careers of Franklin and Nollet are of storybook proportions, many other electrophysicists—and scientists of every kind—rose from humble beginnings. Clearly, scientific accomplishment was an important avenue for upward social mobility in the eighteenth century.

It is true that few women and certainly no slaves achieved recognition for scientific research in the Enlightenment, but the doors to achievement were open surprisingly wide to free men—rich and poor, Catholic and Protestant, peasant and burgher, physician and minister—of most nationalities across Europe.[129] The major prerequisites for conducting scientific research were, of course, curiosity and literacy, but even in the eighteenth century there were many paths to literacy.[130] Largely oblivious to the distinctions that mattered so much in other institutions, scientific societies set an example of tolerance that would eventually become, in America and Europe, the political and moral ideology of equality.[131]

4. Going Public

With effects so startling and unexpected, the use of electrical technology did not for long remain the exclusive preserve of physicists in the laboratory. During the 1740s a sizable and enthusiastic group of disseminators quickly picked up electrical technology and adapted it, giving lectures on electricity and displaying its myriad effects in entertaining demonstrations.[1] A heterogeneous community, disseminators included college teachers, instrument makers, itinerant lecturers, tutors for wealthy families, and members of the clergy. Despite their social and occupational diversity, disseminators all employed equipment that could illustrate electrical effects and principles in a dramatic and captivating fashion.

Among the audiences of disseminators were men and women of varied backgrounds who, while being entertained, learned about electrical phenomena and saw for themselves how educated people could reveal the subtleties of the deity's plan. And thus they absorbed important tenets of Enlightenment elite ideology. At the same time, the bewildering performances of the electrical items themselves, which created curious sights, sounds, and smells—and sometimes instilled pain in participants—called to onlookers' vernacular beliefs from magic and religion.[2] Although eighteenth-century science had explicitly expunged occult and supernatural agents from *proximate* explanations of phenomena, they still resonated in the darkened and liminal setting of electrical displays, a wordless but potent counterpoint to the lecturer's learned banter.

Significantly, a few attendees of electrical lectures pondered whether, and how, electricity might be used in their own activities; sometimes this led to the transfer of electrical technology and the formation of new communities.

—|ı|⊢

Some of the most spectacular demonstrations required only the simplest technologies. By intimately involving people in their displays, electrical disseminators were able to entrance their audiences. Not surprisingly, Gray's demonstration with a suspended boy was copied often and performed with numerous variations. In William Watson's version, the charge was passed through hands from the boy to a young girl; standing on an insulator, she in turn attracted lightweight particles with her other hand.[3]

A charming demonstration of this sort was the "electric kiss." A woman, standing on a resin cake or insulated stool, was connected by an unseen wire to a source of charge, such as a Leyden jar or prime conductor. In its most refined form, an unsuspecting member of the audience was dared to kiss the woman if he could. As his lips approached hers, a spark jumped the gap between them; the painful jolt startled the would-be busser and caused him suddenly to withdraw. Benjamin Franklin found the electrical kiss entertaining and even proposed a way to intensify the shock.[4]

People figured important in another demonstration that merely illustrated the human body's ability to conduct electricity. The lecturer recruited several people and directed them to hold hands in a line or semicircle. When he connected the free hands at either end to a sizable Leyden jar, the participants jumped in unison. This demonstration was sometimes conducted on a large scale before an august audience. Louis-Guillaume Le Monnier, for example, performed the stunt with 140 guards before the French king.[5] Not to be outdone, the resourceful Jean-Antoine Nollet rounded up and shocked 180 gendarmes, whose joint gyrations the king also witnessed.[6]

Occasionally, demonstrations incorporating people made religious references.[7] A favorite was the "beatification," preferably done in the dark. The chosen one was made to stand on an insulator, holding a wire attached to a prime conductor. Thus electrified, the person's body would issue "fire" from ears, fingertips, and hair.[8] In viewing this display, perhaps some attendees were inspired to conclude that the technologies of science gave select humans the godlike gift of bestowing bliss—at least for a few moments.

Other living beings were sometimes used in demonstrations, but they did not live happily or for long. Typically a bird was electrocuted; even Benjamin Franklin killed animals, including a hen.[9] That people could exert their will over defenseless creatures, even electrocute them, resonated with the common religious conviction, made explicit in Psalm VIII, that God had put all things on earth for human use.[10] Audience members could also easily conclude that the electric force had a dark side that, in skilled hands, could be managed safely.

Little more than a source of charge and a few volunteers sufficed for some demonstrations, but most employed an impressive array of display devices. And, to heighten the impact of electrical demonstrations, disseminators invented many hundreds of accessories whose use underscored the lecturer's seeming wizardry. In this chapter I call attention to the most common, significant, or memorable inventions. Let us first examine the use of demonstration devices in college lectures.

–||⊢

With modest beginnings in the Renaissance, the teaching of natural philosophy in universities and colleges in subsequent centuries became increasingly technology-intensive. The growing acceptance of the experimental method not only called into question the received wisdom of Aristotle and other savants of antiquity, but it altered the interactions between teacher and students. No longer could an instructor merely repeat and discuss traditionally revered propositions about the workings of the natural world; he had to illustrate knowledge claims with concrete demonstrations. By the early eighteenth century, the transfer of scientific "knowledge" took place in instructional activities requiring a plethora of demonstration devices.

Demonstrations emphatically showed that experimental laws were manifest only in particular people-technology interactions. Attraction and repulsion, for example, were not just abstractions but phenomena that could be produced at will with appropriate apparatus. Learning science had become more than absorbing and repeating the propositions inscribed in texts: to understand them students had to witness, and sometimes take part in, the manipulation of artifacts. The ability to repeat experiments served to certify the competence of those who would become natural philosophers—a qualification far more important than a pedigree or even a college degree. Texts now took on the important new functions of guiding the conduct and interpretation of demonstrations and of fostering demand for display technologies.

Under the impetus of the Newtonians, especially, natural-science knowledge became embodied in carefully choreographed, technology-rich, demonstration activities.[11] Thus, astronomy lessons incorporated telescopes and orreries (apparatus showing the relative positions and motions of bodies in the solar system); biology had its microscopes, dissection tools, and smelly specimens; mechanics employed inclined planes, air pumps, and pendulums; and electricity required machines, Leyden jars, and scores of accessories. By the middle decades of the eighteenth century, instrument makers had commercialized demonstration technologies numbering in the hundreds.[12]

Not all professors had equal access to state-of-the-art demonstration devices. At the Protestant colleges, for instance, professors were expected to buy their own equipment. The privileged few like Petrus van Musschenbroek could build up a treasure trove of curious artifacts, but professors without independent means struggled to acquire a bare minimum of teaching technologies. In a few cases, wealthy pretenders to a chair of experimental physics beat out better-qualified rivals by bribing the university with a large equipment collection. Sometimes professors sold or willed their collections to the college. The situation was better at many Catholic colleges, as professors' petitions to the church for teaching technologies were often granted.[13]

Happily, basic equipment for demonstrating electrical effects was often less expensive than that needed in other branches of physics. The most costly items were the electrical machine and battery of Leyden jars; the majority of accessories were relatively cheap. Moreover, because they required no optics, precision machining, or exotic materials, the accessories—even large batteries—could be easily cobbled together by a handy and industrious instructor. Despite the barriers that professors often faced in acquiring equipment, many of modest means were able to assemble the minimum needed to give electrical lectures.

Teachers looking to lecture about electricity in their general physics courses could, by the 1750s, obtain guidance from many written sources, including standard physics textbooks and electrical experiment books. By describing experiments, both book genres fostered the teachers' interest in obtaining appropriate equipment.

Many physics texts originated as records of the lectures of natural philosophers, such as Musschenbroek and Nollet, who had offered courses in colleges and to a paying public. Although codifying basic principles, these texts were mainly organized around technology-rich, and sometimes spectacular, demonstrations of effects. Surprisingly, electricity made its way slowly into most physics textbooks. The important two-volume work by 's Gravesande published in 1731 allotted a mere eleven pages to electricity— about 2 percent of the total, and it contained no illustrations of electrical technology.[14] In Benjamin Martin's book of 1743, *A Course of Lectures in Natural and Experimental Philosophy*, electricity was treated in two pages. Musschenbroek's magnum opus was slightly more expansive on electrical topics: in the French edition of 1739, sixteen pages discussed electricity, about 1.8 percent of the book; the 1751 edition differed little.[15] However, the posthumous edition of Musschenbroek's text, published in 1769, at last acknowledged the growth of new effects and principles by including ninety-two pages on electricity—about 6.3 percent of the total.[16] Likewise, William Nicholson's text of 1782, *An Introduction to Natural Philosophy*, devoted seven chapters, seventy-three pages in all, to electricity—or 8.9 percent of the entire book.[17] Because electricity in the late eighteenth century at last comprised a significant part of physics, any teacher who relied on a major text would have incurred some obligation to lecture about and furnish demonstrations of electrical principles. Even instructors previously unfamiliar with electricity would have joined, out of necessity, the evergrowing community of disseminators. Not surprisingly, the number of demonstration devices offered by instrument makers continued to increase.

Electrical lectures were popular. At Göttingen University, for example, Professor Georg Christoph Lichtenberg typically attracted one hundred listeners. Not only did his demonstrations hold the attention of the audience, but others outside the hall were often alerted, by peculiar sounds and smells, to the goings-on. Electrical explosions, for example, sometimes set all the dogs in the parish to barking and startled strangers. But his fellow burghers were tolerant, advising curious passersby that "It's only the professor blowing things up." [18]

There are hints that, by the end of the eighteenth century, electrical demonstrations had reached into schools below the college level. For example, Ladislaus Chernak, a professor of philosophy, reported to van Marum that he had conducted electrical demonstrations before a large audience at a grammar school.[19] As electricity was worked into the curricula of such schools, demand increased for the appropriate display technologies.

It is probably no accident that many physics texts appearing after 1768 had finally bulked up their electrical discussions. In 1767 Joseph Priestley's massive and influential work, the *History and Present State of Electricity*, first saw print. Priestley's well-referenced *History* was an invaluable resource to electrical researchers and writers. One modern scholar, Charles Bazerman, believes that Priestley's book became the paradigm, in all of science, for the discipline-specific textbook.[20]

To put it mildly, Joseph Priestley (1733–1804) led an interesting life.[21] Born in Yorkshire, England, Priestley was the son of Mary Swift and Jonas Priestley, a cloth dresser. He was actually raised first by maternal grandparents and, later, by an aunt in comfortable circumstances. Like many other quick learners, Priestley was encouraged to study for the ministry. He attended parish schools, received private tutoring, and acquired on his own many languages, from Latin to Arabic. Enrolled at the dissenting academy at Daventry, he read widely and began questioning Anglican orthodoxy. After several unsuccessful preaching assignments, Priestley found work as a teacher in another dissenting academy at Warrington, where he taught almost everything, including law and anatomy. He did not, however, give lectures on experimental science or theology, the two subjects about which he was most passionate.

Priestley's marriage to Mary Wilkinson in 1762 also allied him to a family of great ironmasters. Five years later his *History* appeared and Priestley became the minister of a major Presbyterian church at Leeds. Not many years later, the earl of Shelburne, a member of parliament and a friend of Franklin, employed Priestley as his resident scholar and scientist. Wintering in London and repairing to the country in summer with his patron and family, Priestley carried out his most significant scientific work. In 1780, however, Priestley left the earl and moved his family to the major industrial center of Birmingham. There he became a member of the Lunar Society, which brought together learned men of many occupations (who met on the eve of a full moon when travel home on unlighted roads was easiest);[22] he was also appointed preacher at the New Meeting House, a very liberal congregation.

Like many Enlightenment figures, Priestley carried out research in several fields. Perhaps because he rubbed elbows with industrialists in his wife's family as well as in the Birmingham Lunar Society, Priestley nurtured an interest in applying science, especially chemistry, to practical affairs. Employing a systematic research program that involved heating and combining various substances, he succeeded in liberating and identifying many new gases (these were called "airs"), including oxygen, ammonia, nitrous oxide, and sulfur dioxide—all of which eventually found industrial applications. He also invented the first method for artificially carbonating beverages. One of Priestley's most important contributions was to meld electrical and chemical technologies in the study of gases (see chapter 10).

A member of several electrical communities, Priestley was a Franklinian. More than that, Franklin—a quarter century his senior—had been a role model and mentor to Priestley as the latter began his scientific investigations.[23] Not only had Franklin encouraged Priestley to write the *History*, but the Philadelphian also read proofs of this book, which included the first account in the scientific literature of the kite experiment. Beyond keeping up a lively correspondence, Franklin and Priestley were kindred spirits and close friends. In 1771, for example, Franklin journeyed north from London and visited Priestley at Leeds.[24] Priestley's elections to the Royal Society and the French academy were also facilitated by the loyal Franklin, who would press, almost to excess, to secure favors for friends.

Perhaps in return, Priestley argued on the side of the colonists against unfair taxes levied by the British crown.

On the eve of the American revolution in 1775, a teary-eyed Franklin, his diplomacy having ended in failure, spent his last day in London visiting alone with Priestley, commiserating over the looming conflict.[25] Upon returning to Pennsylvania, Franklin learned that, during his six-week voyage home, bloody battles had been fought already between British soldiers and colonists at Lexington and Concord. Nearly seventy years old then, Franklin was immediately drafted by the Pennsylvania Assembly as a deputy to the Second Continental Congress. In letters to Priestley, Franklin furnished some details on the Congress's deliberations and on other events surrounding relations between England and the colonies.

Throughout his adult life Priestley published dozens of theological works, many of them quite controversial. He was an especially ardent supporter of the right to dissent from the Anglican church. Ironically, this deeply devout man, who in public never questioned the existence of a Christian God or an eternal soul, was accused of being an atheist—a common epithet used to tar dissenters. Notwithstanding his affirmation of religious conviction, Priestley's endorsement of the colonists' complaints, early support for the French Revolution, and continued opposition to religious persecution made him a marked man. In 1791 a mob supporting the established church and Crown burned the New Meeting House as well as Priestley's home and laboratory.[26] The Priestley family fled to London but, after continued harassment, left a few years later for the United States. Although Franklin had already passed away, the Priestleys settled in Pennsylvania, where many dissident English families had been welcomed. When Thomas Jefferson assumed the presidency in 1801, the lawyer-politician-scientist befriended the ill and aged Priestley. This friendship was short-lived, however, for Priestley's voice soon fell silent.

—|||—

In the lengthy list of Priestley's books is *A Familiar Introduction to the Study of Electricity*. Published the year after his *History* appeared, this book went through at least four editions. It typified a new genre of science literature—the electrical experiment book—that had arisen in the mid-1740s. Experiment books gave aspiring disseminators (and others) instructions on the conduct of experiments far more detailed than those in physics texts.

Typically an experiment book had two parts. Basic electrical principles comprised the first part, with experimental laws generously represented. The second part presented a description of experiments and usually contained elaborate drawings of equipment. Even in the earliest books, the list of experiments numbered several dozen.[27] Written by—and for—members of many electrical communities, experiment books remained an important genre of literature throughout the eighteenth century, eventually incorporating some two hundred experiments.[28] Because electricity was not yet a quantitative science, even the theoretical sections of experiment books were accessible to interested lay people.

Drawing on experiment books, and building or buying the necessary technologies, disseminators in the tradition of the Hauksbees created instructive and captivating lectures. Technologies employed in the early 1740s, when the disseminator community first began to grow large, merely enabled repetition of experiments devised by Hauksbee, Gray, and du Fay. These items included, for example, glass tube, electrical machine, pieces of metal and sulfur, sealing wax, silk suspenders, and resin cakes. Already in 1746 such technologies were available in England from the instrument maker George Adams.[29] After the appearance of the first experiment books, which took note of the Leyden jar and the antics of the German professors, disseminators created an even wider range of technologies; in due course, many of these were also commercialized.[30]

With the exception of a few people who employed an electrophorus, nearly all disseminators used the core technologies of electrical machine, prime conductor, and Leyden jar as the immediate source of charge.[31] By the late 1760s lecturers could choose from a wide assortment of electrical machines that differed not only in size but also in kind—globe, cylinder, or disk. Surprisingly, disseminators did not always adopt the "latest" machine technologies; even in the later decades of the century, some of them were still using hand-rubbed globe machines.[32] If the relevant performance characteristics were ease of use (no hands), portability, and overall power, such choices make no sense.[33] But lectures were after all performances, and the manner of making charge affected spectators' interpretations: electrical technologies held a symbolic charge.[34]

A lecturer's expectations about symbolic interpretations doubtless influenced his choice of generating technology. For example, newer-style electrical machines would silently tell a savvy audience that the demonstrator was using the latest—that is, "modern"—equipment. In an era when the ideology of progress thoroughly permeated elite culture, the message of modernity might resonate with such audiences. Also, using the latest equipment could help disseminators trying to trump competitors working the same territories. And yet, performing before a relatively untutored audience, the lecturer might wish to trade on his apparent control of occult

Figure 9.   *Left*, dancing figures and *right*, electric carillon. (Adapted from Adams 1799, pl. 2, figs. 13 [left] and 18 [right]. Courtesy of the Bakken Library, Minneapolis.)

powers. Although the use of any kind of generator encouraged this inference (accompanying scientific explanations notwithstanding), hand-rubbed machines in particular allowed spectators to conclude that the lecturer himself possessed supernatural powers as a benign sorcerer or wizard. Perhaps some disseminators used different machines before different audiences.

Apart from the electrical machine itself, demonstration devices often drew on nonelectrical technologies or activities that were probably familiar to spectators. A favorite was the fountain, which issued a spray of electrified water that glowed in the dark. Another water toy, modeled after humorous fountains, consisted of a male child urinating. Waterwheels and windmills were probably the inspiration for a motor that was driven by the electric wind issuing from a point. A motor also powered the electric horse race, which mimicked an existing mechanical toy: paper horses ran on a miniature track, and in some models there were trees and a gun-toting starter. Also imbued with motion were the crowd-pleasing seesaw, dancing figures (Fig. 9, left), and tightrope walker. In the last device, attraction from above moved a paper figure on a wire suspended between the arms of a stand. Another staple in lectures was the electric carillon with small brass bells and clappers, moved by repulsion (Fig. 9, right). A lecturer could also use a doll-like "small head with hair" to illustrate a coiffeur's worst nightmare.[35]

Once experiments showed that a spark could ignite gunpowder, disseminators took the obvious next step: they invented electric guns and miniature cannons (Fig. 10). In Jakob Langenbucher's electric cannon, charge was brought to the "fuse" by a toy soldier in full dress uniform whose outstretched right arm held a ball-tipped conductor. After all that,

Figure 10.   Electric cannon and Leyden jar.
(Adapted from Cuthbertson 1807, pl. 7, unnum-
bered fig. Courtesy of the Bakken Library,
Minneapolis.)

however, the cannon fired only a cork.[36] In later versions of electrical fire-arms, hydrogen gas on occasion took the place of gunpowder as the explosive.[37] Some disseminators placed conductors in the ends of an egg; the discharge of a Leyden jar would cause the egg first to glow briefly, then explode. Filling an ordinary wine glass with wine (or water), introducing two conductors separated by a gap, and connecting them to a large Leyden jar or battery would also explode the glass.[38]

Accessories that gave off a more long-lasting light were incorporated into most lectures. When excited by a nearby charge, virtually any evacuated glass vessel or tube gave off a glow; common configurations included straight and zigzag tubes as well as assorted jars and flasks.[39] Discharges *within* evacuated vessels and tubes produced light too, and many devices with one or two internal conductors were made for displaying this effect (Fig. 11, left and right); they were occasionally accompanied by a small vacuum pump.[40] Compared to the monotony of candles and oil lamps, electrical lights were colorful and intriguing.

A number of charming devices to enlighten an audience rested on a simple effect: charge readily departs from—and thus emblazons—the edges of a metal foil. By pasting pieces of foil on glass, a lecturer could make endless dazzling displays. One popular device appeared at first glance to be a Leyden jar, but instead of a continuous foil coating on the exterior, the "spotted jar" boasted a pattern of small foil circles or diamonds whose edges were surrounded by "electrical fire." Another common variety was the "spiral tube," displaying foil diamonds snaking around a glass tube (Fig. 11, center, below). A more intricate accessory was the "electrical shooter and mark." The framed picture showed a man with a cannon or musket aiming at a target—sometimes a church steeple; the trajectory was

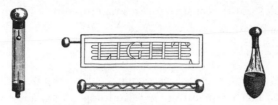

Figure 11.   Display devices that gave off light.
(Adapted from Adams 1799, pl. 3, figs. 49 [left and
right], 32 [center, above], 31 [center, below]. Courtesy of the Bakken Library, Minneapolis.)

illuminated by tiny pieces of foil.[41] Metal foils could also be affixed to
Franklin-style motors (see below), creating a visual extravaganza. For example, the "luminous palace" had eight glass columns, arranged in a circle
and covered with tinfoil diamonds. As the rotor, a raised metal rod tipped
with copper balls, passed a column—also topped with a copper ball—its diamonds became outlined in sparks.[42]

Although modern electric signs did not appear until the end of the nineteenth century, they were foreshadowed in eighteenth-century "luminous
words." Letters of a word were carefully cut from metal foil and pasted on
a piece of plate glass (Fig. 11, center, above). Connected together by small
wires and exposed to a charge, the letters' edges lit up, making the word—
sometimes "love" or the lecturer's name—clearly readable. Numerous
foil-on-glass devices, including luminous words, were available from instrument makers.[43]

Another technology for producing light made use of phosphorescent
substances. Powdered Bologna stone (barium sulfate) was placed into small
glass tubes that had been bent into letters or other shapes. When the ends
of a tube were connected to a sizable charge, the powder inside glowed an
eerie green.[44]

Light-producing displays were ideal for impressing large audiences in
darkened halls, but some accessories were better suited to displays in small-
group tutorials because their effects were not easily seen at a distance.
Charles Rabiqueau, for example, devised a number of displays using liquids, such as wine and water, in standard chemical glassware; these would
have been startling but only if viewed up close.[45]

Because demonstration equipment often had to be moved from place to
place, perhaps carried from home or laboratory to lecture hall, most disseminators eschewed large electrical machines. Some instrument makers

sold "portable" outfits, consisting of a small machine and compact display technologies, in wooden carrying cases.[46]

—|||—

Portable equipment appealed particularly to itinerant lecturers, whose performances reached mainly middle and upper-class audiences.[47] A few established lecturers simply added electricity to an existing physics repertoire, but many others, responding to a seemingly insatiable interest on the part of the public, specialized in electrical shows. They rented halls or homes, attracted audiences through handbills and newspaper advertisements, and skillfully displayed electricity's many wonders.

An electrical demonstration carried out before an audience was a multi-act play in which lecturer, assistants, and equipment all performed their parts. To bring it off, the lecturer had to be a capable producer, director, and choreographer; he also had to have a flair for the dramatic. Eighteenth-century audiences expected to encounter on the scientific stage remarkable—perhaps magical—phenomena. After all, scientific demonstrations competed with countless other amusements—not all of them enlightening—including plays in theaters, fairs and circuses, natural history museums, and freak shows.[48] Those who wished to, could even watch hangings or visit insane asylums to view inmates being tortured. Easily jaded spectators who paid to be entertained by electrical science could not be disappointed.

Among the varied offerings of eighteenth-century amusements, electrical lectures most resembled natural magic, a long-standing tradition of fashioning and displaying elaborate mechanical things, including automatons.[49] Such machines were used not to educate spectators but to amaze them. The most famous of these inventions was Jacques de Vaucanson's duck, which ingested, digested, and excreted food pellets. A more infamous example was a large chess-playing machine that, alas, housed a human motor. Most natural magicians did not engage in such utter deception; rather, they employed varying combinations of springs, weights, hydraulics, and pneumatics to operate their intricate devices.

Both natural magicians and electrical lecturers used technologies to produce surprising and seemingly inexplicable effects. In addition, both employed sights and sounds to stimulate feelings of awe or wonder. And the technologies of both had an otherworldly appearance, unlike the ordinary objects of everyday life; they reinforced a lecturer's claim to esoteric knowledge, intimating that occult forces were at his command. Of course, the in-

ternal workings of natural-magic technologies were usually concealed, enhancing the mystery; in electrical lectures the technology was transparent,
its parts visible to spectators. Yet, while natural magicians jealously kept
their secrets, electrical lecturers attempted to educate, to explain how their
technologies produced specific effects. Notwithstanding the ostensible differences between natural magic and scientific demonstrations, the latter
nonetheless contributed to the belief, perhaps held by the less enlightened,
that manipulations of esoteric technology mediated between the natural
and the supernatural.[50]

What would a modern physicist make of the two-hundred-odd experiments, with their diverse accessories, in a late-eighteenth-century book on
electricity? Although he or she might marvel at the ingenuity of Enlightenment investigators, today's physicist would be puzzled by the huge number of experiments and technologies that redundantly displayed the same
few experimental laws. Several pieces of insulated wire demonstrate the
communication of charge as well as a ring of 140 people. Likewise, that
electricity ignites gunpowder can be shown with a firecracker; it did not require toy guns and cannon. And the sharpened end of a wire is as good as
luminous words to make sparks issue readily from points and other acute
surfaces. In these examples, the underlying science was simple, but the
demonstration technologies were often complex and intriguing, as appropriate for captivating and amusing an audience.

For purposes of demonstrating an effect, different technologies—
though encompassed by the same experimental law—were far from equivalent. The charging of a chicken and a person both illustrated that electricity
could be transferred into a living, conductive body. But for disseminators,
there was a world of difference, aesthetically and theatrically, between the
two effects. It is doubtful, for example, that anyone demonstrated the electric kiss with a chicken. In the eighteenth century, disseminators placed
stress on displaying effects that engaged audiences. In modern physics, pre-
sentations of electrostatic principles in lectures are relatively stark. The
modern way, of course, privileges theory and quantification; the eighteenth
century way privileged the entertaining display of phenomena.[51]

When illustrating scientific effects in a most dramatic fashion, the disseminator brought into the world complex and curious technologies. Spectators viewing these demonstrations occasionally envisioned how the technologies could be employed in different, perhaps more practical, activities.
Beyond simple scientific effects, then, demonstration devices embodied and
displayed emergent *technological* effects, such as electrical weapons and lu-

minous words, which sometimes inspired inventors and entrepreneurs, as we shall see in chapter 11.

—||├—

Along with accounts of the electric kiss, eighteenth-century experiment books indicated, in drawings, that it was acceptable for women to take part in demonstrations as a reservoir, communicator, or conductor of electricity. Describing early demonstrations of the German professors, *The Gentleman's Magazine* asked: "Could one believe that a lady's finger, that her whale-bone petticoat, should send forth flashes of true lightening?"[52] A drawing in Nollet's physics text shows both men and women serving as conveyors of charge. Nollet also depicted a woman generating charge on a globe with her own hands, but the more dangerous role—that of showing the power of the Leyden jar—devolved on a man.[53] In an early Italian work, Eusebio Sguario illustrated a woman—clearly the focal point of the drawing—seated on a board suspended from the ceiling (Fig. 12).[54] Next to her is a large electrical machine being cranked by a man. With the right hand she acquires charge from the machine, and with the left she passes it on to a second man.

Women were also prominent among the spectators at electrical lectures. Indeed, for the elite it was important to be seen at such a place of conspicuous enlightenment where women kindled "fires without poetical figure, or hyperbole."[55] Among the notables noted in early public lectures were a princess of Prussia and the archduchess of the Netherlands.[56] Perhaps France's most famous lecturer, Nollet explicitly encouraged women to enroll in his courses. And he was successful, attracting "the carriages of duchesses, peers, and lovely women."[57] By the same token, in tutorials carried out in homes of the very wealthy, women were among the participants. A drawing in one of Nollet's works on electricity shows a group, including women and children, seated around a large table observing his demonstrations.[58]

According to Benjamin Martin, an itinerant lecturer and instrument maker, it was quite appropriate for British women to learn natural philosophy, particularly since the acquisition of knowledge had *"now become a fashionable Thing."* He labeled as *"monstrous and stupid"* the objection that "Gentlemen will not like them so well for it."[59] As an instrument maker, Martin of course had an incentive to expand the market for his lectures and equipment, but in fact there was no rigid norm against educating women, outside colleges, in matters of science. A later book by Martin, which explicitly targeted both men and women of college age, was perhaps the most creative and charming introduction to physics ever written. Les-

Figure 12. Women sometimes participated in electrical demonstrations. (Adapted from Sguario 1746, frontispiece. Courtesy of the Bakken Library, Minneapolis.)

sons were incorporated into dialogues, including two on electricity.[60] In one illustration, a man, woman, and child are but props behind an accessorized electrical machine sitting on a table.[61] Because even women from wealthy families were barred from attending college in the eighteenth century, they had to resort for scientific education to public lectures, books, and home tutorials.[62]

—||—

Benjamin Franklin, from his earliest involvement with electricity, was a member of the disseminator community. During the months after receiving the glass tube from Peter Collinson, he "had little leisure for any thing else" but repeating experiments for the "crowds" of acquaintances and

friends who gathered at his home.[63] Not surprisingly, Franklin was a prolific inventor of display devices.

Perhaps his most enduring invention was an electric motor. That electricity could produce rotary motion was widely known before Franklin began his researches. During the 1740s, the first accessories that could actually be termed "electric motors" were incorporated into demonstrations.
These devices were often little more than a deformed versorium, merely
an S- or Z-shaped wire rotor with pointed ends, resting horizontally on an
insulated metal pivot. When connected to a Leyden jar's ball or a prime
conductor, the wire's points conveyed charge to nearby air molecules and
then were repelled from them, causing the wire to rotate like a pinwheel.
Instrument makers marketed many versions of this motor, sometimes with
four points.[64]

Motors were occasionally incorporated into more complex equipment,
including electric orreries for illustrating planetary motions and carillons.
In one version of the latter, eight brass bells of varying tones were placed
in a circle on a board; in the middle of the circle stood a small glass pillar supporting a rotor. The clapper, suspended from one arm of the rotor, glanced off the bells as the motor rotated, creating a pleasant sequence
of notes.[65]

These rotary devices perhaps furnished inspiration to Franklin, who invented an electric motor capable of producing considerably more power.[66]
The rotor was a thin wooden disk, about 1 foot in diameter, on whose circumference he arrayed thirty narrow glass strips, each radiating beyond
the disk's edge. On the protruding end of each glass strip Franklin attached
a metal thimble. The shaft, disposed vertically, was made in two parts, one
above, the other below the wooden disk. The lower part of the shaft was a
wooden dowel that could pivot easily on a metal point. The shaft's upper
portion consisted of a metal rod rotating freely in the hole of a brass plate
supported independently. To deliver electricity to the motor, Franklin employed two Leyden jars—one charged normally, the other of reversed polarity—with their balls placed near the wheel's circumference, close to the
thimbles. Set in motion with a small push, the rotor continued revolving
on its own.

Although producing an astonishing effect, the motor's scientific under-
pinnings were simple, based merely on the communication of charge and
attractions and repulsions, as follows. Each thimble, which carried the same
charge as the ball it last passed, was attracted as it approached the other ball
(of opposite charge). As the thimble neared the latter ball, its charges reversed; now being repelled by that ball, it continued the rotor's motion (be-

cause a thimble, with its much smaller surface area, could hold only a tiny charge, it was easily neutralized and overwhelmed as it approached the much larger ball). Many disseminators built and illustrated their own Franklin motors, and it was available from some instrument makers.[67]

Creating motion from an invisible fluid, all electric motors seemed magical, but Franklin's second motor—the "self-moving wheel"—was more magical than most because it rotated in splendid isolation, working without any connection to an external charge.[68] The secret to this motor's motion was simple: the rotor was itself a plate capacitor, furnishing the electricity that kept it moving. The rotor-capacitor was a glass disk, 17 inches in diameter, which had been gilded on both sides except for 2-inch rings adjacent to the edges. On the upper edge of the rotor Franklin placed a lead bullet, connected by wire to the gilding; partway around the rotor he installed a second bullet and wired it to the gilding on the disk's underside. Next, he mounted the rotor in an insulated frame that permitted it to turn freely and put the entire assembly on a table closely surrounded by a circle of twelve small glass pillars, each topped by a metal thimble. Precisely because Franklin understood how his first motor worked, he was able to turn it inside out!

With one charge, Franklin's second motor could spin for about half an hour, making twenty revolutions per minute. When he increased the number of bullet conductors, Franklin found that the motor spun faster but did not run for as long. We can only imagine the delight and sense of wonder that onlookers felt while gazing at this remarkable invention.

Present-day engineers and hobbyists still construct Franklin-type motors, for they are astonishingly easy to make. I did so, using the plastic lid of a cottage cheese container as the rotor; nails, pressed outward from its circumference, serve in place of thimbles. It pivots on a pointed spindle, resembling a giant thumbtack, the kind used by merchants to accumulate receipts by the cash register. Though not very powerful, my motor does reach several hundred revolutions per minute. To supply it with charge, I use a tiny, battery-powered van de Graaff generator—a twentieth-century design—purchased over the Internet.

The plate capacitor of gilded glass was also the basis of the "magic picture," a devilish device actually invented by Ebenezer Kinnersley, one of Franklin's collaborators. He began with a framed picture of the English king (George II), under glass.[69] After removing the picture, he gilded the glass on both sides, except for a 2-inch border, and connected the back side to the frame, which was also gilded. Next, he cut a 2-inch border from the perimeter of the picture. He glued the border to the back of the glass and the remainder of the picture to the front. After replacing the picture-capacitor in the frame and charging it, he put a loose gilt crown above the king's head and invited a naive spectator to remove it. With one hand on the frame, the spectator, approaching the crown with his other hand, "will receive a terrible blow, and fail in the attempt."[70] The magical picture was rather popular: it was copied often and instrument makers brought it to market.[71]

⊣║├

Although Franklin entertained visitors with demonstrations in his home, Ebenezer Kinnersley actually made a living for many years as an electrical lecturer. Kinnersley (1711–1778) was born in Gloucester, England, the son of William Kinnersley, a Baptist minister, and Sarah Turner.[72] The family immigrated to Pennsylvania when Ebenezer was but three, his father serving the Pennepack Baptist church, near Philadelphia. Home-schooled, the young Ebenezer doubtless received a heavy dose of religious education. After marriage to Sarah Duffield and a move to Philadelphia, around 1739, Ebenezer preached against the excessive emotionalism of the Great Awakening, a series of Protestant revivals that occurred during the midcentury. For Kinnersley, God was a "God of Order," not of confusion; he promoted a more "rational Religion" attuned to the growing spirit of enlightenment.[73] He wrote a number of religious tracts, eventually arguing for freedom of religion according to each individual's dictates. Despite a brief falling out with the Baptist church of Philadelphia, Kinnersley was ordained a minister in 1743 but—judged to be too much an individual and a radical—never received a congregation of his own.

Franklin and Kinnersley were neighbors, and the latter had published controversial letters on religious matters in Franklin's newspaper, the *Pennsylvania Gazette*. Their acquaintance led to collaboration on electrical experiments, some of which would form the basis of Kinnersley's lectures. By all accounts, Kinnersley was a gifted orator, a skill not unimportant in delivering science lectures. Some years later, through Franklin's influence, Kinnersley received an appointment as a "professor of the English Tongue and of Oratory" in the College of Pennsylvania, forerunner of the University of Pennsylvania. He is believed to have been the first professor of modern English anywhere, following a plan laid down by Franklin.[74]

According to his biographer, J. A. Leo Lemay, Kinnersley "was the greatest of the popular lecturers in colonial America," but others had preceded him to the podium, and even discoursed on electrical subjects.[75] William Claggett of Newport, Rhode Island, a maker of clocks, learned about the experiments of the German professors and became an adept himself. In late summer of 1747 he initiated lectures on electricity in Boston and other cities. Claggett's first advertisement, in the *Boston Evening Post*, offered to demonstrate electricity on demand to anyone who presented, at the house of Capt. John Williams, the requisite 10 shillings.[76] These performances were immensely successful; by one later account, he earned about £1,500 in a three-week period.[77]

Electrical lecturers turned up in other colonial cities, including New York, Philadelphia, and Charleston, South Carolina. In the last city, Samuel Dömjén, a native of Transylvania, set up shop in 1748 at Blythe's tavern. On Wednesday and Friday afternoons, from 3:00 to 5:00, he demonstrated the "surprising Effects" of electricity for 20 shillings, a far from insignificant sum. Like Kinnersley, Dömjén had learned about electricity from Benjamin Franklin. Apparently he learned it well, for he was able to earn his keep as an electrical lecturer, traveling throughout the colonies and then around the world.[78]

Perhaps buoyed by Dömjén's success, Franklin encouraged Kinnersley, then unemployed, to take an electrical show on the road. Kinnersley followed his friend's advice; clearly, pursuing the occupation of itinerant lecturer was preferable to penury. Drawing on his own experience in furnishing electrical entertainment, Franklin arranged a series of experiments into two lectures and composed the accompanying text. Employing equipment made by instrument makers, Kinnersley began his lecturing career in Annapolis, Maryland, in May 1749.[79] And during a span of twenty-five years he would lecture in every colonial capital city. On one occasion, he even went on a lengthy tour of the West Indies; these latter lectures netted him about £200.

Lectures on electricity were quite popular in the colonies. In the Boston area alone, eight people, including Kinnersley, performed in pre–Revolutionary War times, from 1747 to 1765. And newspaper advertisements for electrical lectures outnumbered those for other scientific demonstrations. The historian William Morse, who tracked down and studied these advertisements, noted that "these lectures . . . must have made a lively appeal to the imagination of the public in the days before the theater came to Boston."[80]

–||–

Regrettably, the text of only the first Kinnersley lecture survives. This 1752 version, entitled "A course of experiments on the newly discover'd electrical fire," has eleven major experiments.[81] Employing a rubbed glass tube, prime conductor, assorted accessories, and a few game spectators,

Kinnersley mainly illustrated attraction, repulsion, and communication. In one demonstration he made use of new technology, an "artificial Spider" invented by Franklin.[82] The spider consisted of a cork body and legs of silk thread. When suspended by a silk line from the prime conductor, the spider jumped back and forth between the knob of a charged Leyden jar and a piece of metal, acting like its living counterpart. A crowd pleaser, the artificial spider became a staple of electrical disseminators (and in later years was available from instrument makers).[83]

In other displays of attraction and repulsion, Kinnersley made use of two metal plates. The first was suspended horizontally by a chain from the prime conductor. Supported from below, the second plate was placed parallel to, but a few inches below, the first one. When a piece of gold leaf, pointed at opposite ends, was placed between the plates with points aimed up and down, it appeared to be levitated, floating in the space between the plates. In contrast, when sand replaced the gold leaf, the particles moved continuously up and down. These displays, with their simple metal plates, were copied often; dancing paper figures were sometimes substituted for the sand and gold leaf. Kinnersley literally ended his first lecture with a bang, saluting the audience "with a Discharge of [the] Cannon from an Electrical Battery by a Spark proceeding from a Person's Finger."[84]

Newspaper advertisements furnish clues to the contents of Kinnersley's second lecture.[85] Apparently, he saved the most sensational experiments—some requiring volunteers from the audience—for this performance. In a display that would have seemed at first incomprehensible, a few drops of electrified cold water were allowed to fall on someone's hand. With nothing but a finger of the other hand, this person kindled a flame (he or she had received the charge from the water and could pass it on to another object). Similarly, the spark "darting from a Lady's Eyes" set alcohol fumes afire.[86] And a Leyden jar's discharge put electricity to work melting metal, ringing bells, penetrating dozens of sheets of paper, and, of course, killing small critters.

The most novel technologies in the second lecture, models of buildings and ships, aimed to illustrate Franklin's conjecture that lightning damage could be prevented by mounting, on the roof of a structure, a pointed conductor for leading the electricity harmlessly to the earth (see chapter 9). When equipped with a lightning conductor, suitably grounded, these models were protected from the simulated lightning supplied by a Leyden jar's discharge. However, when the spark struck an unprotected model, it set off a small amount of gunpowder, causing the house or ship to come apart with some commotion. These displays enjoyed astounding popularity on both

sides of the Atlantic, and similar models were commercialized by many instrument makers.[87] The second lecture, like the first, closed on a loud military note, as eleven guns were set off simultaneously by a spark that had
traveled through 10 feet of water.

In later versions of his lectures, advertised in the early 1770s, Kinnersley incorporated new demonstrations and new accessories. For example,
he showed how a piece of tourmaline could be electrified by boiling it in
water. In an especially dramatic display, Kinnersley placed a piece of iron
wire under water and connected it to a huge battery of seventy Leyden jars;
the iron at first glowed red, then melted.[88] Clearly, successful lecturers, like
Kinnersley, kept abreast of scientific developments and periodically revised
their offerings. A brilliant marketer, he had doubtless learned that audiences would dwindle if the same demonstrations were repeated year after
year without variation. As a man of the cloth, he appreciated that ministers
did not give the same sermon week after week, and even Bach wrote new
organ music for many Sunday services. That the church itself, one of the
most conservative institutions in Western society, expected change and
novelty, suggests that the roots of a novelty-seeking public—prerequisite
for the flowering of modern consumerism—lie deep in the Western world.
In any event, presenting new material along with old crowd pleasers ensured Kinnersley and others a robust audience.

Kinnersley's most dramatic electrical display took place, oddly enough,
in the political arena, as tensions between England and the American colonies increased in 1774 shortly after the Boston tea party. Franklin, then
at age sixty-eight on a long sojourn to England, had been accused of steal-
ing some inflammatory personal letters and making them public. A formal
hearing was held by the Privy Council in London to air these charges. With
Priestley in the small audience, thirty-six members of the council—mostly
friends of the king, now George III—applauded periodically as Alexander
Wedderburn, the chief accuser, unleashed on Franklin an eloquent tirade.
The Philadelphian sat throughout in stoic silence. The council recommended
that Franklin be stripped of his position as postmaster, and it was done.[89]
Upon learning of the indignities suffered by Franklin, Philadelphians rose
in anger, parading on a cart through their streets effigies of Wedderburn
and Thomas Hutchinson, the king's governor of Massachusetts. The pro-
cession ended in the obligatory hanging and burning, but the latter had a
new wrinkle, as Reverend Kinnersley—a lifelong opponent of violence—
himself ignited the effigies with electricity.[90]

Because Kinnersley was the great popularizer of electrical science
throughout the American colonies, members of the public doubtless asso-

ciated his name with electricity more commonly than Franklin's. This situation furnished an opportunity for Franklin's political enemies, including William Smith, provost of the College of Philadelphia where Kinnersley taught. Smith seized the opportunity, accusing Franklin of taking credit for Kinnersley's inventions and discoveries.[91] These charges were refuted by Kinnersley himself, who consistently treated Franklin as his mentor in electricity, acknowledging the originality of his contributions. Kinnersley believed that one day Franklin would be regarded on the same plane as Newton.[92]

In the audience of electrical lectures in America, as in Europe, both men and women were welcomed. Kinnersley even priced his lectures so as to encourage the attendance of women: a solitary person paid five shillings, but "to admit a gentleman and a lady" was only "seven shillings and six pence."[93] One advertisement also counseled male readers to "make my Compliments to Madam your Spouse, and assure her from me, she will be highly delighted with this Gentleman's Experiments."[94] Apart from ensuring a supply of female volunteers, Kinnersley's welcome to women emphasized that this kind of educational activity was suitable for both genders.

Lemay claims that science lectures in America were attended by members of all social classes, including "peasants and craftsmen."[95] Indeed, the many spectators a lecturer attracted in one city, week after week—sometimes month after month—suggests a large and thus somewhat diverse audience. Nonetheless, we should not conclude that this mode of education was entirely democratic; after all, many underclass colonists as well as the urban poor would have struggled mightily to afford one lecture, let alone a series; and it is doubtful that the wealthy brought along their slaves. Yet public lectures—especially on electricity—did instill in middle-class society the desirability of learning about new and often exciting scientific discoveries. At the very least, electrical lectures were a socially acceptable amusement.

-|||-

Although Kinnersley enjoyed great success as a lecturer, religious conservatives occasionally expressed concerns about conducting electrical experiments and, especially, tampering with lightning. Lightning, they held, was a providential sign. If it was God's will to destroy a house by lightning, then man should not interfere (see chapter 9). By this time, however, European physicists had fashioned an elegant ideology for countering religious objections to the pursuit of natural philosophy. Musschenbroek, for example, claimed that science displays, manifests, and celebrates the "infinite wisdom, power, and goodness of the omnipotent Creator."[96] Kinnersley was familiar with this kind of rationale and repeated it in his advertisements: "Knowledge of Nature tends to enlarge the human Mind, and give us more noble, more grand and exalted ideas of the Author of Nature."[97] He went on to claim that the pursuit of such knowledge "seldom fails [in] producing something *useful* to man."[98] And who better than a Baptist minister to assist man in learning more about God and in putting this knowledge to work?

Kinnersley was apparently persuasive, even winning over some rather ardent detractors. One testimonial was especially effusive. An unnamed gentleman, who admitted that he had arrived at the lectures prejudiced against Kinnersley, just as readily acknowledged his conversion afterward: "the Truth of this Gentleman's Hypothesis, appear'd in so glaring a Light, and with such undeniable Evidence, that all my former pre-conceiv'd Notions of Thunder and Lightning, tho' borrow'd from the most sagacious Philosophers, together with my Prejudices, immediately vanish'd."[99] Moreover, that Kinnersley attracted a sizable audience to Boston's Faneuil Hall, for a period of some five months, testifies to his favorable reception in conservative territory.[100]

Religious hostility to natural philosophy, electricity in particular, was generally muted throughout the Enlightenment; the Galileo affair of the previous century echoed only faintly. After all, the findings of eighteenth-century science did not blatantly contradict Scripture and few people pursued an agenda requiring the explicit opposition of science and religion. Despite the apparent materialism of the Newtonians, natural philosophy was simply another way that man could come to understand God's plan. As the prime mover and designer of great systems, God was a remote presence in the background when natural philosophers explained effects occurring in nature or in the laboratory. Thus Paulian affirmed in 1781: "The general laws of nature cannot have God as a physical and immediate cause."[101] And so the scientific works of Catholic and Protestant men of the cloth—Nollet, Priestley, and Kinnersley—who contributed so much to electrical studies rarely mentioned religion or God. Not until the middle of the nineteenth century did science and religion come into conflict, when the geological studies of Charles Lyell, the evolutionary theory of Charles Darwin, and the archaeological discoveries of Joseph Prestwich and Jacques Boucher de Perthes (which demonstrated that humans and extinct animals were contemporaries) decisively challenged the account of creation in Genesis.[102]

In the meantime, men and women throughout Europe and the colonies partook of electrical lectures, which were the most popular form of scientific education and entertainment. And no wonder: electricity exemplified the Enlightenment's emphasis on rational thought and experiment. The arcane technologies and surprising effects of electricity, the newest and most engaging of the sciences, testified articulately to human ingenuity, the possibility of progress, and the ability of people to understand and control new forces.[103] The lecturers' monologue underscored these ideological tenets, but the electrical razzle-dazzle itself admitted other interpretations. Doubtless many audience members attributed the odd sights and sounds produced by the electrical devices to the lecturer's control over occult powers. In the end, electrical demonstrations were profoundly ambiguous happenings.

Although reading about electricity and attending lectures were amusing, some people craved closer contact with these remarkable technologies. After surveying the offerings of instrument makers, these individuals took the next step and acquired electrical systems. In their own homes—or castles, for a few upper-class consumers—they could display the tangible tokens of enlightenment and perhaps advertise their own magical powers by carrying out experiments before family and friends.

5. Power to the People

In Cornelius Tiebout's 1801 engraving of Thomas Jefferson, the new president was placed in the company of a globe, a bust of Benjamin Franklin, and a plate electrical machine.[1] This juxtaposition was no accident, for it alluded to the Jefferson unknown to most present-day Americans. Jefferson was not only the principal author of the Declaration of Independence and the third president of the United States, but he was also a highly educated man who had read Newton's *Principia* in the original Latin and followed the progress of many sciences.[2] In addition, he invented ingenious devices (including a machine for making a letter in duplicate) and was an agricultural experimenter who kept meticulous notes and challenged Buffon's theories on the supposed degeneration of plants and animals in the New World. Surprisingly, Thomas Jefferson was also the father of American archaeology and anthropology, carrying out on his plantation the first scientific excavations and linking the finds to linguistic and ethnographic information. An educated person of the late eighteenth century who spied the curious assemblage in Jefferson's portrait would have appreciated immediately that it alluded to Jefferson's lesser-known side, the learned man of science. Among the elite, especially, probably no artifact was a more potent symbol of science, of modernity, and of enlightenment than an electrical machine.

The electrical machine depicted in Tiebout's engraving was probably just a prop, but did Thomas Jefferson actually own any electrical technology? In fact he did. In 1783 he bought a 15-inch electrophorus from Dr. Bass, a Philadelphia physician and apothecary, but we do not know if, or how, he used it.[3] A few years later Jefferson acquired an amusing device for shocking the unsuspecting: a pocket-sized Leyden jar that could be charged merely by rubbing a ribbon.[4] In addition, Jefferson owned books on electricity, including Franklin's.

Thomas Jefferson was just one of many people in the eighteenth century who had purchased electrical things. Kings, dukes, duchesses, wealthy merchants, even college students bought electrical technology for the home.[5] The fabulously wealthy sometimes hired instrument makers to construct large and ornate machines and accessories. People of modest means could buy less expensive—even inexpensive—equipment from instrument makers or build their own by following instructions in experiment books. And a generation of young enthusiasts grew up, late in the Enlightenment, playing with electrical things. In the hands of this heterogeneous community, consisting mainly of collectors and hobbyists, electrical technology became a diversified consumer product and underwent significant design changes. Fortunately, surviving information enables us to identify a few people who owned electrical equipment and to discuss the place of this technology in their lives.

–|||–

Although the earl of Stanhope and the duke of Tuscany bought electrical technology in anticipation of actually conducting scientific research, many members of the elite merely added electrical machines and accessories to their cabinets of philosophical instruments.[6] Usually, cabinets were entire rooms lined with apparatus-laden bookcases whose main function was to impress visitors. If the instruments in a cabinet were used otherwise, it was for home demonstrations of scientific phenomena. The accumulation of scientific apparatus was one of many collecting passions indulged in by Enlightenment elite, following traditions that reached back to the Renaissance, perhaps earlier, and continue to this day.

Book collections were of course quite common and conspicuously advertised their owners' love of learning. Beyond the display of books, nearly mandatory for Enlightenment elite, collectors had diverse interests. Some focused on fossils, minerals, and other natural curiosities; a few gathered antiquities, from chipped-stone hand axes to the marble facades of Greek temples; others acquired fine art such as porcelain, paintings, and furniture; and some obtained precision instruments of many kinds, for navigation, mathematics, surveying, and natural philosophy. Even today, the elite collect pricey objects, from antique cars to rare postage stamps. But collecting, at least in modern America, is not confined to the very wealthy. Indeed, most Americans collect something, be it CDs, cookbooks, or Elvis kitsch; a few even collect Mickey Mouse memorabilia.[7] Today, collecting has enduring appeal across a broad social spectrum.

During the Enlightenment, ordinary people could build some kinds of collections. Benjamin Franklin, for example, managed to collect books even as a youth with modest funds. However, only the truly wealthy could afford to assemble sizable collections of the best scientific apparatus. Not only were these technologies made in hundreds of types, but some were terribly expensive: fine mechanical orreries cost anywhere from around £100 to £1,500, a large air pump with accessories was £30–60, and microscopes could reach many dozens of pounds.[8] It would have required £400 to buy an assortment of high-grade surveying equipment, including telescopes and barometers.[9] And a collector desiring a top-of-the-line, custom-made electrical machine would have had to lay out hundreds of pounds.

Theories abound about why people, then and now, collect. A few general factors underlie this widespread and persistent activity. For one, friends, relatives, and acquaintances often respect the collector for possessing the expertise needed to obtain and identify the objects and understand their significance. Indeed, curating a collection often requires esoteric knowledge, whose conspicuous display can garner prestige for the collector. And collecting fulfills certain expectations of social class: in the Renaissance and Enlightenment, it was socially appropriate for the elite to amass collections, which signaled—mainly to peers and clients—a person's wealth, commitment to learning, and social power. In addition, the process of assembling a collection, through gifts and other interpersonal transactions, reinforces and expands the collector's social network. Finally, the creation and display of a collection is empowering because the individual exercises exclusive control over a special domain of activity and property.[10] Clearly, collecting is a strongly reinforcing activity that engages all who can afford it—and even some who cannot.

—|ı|⊢

Some of the great private collections of the eighteenth century became the nuclei of museums, from the Ashmolean at Oxford to the British Museum in London.[11] Others were dispersed when the collectors got bored or died or, in the case of French aristocrats, were executed during the Reign of Terror.

The contents of the large collections are fascinating. Fortunately, some collections—or at least inventories of them—do survive and furnish us with glimpses of a collector's interests and electrical holdings. Not surprisingly, the largest collection of scientific apparatus in England was built by George III (1738–1820).[12] It began as the demonstration equipment of Dr. Stephen Demainbray, a lecturer on natural philosophy, who became tutor to the royal family in 1754 when George was in his midteens.[13] Apparently, Demainbray's interests did not extend to electricity; not only did the published text of his lectures fail to mention electricity, but the instruments he sold to the king included only one likely electrical thing—a small globe in which a feather had been suspended.[14] Another portion of the king's collection was apparatus already in possession of the royal family, probably assembled mainly by Queen Caroline, who had an abiding interest in natural philosophy and even had discussions with Newton. When George William Frederick ascended to the throne in 1760, at the tender age of twenty-two, he appointed George Adams to be "Mathematical Instrument Maker to His Majesty."[15] During 1761–1762 the Adams firm crafted many of the stellar items in the king's scientific assemblage. A few pieces were added later, but George III's most active period of collecting seems to have ended around 1769, when he moved the instruments to his observatory in Surrey.

In the nineteenth century, the royal collection was divided and dispersed, but no complete inventory was made. A portion of the collection that ended up in the London Science Museum has been cataloged twice, in 1951 and 1993, and it contains a fair number of electrical things.[16] As sometimes happens in these cases, this remnant of the great collection seems to tell us more about the ravages of dispersal processes than about the king's original electrical holdings.

Like other collectors, George III had a sample of electrical machines. Perhaps the most curious one employed a small globe that rotated on a vertical axis.[17] Another odd machine had two glass cylinders, one for producing positive charge, the other for negative. Adams also supplied the king with a finely crafted cylinder machine whose fittings were all brass. George III owned as well a plate machine; although modest—one 18-inch plate—it was well made and had eight rubbers. (An enormous two-cylinder machine, paid for by the king but not incorporated into his collection, is discussed in chapter 9.)

In the company of the electrical machines were assorted Leyden jars, batteries, and display accessories. The latter included an electric orrery, which showed the motions of the earth, moon, and sun, and a carillon with twelve bells. The king also possessed tubes, flasks, and foil-coated panes of glass that furnished luminous effects. A thunder house, insulated stools, and diverse dischargers and electrometers rounded out his electrical holdings.

Though large and varied, the king's collection of *electrical* artifacts was not very impressive by eighteenth-century standards. Some machines were of fine workmanship, but none was extraordinary, and the list of display accessories is impoverished compared to those possessed by most itinerant lecturers. One possibility is that the king's electrical instruments re-corded in the twentieth century were mainly leftovers. Perhaps, in the nearly two centuries after his collecting interests waned, the choice pieces had been dispersed, doubtless making their way into other private collec-tions and museums.

But there is another possible explanation. The poorly educated George III himself had only a slight interest in science. Indeed, the pub-lished correspondence from his most active period of collecting includes no exchanges with scientists or instrument makers.[18] George's main passions were riding, farming, and procreation—if we may judge by the fifteen chil-dren Queen Charlotte bore in twenty-one years. He also enjoyed the the-ater, especially slapstick and pantomime, but he found Shakespeare's plays too sad.[19] And George III appreciated military uniforms and music. Perhaps the king merely collected scientific instruments because it was an expecta-tion for someone of his exalted social position.[20] That could explain why his collection, though immense overall, was not uniformly distinguished.

The king seems to have relied heavily on the advice of his instrument maker, George Adams, whose specialties were mathematical and naviga-tion instruments, globes and microscopes, and some physics apparatus. In these areas the king's collection contains some extraordinary examples, in-cluding a silver microscope so ornate that it has been deemed "impracti-cal."[21] Because Adams apparently was the exclusive supplier of instruments to the king, the collection's emphasis on very pricey nonelectrical technol-ogy is understandable.

Perhaps both factors—indifferent collecting channeled by one instru-ment maker and dispersal processes—account for George III's assemblage of electrical things as cataloged in the twentieth century. One conclusion seems quite likely: the king lacked the know-how to put on his own elec-trical demonstrations.

—| | |—

Fortunately, some collections have been inventoried in a timely manner. Among the most noteworthy was the collection put together by John Stu-art (1713–1792), third earl of Bute.[22] Born in Scotland to a powerful fam-ily that had supported union with England in 1707, Bute was raised in En-gland. In contrast to George III, he was well educated. He attended Eton and then received a degree at the University of Leiden in 1732, some years be-fore Musschenbroek joined the faculty, but perhaps he had witnessed lec-tures by the great Newtonian s' Gravesande. He later became a member of the House of Lords and secretary of state, and he was a close friend and

influential confidant of the much younger George III. Bute was deeply interested in science, especially botany, and made a number of published contributions; he was as well a friend of Peter Collinson, with whom he corresponded on matters mainly botanical.[23]

In addition to being, as the historian Pain so bluntly put it, "vain, pom-
pous and self-important," Bute was exceedingly rich and so was enabled
fully to indulge his collecting passions, which seemingly knew no bounds.[24]
He gathered together minerals and fossils, books on botany and natural
history, and prints. And in England his collection of philosophical instruments was second in size only to that of the king. When Bute died in 1792,
his instrument collections were put up for auction; not only does the auction catalog survive, but so does a list of the buyers and prices fetched.

Over a three-day period in February 1793, 255 lots of optical, mathematical, and philosophical instruments were sold to seventy different buyers, yielding a total of £1,337, a tiny fraction of their cost when new. Among
the buyers were the electrical experimenters Abraham Bennet and Tiberius
Cavallo; the latter purchased one of the most expensive lots, consisting of
five Dollond telescopes—one 10 feet long in brass and mahogany—for
only £42. The longest telescope—with a 10-foot tube—was custom-made
and had originally cost many hundreds of pounds.[25] Instrument makers
were also in attendance, including Edward Nairne and a representative of
W. & S. Jones (successor to Adams). Most buyers were probably not other
wealthy collectors; perhaps it would have been unseemly for the elite to be
seen picking at the bones of one of their own. Rather, many buyers were
members of professions, including architects, teachers, and scientists, who
presumably could actually have used the instruments in their work. And of
course dealers as well took part in the bargain hunting.

Of the 255 lots, only eleven were electrical apparatus. However, these
included five electrical machines, which ranged from a globe machine, to
a 10-inch cylinder machine, to a Dollond machine with two 2-foot glass
plates. In addition, Bute possessed four batteries, including two of nine
jars each in mahogany cases. Accessories abounded, but they were usually
lumped together with a machine in a lot, and so we cannot know their identity. For example, the 10-inch cylinder machine came with "an extensive
variety of well chosen and useful and entertaining apparatus," but that is
the only description supplied.[26] Separately listed, however, were an atmospheric electrometer; Ferguson's cardpaper models of clock, orrery, and
grist mill (see chapter 11); a double electrophorus; and the cord for an electrical kite and associated pith-ball electrometers. Bute's main collecting interests apparently did not reside in electrical apparatus; nonetheless, he had

accumulated a more than respectable assemblage of machines and accessories. Sources are regrettably silent on whether any of the machines had ever been set in motion.

Perhaps reluctant in their new republic to emulate the ostentatious consumption patterns of wealthy Europeans, and lacking wealth on that scale anyway, Americans tended to build more modest collections. Like Benjamin Franklin, Thomas Jefferson was a passionate collector of books, and he bought thousands of them, which eventually brought him to the brink of bankruptcy. He was bailed out by the new Library of Congress, whose first purchase was Jefferson's book collection. These works, which inscribed the accumulated knowledge of science along with the ideas of great thinkers, were of far more interest to Jefferson than experimental apparatus. Compared to the earl of Bute and other members of the old-world elite, Jefferson was neither rich nor a serious collector of philosophical instruments. Yet as an educated man of the late eighteenth century, he owned a few tangible tokens of the human ability to create new knowledge.

—||—

Surprisingly, collectors had a profound influence on the design of scientific instruments, including electrical ones. Since collectors at that time comprised the largest market for scientific apparatus, instrument makers naturally catered to them.[27] They affected the commercial manufacture of electrical things in several ways.

Because the items in many collections were primarily for display, not use, visual performance loomed large. For their wealthiest clients, instrument makers fashioned highly appealing apparatus of brass and glass and wood, for which they could charge extravagant sums. Many of the large and imposing electrical machines on exhibit in today's science museums were custom-built for wealthy collectors. Even now their appearance is striking, almost otherworldly, as it would have been in the eighteenth century. One telling example is the enormous plate machine that the duc de Chaulnes commissioned in the late 1780s, modeled after the very large van Marum machine. Its one glass plate measured 63 inches in diameter, with the prime conductor proportioned accordingly.[28] This machine, whose presence would have dominated many a room, silently testified to the duke's discerning taste as well as to his commitment to learning. More than a generator of electricity, it was an instrument of social and political power.

Given the existence of this sizable and demanding market, instrument makers were constantly on the prowl for new devices to make. In the absence of collectors, the variety of electrical accessories—perhaps even machines—actually brought to market would have been far more limited.[29] Several new kinds of electrical machines may have been commercialized chiefly for their novelty, and thus their potential appeal to collectors.

Finally, the market of collectors led to an increase in both the number of instrument shops and the number of employees. In the larger shops, methods of production underwent changes, especially as artisans produced standard designs in greater quantities. Possible economies of scale, as well as competition among the many makers, also permitted machines and accessories to be sold at lower prices.

Smaller, less expensive electrical machines along with relatively cheap display accessories were within the limits of many middle-class budgets. Thus, I suspect that electrical technology in the eighteenth century reached a market far larger than elite collectors. However, whether ordinary people in nontrivial numbers purchased electrical technology is difficult to demonstrate because their lives are poorly documented. When scholars want to learn about the mostly anonymous people of the past, they sometimes turn to evidence unearthed in archaeological digs. So I consulted a few colleagues, eminent archaeologists who specialize in the eighteenth century, and asked if they had found any electrical technology. Uniformly they replied: "I didn't know there were any electrical things at that time." I wonder if archaeologists have occasionally found the remains of electrical technology but have failed to identify them as such. Perhaps shards of glass globes and cylinders have ended up in the category "glass, misc., unidentified." We can hope that someday an archaeologist might uncover—and recognize—the remains of electrical artifacts in the trash areas of a modest eighteenth-century residence.

Lacking relevant archaeological evidence, we must substitute other sorts of information and arguments to support the claim that electrical technology had become a consumer product not confined to a tiny elite. The strongest evidence is instrument catalogs from the late eighteenth century, most of which list electrical machines in several sizes and prices. For example, a 1799 catalog of W. & S. Jones offered an assortment of cylinder and plate machines, including prime conductors and Leyden jars, which ranged in price from less than £3 to £10 10s.; "New and much improved" models were available for £4 up to £42.[30] In Germany, Gütle made standard globe and cylinder machines in prices that went from 4 to 12 taler.[31] Although Nairne's "Patent Electrical Machines" were pricey—£16 16s. to £168—he also sold a modest machine for "common Electrical Experiments" at £4 4s.[32]

That most instrument makers offered an array of relatively inexpensive machines for many years appears to indicate the existence of an appreciable demand for low-end equipment. Although itinerant lecturers might have favored small machines for their portability, and scientists and wealthy collectors might have bought some examples, it is doubtful that such people comprised the entire market. I suggest that more ordinary people also bought these inexpensive products, perhaps alerted to electrical wonders by lecturers and experiment books.

But could such people actually afford £4 or so for an electrical machine and another pound or two for accessories? Let us first compare the cost of electrical technology with other eighteenth-century indulgences and then consider incomes. A pit seat in a London theater cost 2s. 6d., but box seats cost more; and all prices were doubled for premieres.[33] Nonfiction books ranged around 5s. to 15s. A coach trip from Oxford to London, only 100 miles away, cost £4 8s. in 1774.

Like today, the eighteenth century was a time of vast social inequality. In England, the four hundred wealthiest families of the landed nobility could count on annual rental incomes of £10,000 — roughly £600,000 in today's currency. In contrast, a poor prostitute might make £50 a year. Even worse was the plight of teachers, some of whom earned just £20 or £30. Although bishops and other high clerics were handsomely remunerated — sometimes receiving from the Crown thousands of pounds annually — the bottommost rung of the clergy sometimes flirted with penury at wages little better than teachers' and had to supplement their incomes by tutoring or hauling coal to market.[34]

Between these extremes, however, resided the vast majority of the English. Below the wealthiest landowners, for example, were fifteen or twenty thousand owners of estates having annual incomes above £300. But agricultural land was no longer the only fount of wealth. Industrialization was already well under way in Britain by the middle of the eighteenth century, and was creating new classes of people not tied to a farming lifeway.[35] Mines, foundries, mills, and shipyards, as well as factories — large and small, which churned out everything from dinner plates to umbrellas to beer — were already generating wealth on a vast scale. The owners and managers of these industries were doing quite well, thank you, even if their laborers — whose annual incomes ranged between £20 and £150 for seventy- or eighty-hour weeks — were not.[36] Many traders and shopkeepers earned hundreds of pounds per annum — as did college professors paid by the Crown — and like Benjamin Franklin's father, could support a large family with money to spare for the occasional luxury.[37]

In England at least, tens of thousands of families could easily afford to bring electrical technology home. Even the poorest parson could scrimp and save if he wanted to buy an electrical machine. It was simply one kind of consumer product that competed, along with countless other hobbies and amusements, for discretionary income. Just how far electrical technology trickled down below the very wealthy remains to be learned, but I am confident that many thousands of ordinary people in England, on the continent, and in the colonies found this fascinating symbol of enlightenment irresistible.

—|ı|—

By considering the amusements of youth, we can find still more grounds for believing that electrical technology had already become, by the late eighteenth century, a middle-class consumer good. Inevitably, experiment books and electrical technology would occasionally fall into the hands of inquisitive youth. And the dozens of experiment books written in German, English, and French often gave little more than lip service to electrical theory; rather, they emphasized the apparatus and manipulations needed to produce specific effects. That many experiment books were inexpensive and written in plain language, lacked mathematics and arcane symbols, and could be put into practice without great expense, suggests that they were accessible to literate young people. These books, along with the spectacular effects that could be created if their instructions were followed, rendered electricity an appealing pastime for hobbyists and amateur scientists.

If young people, doubtless mostly male, were experimenting with their own electrical equipment, we might expect some enterprising authors to write books aimed specifically at a youth market, offering do-it-yourself projects.[38] Indeed, Priestley's *Familiar Introduction to the Study of Electricity*, first published in 1768, was written for "beginners," and he was especially diligent in alerting "all young electricians" about the dangers of large batteries.[39] In addition, Priestley's masterful *History* already contained by 1775 a lengthy section on "Practical Maxims for the Use of Young Electricians."[40]

A German author of electrical experiment books, Georg Heinrich Seiferheld (1757–1818), not only targeted his how-to books at young people but also invented dozens of electrical accessories that children could make at home, transforming common materials into utterly new—and wondrous—configurations.[41] Between about 1790 and 1805, Seiferheld published more than a dozen sequentially numbered experiment books, all bearing the title *Collection of Electrical Amusements for Young Electricians*, which were sometimes bound and sold together. He declared that he had written his books not for scholars or historians or mechanics, but for "friends and beginners." [42] In the earliest numbers, Seiferheld, who transitioned from a career as lawyer to that of physics teacher at a secondary school, merely copied technologies and experiments from other books.[43] Thus he included the entire gamut of display-oriented accessories, from carillon to electric fountain.

By the late 1790s, however, Seiferheld had begun to exhaust the possibilities of copying. In search of new projects, he struck out on his own, revealing an inventive streak that in the Enlightenment was nearly unmatched among those working with electrical technologies. Today Seiferheld—who made no significant contributions to electrical science—is all but forgotten; his creations were, in a word, only toys. Most of the later ones were based on rather simple effects, some known in Hauksbee's time. Yet what Seiferheld showed above all else was that electricity could be incorporated into countless complex devices, from card game to fortune-teller, which could surprise and delight young people.

Among the many dozens of Seiferheld's electrical games and amusements, the present chapter mentions just a few that capture the spirit of his efforts (for several additional examples, see chapter 11). Perhaps the most fascinating was his "electric oracle." [44] Powered by a Leyden jar, this game was built into a small box with a display on top. Play was simple: a person merely chose one of six available questions; the choice was indicated by closing the appropriate switch. Immediately a spark lit up the area of the display holding the correct answer and a bell also rang.

Like most of Seiferheld's amusements, the electric oracle could be built with easy-to-obtain materials, including wood, several kinds of paper, metal foils (tin or silver), brass wire, and glue. Of course the youthful experimenter also needed an electrical machine and a Leyden jar. Not surprisingly, Seiferheld supplied instructions for assembling a Leyden jar, but the machine presumably had to come from an instrument shop.[45] What the making of Seiferheld's games and toys required, above all, was manual dexterity, a few common tools, and patience.

The electric oracle was just one of several amusements having occult overtones. For "ghostly appearances," the experimenter built a small temple having paper walls, on the inside of which he drew human figures that materialized when backlit by sparks.[46] The electric fortune-teller was an especially elaborate game (Fig. 13).[47] The fortune-teller was first painted on cardboard, cut out, and then placed in a small shrinelike enclosure, open at the front. In the fortune-teller's hands was a book, on which the fortune magically appeared in response to a player's question. There were eight

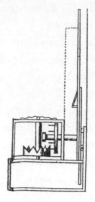

Figure 13.   Seiferheld's electric fortune-teller.
(Adapted from Seiferheld 1796, tb. 2, figs. 3 [left]
and 4 [right]. Courtesy of the Dibner Library,
Smithsonian Institution.)

permissible questions, and each had to have a precise number of words,
from four to eleven; it was these numbers that the operator dialed on a
small box in front of the fortune-teller. Through an unseen mechanical
linkage from the dial (Fig. 13, right), the appropriate answer appeared on
the book, illuminated by a spark. To youngsters, the materialization of the
lighted answer must have seemed quite magical. Even more sophisticated
players, perhaps not impressed by the electrical display, would have been
mystified until they figured out the numerically based system. This was
one of many games in which electricity served merely to light a display.

Finally, although disseminators devised a number of electrical weapons,
Seiferheld encouraged his young readers to make an entire arsenal, including flame throwers, pistols, flintlocks, mortars, and cannons.[48] The guns
achieved their effects by setting off, with the spark from a Leyden jar, either gunpowder or a hydrogen-air mixture. The electrical flintlock was especially ingenious, for it contained a built-in Leyden jar; pulling the trig-
ger moved the jar far enough forward to close a circuit, which sent a spark
into the combustion chamber. When the gaseous mixture exploded, the
barrel expelled a projectile. He even supplied instructions for building a
hydrogen generator and furnished a method for obtaining a combustible
air-hydrogen mixture.[49]

That Seiferheld's singular creations might have captivated a small, but
enthusiastic, youthful audience is beyond doubt. By building and showing
off these games, the young experimenter, no less than the professional lec-

turer, displayed arcane knowledge and a power over unseen forces. Friends and family would have been suitably impressed.

In the twentieth century, tinkering with electrical technologies often inflected a young person's adult activities, perhaps even career choice. Many an electrical engineer—or even writers about electrical technology, such as I—built kits and played with electrical things when young. Such experiences can foster a more general enthusiasm for gadgetry and new technologies.[50] I suspect that this process had begun by the late eighteenth century, given the widespread availability of youth-oriented experiment books and the ready accessibility of electrical technology as well as the materials needed for making it at home.

—|ι|⊢

In at least one case, a young man's engagement with electrical technologies would, through his wife, have a profound influence on literature and popular culture for the next two centuries. Percy Bysshe Shelley was born in 1792 to a long line of English country gentlemen.[51] As a child, he enjoyed regaling his four younger sisters with tales he had concocted of monsters and such. Timothy Shelley wished his son to become "a good and Gentlemanly Scholar" and so enlisted the vicar of Warnham to tutor the child.[52] Apparently the vicar was a good teacher, for soon Percy had authored a play and some poetry, which was published at family expense.

At age ten Shelley was dispatched to school at the Syon House Academy, where for several years he suffered the cruelty of his peers, gradually becoming radicalized. But he also found time to attend Adam Walker's science lectures and read widely, including Gothic mysteries.[53] At his next school, the prestigious Eton where Walker also lectured, Shelley fared no better at the hands of fellow students, especially when he tried to overturn long-established traditions. Although the socially isolated Shelley spent much of his time reading, he once put together a steam engine, but it exploded; he also tried "rais[ing] ghosts at midnight and vigils in deserted graveyards."[54]

His stint at Eton over, Shelley matriculated at Oxford University shortly before his eighteenth birthday. A tall lad with long, bushy hair, he was said to have eyes "large and animated, with a dash of wildness in them."[55] Shelley did, however, make one close friend at Oxford, Thomas Jefferson Hogg, and they would argue interminably about all matters that absorbed undergraduates, from literature to religion. Shelley's interest in religion culminated in 1811 with his publication of a seven-page pamphlet called *The Necessity of Atheism*.

To say that Oxford then was a scientific backwater would be generous, for science was regarded by most literary dons as an alien beast, best slaughtered or chased away. Like many of his teachers, the misnamed Thomas Jefferson Hogg hated natural philosophy and tried to wean his friend Shelley from its delights. Nonetheless, Shelley continued his reading, becoming especially enamored of the writings of Erasmus Darwin (grandfather of Charles), who was the cofounder of the Birmingham Lunar Society, physician, biologist, geologist, poet, and electrical experimenter.[56] Although Darwin's most important works were in botany, medicine, and agriculture, he was also a visionary who saw, especially in science, glimmerings of new technologies. And he penned plans for inventions, some seemingly outlandish in their time, such as a rocket motor powered by oxygen and hydrogen, steam turbine, multiple-mirror telescope, and steam-powered carriage. In the tradition of natural magic, he even constructed a bellows-powered talking machine that, with leather lips and a ribbon vocal cord, could mimic human speech. A friend of Franklin, Darwin regarded Nollet's attacks on the Philadelphian's electrical theories as "weak and impotent."[57]

Following Erasmus Darwin, Shelley had grand visions of society improved by science-based inventions, such as deserts turned into fertile fields by chemical discoveries, transportation revolutionized by aerial machines using a new mechanics of flight, and the limitless power of electricity harnessed for human good.[58] But Shelley would not live long enough— he died in 1822 at the age of thirty—to see any of these visions realized.

In the meantime, Shelley turned his quarters at Oxford into a laboratory littered with apparatus. His chemical experiments often ended with explosions or, at best, new stains and holes in the carpet. Hogg, who wrote a biography of Shelley, furnished a memorable description of his residential laboratory. In addition to a dense scatter of clothes, books, and beverage bottles, Shelley's quarters contained an Argand lamp for heating concoctions, assorted glass receivers, and a vacuum pump. Also among the scientific bric-a-brac were an electrical machine and Volta's recently invented galvanic trough, the first electrochemical battery (see chapter 6). Hogg also recounted Shelley's antics:

> He then proceeded, with much eagerness and enthusiasm, to show me the various instruments, especially the electrical apparatus; turning round the handle very rapidly, so that the fierce, crackling sparks flew forth; and presently standing upon the stool with glass feet, he begged me to work the machine until he was filled with the fluid, so that his long, wild locks bristled and stood on end.[59]

A more recent biographer has seen in Shelley's dabblings in experimental science a "Curiosity at Nature's marvels and the desire to control them"; and there was also the "spice of danger or the charm of uncertainty." [60] These were the attractions that chemistry and electricity held for this most gifted student, whose literary offerings, especially *Prometheus Unbound*, would contain a wealth of scientific allusions.

Although Shelley was far from being a typical Oxford undergraduate, much less a typical English youth, he does represent that tiny fraction of late-Enlightenment teenagers enamored with electrical apparatus and experiments. Unlike others in this group of enthusiasts, Shelley's brief life has been documented in rich detail. But we may be confident that his passionate interest in electricity and in the purchase of electrical things was far from unique.

Clearly, Percy Bysshe Shelley exemplifies a consumption pattern that persists to this day: bright and curious youth—only a handful of them, mostly males whose intellectual interests (and perhaps social skills) put them outside their generation's mainstream—develop an affinity for electrical tinkering and buy appropriate parts and products. In my generation, these hobbyists played with radios, televisions, and audio equipment; today they explore computers and robots. [61]

—||—

Shelley's troubled first wife eventually committed suicide, but he did find another woman, a kindred spirit, to share his life. The daughter of the literary giants Mary Wollstonecraft and William Godwin, Mary Shelley grew up in the country, mostly near Dundee, Scotland. [62] There she began to write, putting words to "the airy flights of my imagination." [63] Although born to writing, she had never thought of recognition on her own. Percy helped change Mary's attitude: not only did he encourage her to establish a literary reputation worthy of her parentage, but he also recommended the reading of science and acquainted the young woman with some of his favorite authors, including Erasmus Darwin.

In 1816 Mary and Percy, not yet married, spent the summer together in Switzerland. One of their neighbors was George Gordon, Lord Byron, who was working on the composition of *Childe Harold*. The summer was rainy and dreary, confining them to the house for long periods; for amusement they read German ghost stories that had been translated into French. Mary, who was only nineteen years old, had little success in putting pen to paper until Byron suggested that they "each write a ghost story." [64]

Neither Percy nor Byron completed the assignment, but Mary persisted, hoping to produce, she said, "a story to rival those which had excited us to this task." [65] It would have to be, she insisted, a tale that spoke to "the mysterious fears of our nature, and awaken thrilling horror—one to make the reader dread to look round, to curdle the blood, and quicken the beatings of the heart." [66] In this she succeeded brilliantly. Moreover, the novel that emerged—Mary Wollstonecraft Shelley's first—was also fascinating and deeply provocative. Although she would write many more books, none did more to establish her reputation as a literary giant than *Frankenstein.*[67]

The crafting of *Frankenstein* reflected its author's familiarity with electrical technology, which stemmed mainly from conversations with her husband. Yet the idea that dead people, even body parts, could be reanimated by electricity was the product of neither Shelley's fertile imagination. Rather, it emerged from experiments performed by members of another electrical community, the electrobiologists. Before discussing the influence of Enlightenment electrical technologies on *Frankenstein,* then, I turn in the next chapter to the electrobiologists' activities and inventions.

6.   Life and Death

Beginning in the mid-1740s, public lectures as well as reports in newspapers, magazines, and society journals had alerted scientists of all stripes to the novel effects produced in electrophysics laboratories. Not surprisingly, a handful of people interested in botany, physiology, and chemistry saw in electrical technology a promising research tool; likewise, electrically savvy scientists, including Franklin and Nollet, appreciated that their technology allowed them to engage new subjects. And so arose several science-oriented electrical communities whose beginnings coincided with the widespread commercialization of electrical technology. Members of one overarching community, which I call *electrobiologists*, applied this technology in studies of plants (electrobotany) and animals (electrophysiology), with sometimes startling results. By the time Mary Shelley wrote *Frankenstein*, in 1816, more than a half century of experiments had already shown how electricity affects organisms—living and dead.

Electrobiologists investigated everything from seed germination to the very nature of life itself. Significantly, apparatus such as the electrical machine and Leyden jar became models for understanding physiological processes, including the operation of the nervous system. The purported discovery of "animal electricity" reverberated throughout the literate world and even affected popular culture. What is more, studies showing the influence of electricity on human physiology and anatomy laid a foundation for new medical therapies, some still used today (see chapter 7).

In adapting electrical technology for plant research, electrobiologists fashioned the first electrical machines that ran without human power. As an outgrowth of studies on animals they also invented an entirely new kind of technology—the electrochemical battery. Let us examine the activities and technologies of the electrobiologists, beginning with their work on plants.

⊣ı∣⊢

Jean Jallabert (1712–1768), a Swiss investigator, was one of the first to conduct electrobotanical research.[1] Like his close friend Nollet, Jallabert had a clerical background; later he attained professorships in physics, mathematics, and philosophy. Inspired by Nollet, Jallabert initiated electrical studies in the mid-1730s but at first merely repeated and refined earlier works.[2] However, when at last he turned his attention to plants and animals, Jallabert ventured into little-known territory.

Like others working at about the same time, Jallabert found that plants could be electrified, and the effects were sometimes dramatic.[3] For example, after electrifying a flowering plant in a darkened room, he could see light issuing from the tips of its petals and leaves, and he also showed that electricity could cause a drooping flower to revive.[4] Others discovered that touching certain charged plants immediately caused their leaves to wilt or curl; this became the celebrated phenomenon of the "sensitive plant."[5] If electricity could thoroughly permeate a plant's substance, as these experiments indicated, then, wondered Jallabert, might electricity enhance plant growth? To answer this question, he compared the growth of charged specimens with uncharged controls subjected to identical conditions of soil, moisture, and light.

Jallabert began with studies of seed germination, which required no elaborate apparatus. In one experiment, he merely filled two porous pottery vases with water and placed mustard seeds on their sweaty outsides; one vase was electrified for eight or nine hours per day, the other was not electrified at all. After just two days, mustard seeds on the charged vessel began to germinate, but those on the control showed no signs of life. The electrified plants were also first to boast blossoms and leaves. In a similar experiment employing two metal vases filled with the same soil, he found that electrified mustard plants grew much more rapidly.

Regrettably, Jallabert did not describe how he applied charge to his pottery and metal vessels. However, in experiments with hyacinth, daffodil, and narcissus bulbs, he reported conveying charge from his electrical machine with a brass conductor. The bulbs had been placed in water-filled carafes, one of which rested on a resin cake and was electrified for eight or nine hours a day for eleven days. Consistent with Jallabert's earlier findings, the electrified bulbs enjoyed more vigorous growth than the controls. About the same time, Nollet carried out similar experiments on several species

Figure 14. Nollet's apparatus for electrobiological experiments. (Adapted from Dibner 1957, 29. Courtesy of the Burndy Library, Dibner Institute, MIT.)

with the same result: compared to controls (Fig. 14, on table), seeds in electrified metal containers germinated more rapidly (Fig. 14, above table).[6]

—|ı|ⵊ—

Picking up where Jallabert and Nollet left off, Pierre Bertholon (1741–1800), the abbé of St. Lazare and a professor of experimental physics at Montpellier, undertook many electrobotanical experiments.[7] A priest until his renunciation in 1797, Bertholon was a member of several electrical communities, and we shall encounter him again (in chapters 7 and 8). Bertholon's major electrical works deal with plants, earth science, and human disease; all advance comprehensive systems for understanding both diverse natural phenomena as well as the effects revealed by experiment. A Franklinian, he himself was an accomplished experimenter and sometimes built original apparatus.

Of special interest in the present chapter is Bertholon's theoretical system for explaining how, in nature, atmospheric electricity influences germination and growth as well as flowering and fruiting, for it gave meaning to his interesting experiments and new technologies.[8] This work was reported in a sizable book entitled, not surprisingly, *De l'électricité des végétaux*, published in 1783. Bertholon built this theoretical system meticulously, premise by premise. The result was a grand edifice, parts of which now seem far-fetched, perhaps kooky. Yet in the late eighteenth century it was serious science; moreover, several of its seminal ideas are still accepted today.

Bertholon believed in a great continuum of life, from the lowliest plants to the highest animals. He even enumerated, in some detail, many parallels between plants and animals in structures, functions, and life histories, suggesting that they indicated "a similarity more complete than anyone could imagine."[9] Arguing by analogy, Bertholon proposed that because electricity had so many effects on animals (see below), it must also have equally far-reaching—and largely beneficial—effects on plants.[10]

But how do plants in their natural environment receive its effects? They obtain it, Bertholon argued, from atmospheric electricity.[11] Specifically, electric fluid in the upper reaches of the atmosphere attaches itself to water molecules, which rain brings downward; snow, hail, and fog also carry electricity to the earth's surface. Entering through roots, the charged water circulates in the plants and thus promotes more rapid and greater growth. But Bertholon covered his bets by also positing that plants acquire atmospheric electricity directly from the air, which enters through pores in the leaves.

To shore up his argument, Bertholon studied the conductivity of different plants, employing apparatus that included an electrical machine and Leyden jars. Also a part of the apparatus, two people judged the strength of the shock they received in a circuit that included the plant being tested.[12] Aware that his human measuring instruments were imprecise, Bertholon repeated the experiment many times and obtained "true results."[13] He found that the tested plants could be grouped into best, moderate, and worst conductors. Among the best were cactus and succulents; moderate conductors included many herbaceous plants; and some of the worst were trees.[14] These differences in conduction Bertholon attributed to differences in water content. He also observed that dried plants were poor conductors but "transmit the shock" well when moistened.[15]

Bertholon noted that plants prosper in regions having frequent thunderstorms—the times when atmospheric electricity is at a peak.[16] His own annual records of storms and plant growth seemed to confirm the correlation: "the stormiest years, the most electric, were the most fertile."[17] Clearly, there must be a link between plant growth and atmospheric electricity. But a more direct experiment would be needed to demonstrate that rainwater, electrically charged, was the actual growth-enhancing mechanism. So Bertholon placed identical plants in two pots, watering one with rainwater, the other with ordinary garden water.[18] He was doubtless delighted to find that the plant given rainwater grew more vigorously (over two centuries later, we recognize that rainwater carries a charge and that the nitrates in it—formed by lightning—probably accelerate a plant's growth). Bertholon augmented these findings by repeating the experiment with different species. In further experiments, Bertholon found that he could simplify the apparatus by placing seeds directly on a prime conductor or on a magical table—a version of Franklin's plate capacitor. These seeds, along with unelectrified controls, were then planted in identical containers with all other factors—soils, water, light, and temperature—held constant; the treated seeds sprouted faster.[19]

Bertholon carried out additional studies on seeds of poppy and tobacco. This time, he allowed the plants—periodically electrified—to mature. According to his effusive report, electrified plants grew far more luxuriantly than controls. Drawing also on the findings of other investigators, Bertholon asserted that, in nature, atmospheric electricity beneficially affected all parts of plants at all stages of growth.[20] He even went so far as to suggest that plants grow perpendicularly to the horizon because they move toward the highest concentration of electric fluid.[21] Not content only to experiment and theorize, in the last part of *De l'électricité des végétaux* Bertholon developed ingenious technology for applying electricity to cultivated plants (see chapter 11).

—|ı|⊢

Electrobotanists who performed long-term laboratory experiments had to solve an immediate problem: how could they keep an electrical machine in continuous operation, hour after hour? One solution, so obvious to many researchers that it required no mention, was to have servants and assistants turn the crank. Doubtless the studies of Nollet, Jallabert, and Bertholon were conducted with the aid of anonymous helpers. But a few investigators framed the problem explicitly and in narrower technical terms, which led them to propose new kinds of electrical machines needing no human prime mover. Priestley, for example, suggested that a "machine for perpetual electrification" could be powered by water or wind, but this vague idea was nei-ther developed further nor realized in hardware.[22] Others built prototypes and reported them in the scientific literature.

In the 1770s, many years before he commissioned the monster machine at the Teyler Museum, Martinus van Marum created a radical electrical machine suitable for botanical research while trying to solve another nag-ging problem.[23] One difficulty most experimenters faced was that the glass used in machines collected moisture, especially during humid summer days. That is why so many investigators covered their insulators with sealing wax; some even coated the *insides* of their cylinders and globes. But this fix was not workable for glass plates, since both sides, in contact with rubbers, had to remain bare. Evidently, another approach was needed. Trying out different materials, van Marum learned that he could eliminate the buildup of moisture by using a plate made of shellac (such plates resembled the thick phonograph disks sold in the early twentieth century). But that was only the beginning.

In pressing against a machine's plate, be it glass or shellac, the rubber created friction that not only generated charge but also increased the amount of work required to rotate the disk. Seeking to reduce this friction, van Marum came up with an alternative to the traditional amalgam-coated leather cushion: he simply immersed the rotating disk in a trough of mer-cury. And since the mercury was a better conductor than the conventional rubber, this design conveniently generated positive and negative charges. Involving so little friction, van Marum's machine could be operated by a simple weight-and-pulley system and so was ideal for carrying out elec-trobotanical experiments. Although reports of van Marum's shellac-and-mercury-trough machine were published in French and German as well as in his native Dutch, few outside the Netherlands seem to have taken much notice of it.

The Italian Giuseppe Toaldo, a cleric and professor of physics at Padua, in 1782 took a rather different tack to building a machine for electrobotani-cal research. He designed a compact, spring-driven machine that could run for four hours after one winding.[24] The clocklike mechanism, consisting of wheels and springs, was housed in a small brass box, 10 inches square and 4 inches tall. In one configuration, it spun an 8-inch glass disk horizontally; with a slight adjustment, it could also drive the disk vertically or rotate a small cylinder. Despite its diminutive dimensions, this machine was capa-ble of charging a tiny Leyden jar sufficiently to deliver a smart shock. Toaldo also pointed out that the basic design could be scaled up somewhat, depend-ing on the researcher's needs.

Franklin's friend Jan Ingen-Housz devised another solution to the prime-mover problem: he dispensed with the electrical machine as the immediate source of charge.[25] To supply power for his botanical experiments, Ingen-Housz used a Leyden jar, which he kept constantly charged. The Leyden jar also served another purpose, for inside he had placed a cup of water. Floating on the water was a piece of cork on which he could put some seeds, such as mustard; this arrangement he wrapped in blotting paper to keep the seeds moist. With a conductor, he connected the water in the cup to the jar's inside surface. He also established a control, identical in all respects, except that the second Leyden jar was not charged. In contrast to Nollet and Jallabert, Ingen-Housz found that the seeds in both Leyden jars germinated at the same rate.

Apparently, the new technology of electrobotanists remained confined to this tiny group of researchers. Despite their promise, neither van Marum's nor Toaldo's machine was commercialized. Did instrument makers judge the anticipated market—a handful of electrobotanists and wealthy collectors—to be too small to warrant the development costs? After all, these machines were so different from conventional models that much trial and error would have been required merely to arrive at marketable designs. Incidentally, none of the clever accessories invented by electrobotanists were commercialized either, perhaps because they could be easily assembled from common components.

Electrobotanical experiments have fallen in and out of favor several times during the past two centuries; often they have been mired in controversies over the comparability of different experimental designs.[26] But, despite Ingen-Housz' negative findings, there seems to be a consensus that electricity enhances plant growth. Equally surrounded in controversy, eighteenth-century electrophysiological experiments, especially on humans, also furnished fascinating findings and new electrical technologies.

—∣∣⊢

The electrophysiologists, like the electrobotanists, got off to an early start. By the mid-1730s, for example, Stephen Gray had shown with poultry and a small boy that living animals could both conduct and store electricity, and countless experimenters had also learned that a Leyden jar's discharge could kill small animals. Nollet, who himself dispatched birds, cautioned that these experiments should not be done before "delicate persons, especially pregnant women."[27] Soon investigators were exploring the possibility that electricity could affect a living body's vital functions.

There were several theoretical reasons to expect that a charged animal would have, for example, a more rapid pulse or perspire more profusely. In the first place, electricity had been shown to accelerate the flow of liquid in capillary tubes as well as from funnels and watering cans; these effects were caused by the repulsion of water molecules carrying the same charge. In the second place, the famed physiologist William Harvey had demonstrated in the previous century that blood, pumped by the heart, circulates through arteries and veins. By the time electrophysiologists began work in the mid eighteenth century, investigators took it for granted that, like blood, other liquids circulated in the animal body, and that these flows helped sustain life. If electricity sped up the movement of all these liquids, then this accelerated motion should be evident, for example, in more rapid pulse and more copious perspiration.

In the late 1740s Nollet carried out several experiments on animal perspiration that supported these expectations.[28] His basic plan was to take two animals of one species, such as cats, and electrify one for five hours, using the other as a control. Their weights would be measured before and after, and any differences—attributable to varying amounts of perspiration—noted.[29] Appreciating that the creatures might be loathe to sit still for this study, Nollet, the consummate instrument maker, devised cages of wooden slats and sheet-metal bases to confine them. He placed the cage with its experimental subject on a metal shelf suspended by chains from a silk cord (Fig. 14). By connecting the chains to his machine, Nollet could pass charge to the cage and thus the animal inside.

In his first trials with cats, Nollet found, as predicted, that the electrified animal lost more weight. Yet he worried that this effect might have arisen if the cats had intrinsically different temperaments. To rule out this cause, he repeated the experiment with one alteration: the former control cat was now electrified, and the previously electrified cat became the control. The results were the same. He next tried different animals, including pigeons and other birds, and again found that electricity accelerated weight loss. By eighteenth-century standards, Nollet's experiments were elegant and definitive; not surprisingly, later investigators cited them often.[30]

Using human subjects, several investigators extended these findings. A common experiment was to take a person's pulse before and during electrification. One of the most comprehensive studies of this kind was conducted by the Dutch investigator, Wilhelm von Barneveld. In a data table strikingly modern, Barneveld reported 169 trials on more than a dozen people who ranged in age from nine to sixty.[31] Aware that fear or anxiety could also accelerate the pulse, Barneveld strove to put his unnamed subjects at ease. Even so, in almost every trial—first with positive, then with negative electricity—the pulse speeded up, sometimes rising 10 percent above the individual's "natural" heart rate. Barneveld's apparatus for these experiments was simple: the person to be electrified merely sat on an insulated chair and held a metal rod connected to the disk machine.[32]

Electrophysiological research produced even more dramatic findings. By analogy with the electrified watering can, some investigators conjectured that a punctured person would, during electrification, bleed more profusely, a hypothesis that was soon verified. Indeed, already in 1746 Benjamin Martin included such an exercise in his experiment book. He recommended placing the subject on a resin cake and opening a vein in his arm—a common medical procedure at the time (see chapter 7). With a hand grasping the prime conductor, the person could be charged or not at will. Martin reported that when he electrified the subject, blood spurted out "with a much greater Velocity, and to a much greater Distance."[33] Please don't try this at home.

⊣‖⊢

Enormous batteries of Leyden jars gave electrophysiologists a powerful technology for exploring electricity's dark side; these experiments in dealing out death also reveal to us today an unsavory side of science. One of the most avid experimenters in this area was none other than Joseph Priestley, dissenting minister.[34] Familiar with earlier reports on electrocuting small animals, Priestley believed that many questions remained about how, exactly, electricity did its gruesome work. Did electricity, for example, really burst all the blood vessels? His basic strategy was to administer a large shock and observe the result; if the unfortunate creature died, he dissected it in search of internal effects. Priestley's subjects included a mouse, rat, frogs, large kitten, adult cat, and a dog. In general, he found few internal traces of the massive shocks, but then again he was not an experienced anatomist.

The dog, "the size of a common cur," provided an interesting case.[35] After assembling an enormous battery, Priestley placed its conductors on the hapless dog's head. Immediately after receiving the shock, the dog's limbs extended fully and it fell backward, lying motionless for around a minute. After some convulsions and rattling in the throat, the animal began to salivate profusely; a half hour later it started to walk. The next day Priestley observed that the dog was blind, perhaps because the shock had been delivered so close to its eyes. Curious about the internal effects, Priestley shot the dog in the head and then dissected the eyeballs. The cause of the blindness was soon revealed: the shock had cooked the corneas, turning them white and opaque, so they no longer passed light.

In the name of science, even larger animals were sometimes sacrificed on the electric altar. On one occasion, members of the Royal Society assembled, on March 12, 1781, to witness the spectacle. Using his own very large battery, Reverend Abraham Bennet electrocuted a sheep.[36] It is doubtful that this demonstration yielded any new science; that electricity, in the form of lightning, could kill large animals was already appreciated.

No doubt the biggest batteries, such as the installation at the Teyler Museum, were also capable of doing serious harm to humans. In experimenting with these apparatus, investigators flirted with disaster; even the most careful worker occasionally got a nasty shock. And Benjamin Franklin was no exception. Two days before Christmas 1752, he was preparing to dispatch a turkey with a battery of two enormous Leyden jars, fully charged. Distracted by conversation, Franklin managed to touch the wire connecting the balls with one hand and the chain joining the jars' exteriors with the other. As the battery discharged through the great man, there was a flash of light and a "crack as loud as a pistol."[37] Franklin testified to having felt a "universal blow throughout my whole body from head to foot," followed by gradually diminishing convulsions.[38] Although the back of his neck and arms remained numb until the next morning, he apparently suffered no long-lasting effects (a 10-lb. turkey, finally shocked, was less fortunate; it was, insisted Franklin, "uncommonly tender").[39]

Receiving the occasional jarring jolt came with the territory of electrical research. Priestley too was once zapped by his own battery.[40] Stories of such mishaps, told many times and doubtless embellished in each retelling, were obvious warnings to the unwary, but they also served as badges of honor for researchers, not unlike a soldier's tales of battle.

—|¡|—

A battery of Leyden jars also turned out to be a useful tool for modeling the process that enabled certain "fish," particularly the torpedo (*Gymnotus*), to numb prey as well as humans in close encounters.[41] The torpedo's remarkable ability had been known since antiquity—reported by both Aristotle and Pliny—but its nature remained a mystery. After the florescence of electrical research in the mid eighteenth century, some investigators began to suspect that these fish were in fact electric.

A particularly noteworthy paper by John Walsh, relayed to the Royal Society by his friend Franklin and published in 1773, reported experiments with freshly caught torpedos.[42] His conclusion was that their numbing effects were achieved electrically; the sensation they produced in people was identical to the familiar shock of the Leyden jar. Indeed, the fish's electric organs were in effect miniature Leyden jars, which could be discharged in various modes. This appealing analogy was beset, however, by some perplexing anomalies. The torpedo could discharge dozens of times with no apparent diminution of the shock. But try as he might, Walsh could not cause the fish's electric organs to yield any sparks. These and other little glitches gave skeptics an opening.

The torpedo problem was taken up by one of the most respected and enigmatic figures of eighteenth-century British science, Henry Cavendish (1731–1810).[43] Born into a very prosperous family, he was descended from dukes on both sides. His mother, Lady Anne Grey, died just two years after Henry's birth. Lord Charles Cavendish, his father, sent him off to Dr. Newcome's academy at age eleven, where he associated with other scions of the upper class. Next he matriculated at St. Peter's College, in Cambridge University; although remaining for four years, he took no degree. Many men in Henry's social position would have prepared for a career in law but—following his father—he took an interest in natural philosophy. However, unlike other scientists of the eighteenth century, Henry did not have to squeeze his research between other responsibilities; and he never married. With his father's indulgence, and later inheritances, Henry could devote all of his energies to scientific activities, including distinguished service to the Royal Society. As one of England's wealthiest—and most antisocial—men, Henry had the luxury of pursuing his scientific passions wherever they led. And they led in many directions, including geology, mechanics, mathematics, chemistry, astronomy, and electricity.

Unfortunately, Cavendish was a perfectionist. Unsatisfied with the products of many projects, he published no books and only around twenty papers. Several papers were significant and influential, especially his work on the composition of water (see chapter 10), but Cavendish's legacy might have been far greater had he accepted mere progress instead of pursuing perfection. For what he regarded as imperfect was often judged by others to be an appreciable advance. Not surprisingly, Cavendish left behind a treasure trove of notes and manuscripts, which revealed to later investigators that he had performed a Herculean amount of valuable research that never reached print. Nonetheless, his published papers along with an encyclopedic knowledge of science enabled Cavendish to achieve a stature in British science unsurpassed in his lifetime. Even today, he is usually mentioned in the company of Isaac Newton.

Cavendish's research on the torpedo, which *nowadays* seems trivial in comparison to some of his unpublished work, was fully reported in the *Philosophical Transactions*. Understanding the torpedo's mechanism was, in the late eighteenth century, a pressing problem. Although convinced by Walsh's research that the torpedo's power was electric, Cavendish decided that the matter could be laid to rest only by more definitive experiments. These needed to be conducted, not on the torpedo itself, but on apparatus that enabled a researcher to model particular effects.

Cavendish first took up the matter of the missing spark. His experiments, elegant by any standard, employed a number of Leyden jars along with an ultrasensitive electrometer he made from gilt straws. By passing charges between and among jars in various circuits, Cavendish was able to reduce the "force" of the electricity—indicated by spark length or the separation of the straws—while increasing its "quantity." [44] On the basis of the crucial distinction between force and quantity (which correspond, respectively, to the modern concepts of voltage and current), Cavendish—taking the shock himself—showed that "the strength of the shock depends rather more on the quantity of fluid which passes through our body, than on the force with which it is impelled." [45] Thus, the torpedo, with its force too small to create a spark can, if the *quantity* of electricity is sufficient, still deliver a memorable jolt. Of course Cavendish was right, but this question would arise again, a few decades later, in connection with the electrochemical battery (see below). [46]

In focusing on another anomaly, Cavendish also established a new principle that contradicted one of Franklin's frequent claims. How, Cavendish wondered, could a torpedo's discharge travel some distance to a prey animal or person, without first being totally dissipated in the surrounding saltwater—a good conductor? Cavendish offered an argument based on his experiments in conduction using wires and Leyden jars. [47] The gist was that electricity in a circuit follows not only the path of least resistance—as Franklin had incorrectly maintained [48]—but divides up, passing through all conductors (the quantity going through any one depending, inversely, on its resistance). To wit, a long piece of very thin wire carries a smaller quantity of electricity than a thick brass rod when both simultaneously discharge the same Leyden jar. Thus, while much of a torpedo's discharge between its electric organs passes through the immediately adjacent water, at greater distances the water still conducts enough electric fluid to deliver a shock (Fig. 15, bottom). We can regard Cavendish's principle as the earliest formulation of what would become known, a half-century hence, as Ohm's law.

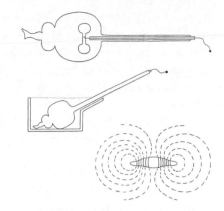

Figure 15. *Top*, Cavendish's model torpedo; *middle*, in its tank; and *bottom*, the fall-off of charge with distance from the fish. (Adapted from Cavendish 1776, figs. 3 [top], 4 [bottom], and 1 [bottom].)

Cavendish could have stopped there, having strengthened the case that the torpedo's strange power is electric, but he went further by experimenting on a model of the torpedo made from leather (Fig. 15, top).[49] He gave his fish a backbone of sorts by cutting and layering thick shoe leather in the shape and size of a torpedo. The electrical organs were represented by barbell-shaped pieces of pewter attached to each side of the leather backbone. To link the pewter organs to an external charge, Cavendish connected them to wires passing through long glass tubes along both sides of the torpedo's tail. Finally, the torpedo received a sheepskin covering. The artificial torpedo was then given its habitat: a wooden trough filled with saltwater (Fig. 15, middle). In electrifying his banjo-shaped fish effigy, Cavendish employed a large battery; the forty-nine jars could be connected in various combinations to the wires protruding, above the water, from the ends of the glass tubes.

Taking the charge from the artificial fish himself, Cavendish carried out many trials, often comparing the shock to that of a real torpedo. As expected, he could feel the electricity with his hands in the water, a few inches away from, but not touching, the leather look-alike. He even obtained the assistance of John Walsh, who acknowledged that Cavendish's "artificial torpedo produced just the same sensation as the real one."[50]

Although the sensation might have been the same, the magnitude of the shock was not, and this especially troubled the compulsive Cavendish. Assuming that the cells of real electric organs were miniature Leyden jars, he calculated their capacity based on anatomical measurements. His surprising conclusion was that an actual torpedo's battery held a quantity of electricity fourteen times greater than that of his Leyden jars.[51] Presumably, with a larger battery, he might have more closely mimicked the torpedo's electrical discharge.

Although his paper on the model torpedo, delivered to the Royal Society in January 1775, was itself a model of persuasive scientific discourse, Cavendish was aware that words alone cannot convey sense impressions.[52] To convince others that the shock of the torpedo was electric, he invited several distinguished men of electrical science to his laboratory to feel, as well as see, his model in action. The group, consisting of Joseph Priestley, Timothy Lane, and Edward Nairne, also included Thomas Ronayne, an outspoken skeptic who had challenged John Walsh's conclusions. On Saturday, May 27, 1775, they assembled to take part in performances of Cavendish and his facsimile fish. Apparently, all became believers.[53] Cavendish's report along with later testimonials from the witnesses at last laid to rest questions about the nature of the torpedo's awesome power. The finding that living organisms can produce electricity without friction provoked even more momentous discoveries in biology.

—|ı|⊢

The works of Walsh and Cavendish were not the first to implicate electricity in physiological processes. Investigators much earlier in the century, including Jallabert and Nollet, had noticed that a shock applied to the skin caused contraction of the underlying muscles; and, as Franklin and others learned, large shocks could also lead to trembling, numbness, and convulsions. Some investigators had dissected animals and applied electricity directly to their muscles, thereby producing motion.[54] These provocative findings contributed to new understandings of the animal nervous system.

That the brain is the seat of intelligence, that it integrates information from sense organs, and that it directs the voluntary movement of muscles—all through connections made by nerves—came to be accepted by late-eighteenth-century anatomists and physiologists. The burning question was, how does this system work? In particular, what passes along nerves that enables the system to translate volition into motion?[55] Many believed that some subtle fluid must circulate throughout the nervous system, carrying information from sensory organs to the brain and from the brain to all muscles. But the postulated "nervous fluid" would be unlike any other bodily fluid because it moved so quickly. After all, the time lag between wanting to move a leg and its actual movement was imperceptible. Clearly, this fluid could not be like blood or lymph, tears or digestive juices; nor was there an obvious mechanism for pumping the fluid from place to place or for rapidly reversing its direction.[56]

Isaac Newton was actually the first to propose that the nervous system operated electrically.[57] This conjecture was taken up, decades later, by several Italian investigators, including Giambatista Beccaria, who argued in 1753 that electricity could account for the "speed and changes in animal sensations and motions."[58] His countryman, Felice Fontana, suggested that the nervous system was analogous to an electrical machine; electricity passed through nerves, which perhaps were covered with an insulating substance.[59] Although many investigators came to believe that electricity was the subtle fluid that made the "connection between mind and matter," a chorus of critics ensured that the mechanism of nervous action remained in question.[60]

This issue was addressed in the 1780s by the distinguished Italian physician and anatomist Luigi Galvani. Little did Galvani know that his research would become the focus of one of the most contentious debates in all of science and that its outcome would include a new electrical technology that captivated experimenters for decades.

-||||-

Luigi Galvani (1737–1798) was the third of four children born to a well-off Bolognese family.[61] After deciding to become a medical doctor, Galvani trained at the University of Bologna—the oldest university in the world—and received his degree in 1759. He took up the practice of medicine, gave public lectures, and carried out anatomical research. Among his early publications were treatises on the kidneys and ears of birds.

In 1762 Galvani married Lucia Galeazzi; she took an avid interest in his work, much of which was actually done in a room of their spacious apartment.[62] Although Galvani was hardly unique in having a home laboratory, his was somewhat unusual because it contained, among other apparatus, dissected animals and parts of human cadavers; these he used in honing his surgical skills and in giving lectures. Such a domestic scene was familiar to Lucia, for her father, Gusmano Galeazzi, was also an anatomy professor who brought his work home.

A few years after their marriage, Galvani obtained a paid lectureship at his alma mater and was eventually elected professor of obstetrics. Throughout his life, Galvani continued to practice medicine and to conduct research in anatomy and physiology; in the latter activities he adopted insights and technologies from diverse sciences, including chemistry and physics.

Following the lead of Giuseppe Veratti and assisted by a nephew, around 1780 Galvani began to look into the effects of electricity on muscular motion.[63] By achieving a better understanding of this phenomenon, he hoped to devise new treatments for diseases, especially of the nervous system. Galvani's apparatus consisted mainly of a plate electrical machine of modest size, plate capacitor, Leyden jar, various conductors, and assorted animals; his favorite laboratory animal was the frog.[64]

Galvani began many experiments by partially dissecting a frog, keeping only its legs with the crural nerve. The nerve, which he wrapped in metal foil, could be excited by the disk machine or Leyden jar. Galvani quickly discovered that the frog's legs moved when the machine sparked, even when they were not connected to the machine. It was in further exploring this bizarre effect—whose cause remains controversial to this day—that Galvani made his greatest discoveries.

In playing around with scalpels and other metal objects, bringing them into contact with his fragmentary frogs, Galvani found that they too could excite contractions, even when the electrical machine stood motionless. Galvani, a physician and anatomist influenced by Fontana's views, had a ready explanation for this curious effect. The muscles, he claimed, were miniature batteries of Leyden jars that stored electricity. The metal conductors, like nerves in living animals, merely discharged the muscle batteries.[65] But what was the origin of the batteries' electricity? Galvani proposed that the immediate source of this animal electricity was the cerebrum, which extracted the electricity from the blood and, in turn, communicated it through the nerves to muscles.[66] Galvani reported his findings and speculations in a 1791 work published in Latin, which was soon being avidly read in the original and in a 1792 Italian translation.

—|||—

Investigators throughout Europe rushed to repeat Galvani's experiments. Following the introduction of Galvani's work to England by Tiberius Cavallo, who read two letters by Volta on the subject to the Royal Society, several British scientists undertook extensive researches, including Richard Fowler and John Robison.[67] In France, scientists were at first so incredulous that the French academy appointed a series of commissions to ascertain whether the effects claimed by Galvani were real.[68] Although a French commission conceded eventually that animal electricity was a phenomenon to be taken seriously, few in France pursued the subject until after 1801. However, in Italy, Britain, and Germany, many investigators rapidly refined Galvani's findings.

In general, these researchers employed apparatus that were endless variations on the basic Galvanist pattern. The work of Eusebio Valli, a physician, was typical. After repeating Galvani's experiments on frog legs, Valli extended his studies to other animals similarly prepared, including lizard, mouse, rat, eel, lark, rabbit, kitten, dog, and horse; occasionally he also experimented with living animals.[69] On the basis of these studies Valli showed that animal electricity was present throughout the animal kingdom. And, in experiments with different metals, such as lead, antimony, and gold, he found that all could conduct animal electricity; even water and other fluids could play this role.[70] Finally, he varied the metal foils used to wrap the nerves and ranked them according to the magnitude of muscular movements.[71] These extensive substitutions for the basic elements of Galvani's apparatus enabled Valli to support broad generalizations about animal electricity.

Curiously, Valli was sensitive to the possibility that his manipulations of animals, living and dead, would arouse charges of cruelty.[72] His experiments not only accumulated a sizable body count, but his methods sometimes inflicted prolonged suffering. For example, he killed one dog with arsenic, taking in stride its nine hours of painful demise, and starved another for twenty-three days before dispatching it.[73] In his own defense, Valli acknowledged that, in the dogged pursuit of science, he had also suffered: "I have sacrificed every comfort and convenience of life, with all the pleasures of society . . . [and] experienced every possible hardship and fatigue."[74] This defensive posture indicates that, in the last decades of the eighteenth century, there was some opposition to torturing creatures. Not until the early nineteenth century, however, would societies be founded to prevent cruelty to animals.[75]

Abject cruelty aside, Valli's studies enabled him to justify explicitly the idea that electricity is the basic principle of life, constituting its "vital power."[76] As the nervous fluid, not only is electricity involved in volitional behavior, but it also makes possible involuntary motions such as digestion and yawning. Thus, when all electrical activity ceases, death is irreversible; however, the decline of electrical activity after death occurs gradually.[77] Tantalizingly, Valli also alluded to research by one Dr. Abildgaard, who administered a "violent shock" to the head of fowls, which rendered them seemingly lifeless and then "re-animated" the birds with gentle shocks to the heart and lungs.[78] Studies such as these would give substance to the concept of "apparent death" and lead to electrical technologies for reviving human accident victims (see chapter 7).

—⊣||⊢—

Despite painstaking research by Valli and others, understanding the *mechanism* of animal electricity remained out of reach. Indeed, many researchers disputed Galvani's model for the electrical operation of the nervous system and offered their own. Galvani's most persistent and successful critic was the University of Pavia's Alessandro Volta, who learned of Galvani's research in March 1792 and swiftly began his own investigations. Although at first entertaining the possibility of a peculiar animal electricity, Volta—always the electrophysicist—rapidly rejected it.

One of Volta's most telling experiments was performed with the simplest of apparatus, which nonetheless disclosed new and exciting effects.[79] He took a dead frog and, without dissecting it, placed a silver coin on one of its thighs while wrapping the other in tin foil. When he connected the dissimilar metals with a conducting thread, the frog's legs convulsed strongly. In pondering this strange phenomenon, Volta began to formulate new physics. Electricity can be generated, he suggested, at the junction of two different metals. To convince himself that such a simple device produced electricity, he engaged a most sensitive instrument: his tongue.[80] When Volta placed two pieces of metal on his tongue and curled it, bringing the metals into contact, he experienced the same sort of "acidulous" taste caused by any electrical discharge.[81] To assure himself that he was not merely tasting the metals, he put them in a glass of water and, placing his tongue on just one piece of metal, still sensed electricity. Later in 1792 Volta published the first version of his new theory in which dissimilar metals in contact become "true motors of electricity."[82] Volta's theory is known as the "contact" theory of electricity.

Other experiments done with dissected frogs, lambs, and Leyden jars helped Volta conclude by the end of 1792 that the concept of "animal" electricity was superfluous. In the living animal, he suggested, muscular contractions resulted from an externally caused imbalance in electrical fluid.[83] Animals, especially frogs, were merely sensitive electrometers. Thus, there was no need to postulate a peculiar kind of electricity inside animals, for the force that causes contractions in all cases comes from the outside.

From this point onward, hostile camps formed around the theories of Galvani and Volta. Among the supporters of Galvani's views were distinguished scientists, including Lazzaro Spallanzani and Alexander von Humboldt. Galvani's most ardent supporter was Giovanni Aldini, another nephew and a research assistant, who took up the standard, as Thomas Huxley would for Charles Darwin decades later. But the Volta camp also had doughty followers, including respected physicians and chemists. With one early exception, Galvani himself did not respond to Volta's critiques; others took up the charge on his behalf. But Volta's own volleys kept coming.

As Marcello Pera points out in his aptly titled book on this controversy, *The Ambiguous Frog,* both camps could explain some effects but had to ignore or finesse others. The frog experiments in particular admitted of many interpretations. Aldini noted, for example, that a single piece of metal connecting nerves and muscles caused the frog legs to quiver; but this effect was not allowed in Volta's new physics, since two *different* metals in contact were needed to generate electricity.[84] To clinch the point, Aldini constructed some simple but ingenious apparatus—using glass and wooden vessels, mercury, and dissected frogs—to show contractions in the absence of Volta's contact electricity; he also demonstrated contractions using only a rod of carbon—a nonmetallic conductor.[85] Eusebio Valli, another Galvanist, delivered the coup de grâce: using himself as a conductor, linking the nerves and muscles of a dissected frog, he managed to make its legs move.[86]

This apparent proof of animal electricity won over many doubters, but it had no direct bearing on the nature of Volta's contact electricity. Nonetheless, in pursuing their case against Volta's theory, Valli and other Galvanists denied altogether that contact electricity existed. Calling into question the experimental findings of the foremost living electrophysicist was not a smart move. Volta would respond forcefully with new experiments, but first he gave a little ground by graciously conceding that he had gone too far in claiming that two different *metals* were needed to excite muscular action; and he eventually abandoned his belief that electrical imbalances in animals always had an external origin. However, Volta did not embrace the idea of a unique *animal* electricity. More important, this careful experimenter insisted that no one had refuted his discovery of contact electricity.

In seeking to show the reality of contact electricity, Volta decided that animals added unnecessary complexities to an experiment. Instead of using a frog or his tongue as the electrometer, Volta tried William Nicholson's doubler and his own condensing electrometer (a combination Leyden jar and electrophorus), both of which could amplify minute charges. With the charge increased, the electricity—formed by numerous combinations of metals—registered on Volta's thin-straw electrometer.[87] Published in 1797, these findings vindicated Volta's theory of contact electricity but had nothing to say about *animal* electricity per se.

Both Galvani and Volta had discovered important new effects, but neither took advantage of opportunities to compromise on theory. In the meantime, political events, in the form of Napoleon's invading army, overtook both protagonists. Galvani publicly refused to swear allegiance to the new republic and, on April 20, 1798, was stripped of his professorship; he died a few months later. Volta also objected to the new regime, but nonetheless made his accommodation with it and continued his research.

—|ı|—

Although Galvani himself was out of the picture, the Galvanists were not, and so Volta carried on the battle. In 1800, an auspicious year for a new technology and the new science it would beget, Volta delivered a bombshell. Building on his earlier research, he reported two new apparatus that generated electricity from metals in contact.[88] The first was the pile, known almost universally afterward as the "voltaic pile."

The pile was literally a stack of small disks, alternating silver *(argento)* and zinc *(zinco)*, each pair separated by a piece of pasteboard soaked in saltwater. With only twenty silver-zinc pairs, their force augmented by his condenser, the pile registered a strong reading on a gold-leaf electrometer (see chapter 8) and, more impressively, yielded a spark. When Volta reconfigured the pile so that its ends were conducted into cups of saltwater, the pile generated still more effects (Fig. 16).[89] Indeed, Volta sensed a slight shock after he immersed several fingers in one cup and placed fingers of the other hand at different places along the pile. Volta even noted a slight prickling feeling when he was in a circuit with only three or four pairs.

A conclusion obvious to Volta was that the pile's force could be intensified by adding more metal pairs, apparently without limit. To facilitate this piling on, Volta made a version of the device with glass or wooden rods to hold the disks in place, which easily accommodated eighty or a hundred pairs.[90] He also tried out different metals and found that many combinations yielded the same effects; similar experiments in the next century would eventuate in the electromotive series.[91]

Intent on expunging from science all vestiges of "animal" electricity, Volta employed his pile as a model for the torpedo's electric organs. In these organs, he suggested, resided small columns with thin conducting membranes that generated current like the pile's metal disks.[92] The torpedo's power came not from an animal electricity stored in miniature Leyden jars, but from ordinary electricity produced by contact among different conducting tissues. It was a purely physical process involving nothing unique to animals.

Figure 16.   Two versions of Volta's pile. (Adapted from Volta 1800, pl. 17, figs. 2 [right] and 3 [left]. Courtesy of the Bakken Library, Minneapolis.)

Volta's second apparatus was no more complicated than the first. His "crown of cups" was just a line of saltwater-filled cups; these could be made of almost any material, such as wood or clay, but crystal was best.[93] The cups were connected by metal arcs, strips of silver and zinc joined in the middle, which dipped deeply into the saltwater. A device of greater force (or intensity) could be made by chaining forty or even sixty cups in a row. In testing this invention, Volta felt a shock simply by dipping fingers of each hand into cups some distance apart; the farther the cups were separated along the chain, the stronger the shock. Evidently, the effects produced by the pile and crown of cups were identical.

Volta also discerned the influence of his new apparatus on other human senses. Using his body as an instrument, he played around with various circuits that included parts of his head. He found, for example, that when he electrified an eye, he saw a bright flash.[94] In one experiment, he was able to produce—all at once—a flash of light, convulsions in the lips and tongue, pain at the tongue's tip, and a sensation of taste. But Volta had still more effects to probe. He took two metal rods, which were connected to a pile of thirty or forty pairs, and introduced them into his ears. At the instant the circuit was completed, he felt a shock to the head and, simultaneously, a sound he had trouble putting into words.[95] Fearing that shocks to the brain might be dangerous, however, Volta declined to repeat the exercise. On the basis of these somewhat uncomfortable experiments, Volta concluded that the nervous system operated electrically (as Galvani and countless others had insisted) but with contact—not some special animal—electricity.[96] More important, he claimed that the pile, not the Leyden jar, was the most appropriate model for how the nervous system got its juice.

In discussing his new apparatus, Volta assigned no fundamental role to the saltwater or other conducting medium—today called an "electrolyte"—

that connected the metal pieces. And he did not observe or report effects suggesting that his pile and crown of cups involved chemical processes. But several investigators suggested almost immediately that Volta's inventions were the first *electrochemical* batteries; their views did not immediately prevail.

Volta gave the name "artificial electric organ" to his inventions, but that cumbersome label failed to stick.[97] Rather, in an ironic twist, the technologies (and the electricity they generated) were universally called galvanic or galvanism, a term Volta himself occasionally used.[98] But for Volta, this term hardly paid homage to Galvani; rather, it underscored the identity between the electricity in Galvani's frogs (and other living creatures) and in Volta's new laboratory apparatus. Although Volta did not grasp the electrochemical basis of galvanism, his stunning new technologies and sterling reputation carried the day; Volta was victorious. And, for several decades, "animal electricity" became a taboo term.

Although "animal electricity" had been consigned to scientific oblivion, the Galvani-Volta conflict in its later years was not about *whether* animal nervous systems functioned electrically but about *how* they generated that electricity. Neither man got it quite right. But, in inventing new laboratory apparatus to explore this question, and in advancing one-sided theories, both physician and physicist established beyond any doubt the electrical basis of the nervous system.

Published in 1800 in the *Philosophical Transactions*, Volta's first paper on the invention of the pile and crown of cups stirred great interest. Using common materials—a pile could even be made with ordinary coins—many people throughout the scientific world, including Joseph Priestley and Martinus van Marum, built their own versions of Volta's inventions and rapidly published the results of new experiments.[99] Predictably, instrument makers brought them almost immediately to market.[100] Not since the appearance of the Leyden jar a half century earlier had a new electrical technology fomented so much scientific activity. No longer could anyone doubt Volta's claim that certain configurations of metals in contact, perhaps all conductors, could generate electricity.

Compared to high-tension, low-current electrical machines, galvanic devices—our present-day "batteries"—produce electricity of low tension and high current, a pattern that Volta himself pointed out in 1802.[101] This technology furnished an exceedingly convenient and versatile tool for performing new experiments in physics and chemistry (see chapter 10). Because experiments with galvanic apparatus led to Hans Christian Ørsted's recognition of electromagnetism in 1820 and, soon thereafter, to Michael

Faraday's development of new kinds of motors and generators, Volta's inventions are usually regarded as marking the beginning of the modern electrical age. Yet Volta's pile and crown of cups arose originally, not because he wanted to create new equipment for physical science, but because he was obsessed with solving intriguing problems in electrophysiology.

⊣|⊢

True to its beginnings, galvanism became the favorite source of electricity for physiological investigations. Among the people who quickly entered this research arena was Galvani's standard bearer, his nephew Giovanni Aldini (1762–1834), who in 1794 had become professor of natural philosophy at the University of Bologna.[102] Claiming an interest in the possible use of galvanism for reviving the "apparently dead," Aldini undertook research on dead chickens, sheep, and oxen. These provocative studies were a mere warm-up exercise for his peculiar experiments on human bodies. The latter research occasioned much criticism, yet Aldini nonetheless persisted for a few years.

In one series of studies, Aldini took the hand of a corpse and connected it to one end of a voltaic pile; on the pile's other end he ran a conductor to the corpse's ear. With the circuit thus formed, Aldini observed some surprising effects, including "various contractions, sometimes of the fingers, sometimes of the hand, and sometimes even of the whole arm. The fingers bend and unbend very visibly, and sometimes the whole fore-arm is carried towards the chest."[103] On the basis of these findings, Aldini speculated that different patterns in muscular contractions might correlate with the deceased's sex, age, and temperament.

These experiments were performed on the bodies of people who had died of diseases. Noting that such bodies might have wasted away somewhat, their vital fluids perhaps spoiled, Aldini sought a source of fresher specimens. In 1802 the government of Bologna granted him permission to "galvanize" the bodies of executed criminals. The corpses were not entirely intact, for the act of execution had severed the head; but this major mutilation did not deter Aldini. In one experiment with a dead head, he first moistened the ears with saltwater and then inserted the wires coming from a voltaic pile. According to Aldini's account, "All the muscles of the face underwent frightful contractions" and the eyelids in particular became quite animated.[104] After connecting the pile to the tongue and one ear, Aldini noted still more movement, including withdrawal of the tongue.

Bologna's executioner was able to produce a second specimen on the day of Aldini's experiments. After repeating the earlier experiments, Aldini had an inspiration: he would make the two severed heads perform in tandem. So he pushed them together on a table and connected the voltaic pile to the left ear of one and the right ear of the other. The result even animated Aldini: "it was surprising and even frightful to see these two heads making at the same time horrible contortions, as if at each other, so that some of the spectators who were not prepared for such results, were exceedingly terrified." [105]

For a time, studies like Aldini's on human bodies were pursued in several European nations.[106] The French even appointed a commission of distinguished scientists to learn whether galvanism could be used to revive battlefield casualties. A German physicians' club experimented on twenty freshly executed brigands, comparing the stimulative effects of galvanism to that of electrical machines. And in 1803 Aldini himself conducted experiments on an executed criminal at the Royal College of Surgeons in England. Writing more than a dozen years later, Aldini acknowledged that additional experiments along these lines yielded no new knowledge. Continuing to torture the dead in this way was now regarded as "unjust and immoral." [107] He even expressed concern that public lecturers might adopt such antics, which would be above all a "prostitution of galvanism." [108]

-|||-

In the first years of the nineteenth century, galvanism was all the rage, not only in scientific books, journals, and society publications, but also in magazines and books that catered to less specialized readerships. Even *The Gentleman's Magazine*, which decades earlier had lost most interest in electrical technology, contained an article on galvanism in 1803.[109] And, doubtless, the ghoulish experiments carried out on cadavers throughout Europe, often in anatomical theaters, had become common knowledge.

Galvanic apparatus, like electrical machines and Leyden jars, was acquired by hobbyists and youthful enthusiasts of science. Percy Shelley, among his other scientific bric-a-brac, had a galvanic device in his college residence.[110] Perhaps he had even read about the experiments of Aldini and others on severed heads; surely someone as scientifically literate as the young poet could not have been unaware of these sensational happenings. Apparently, he passed along these little knowledge nuggets to Mary Shelley; she may also have learned about Aldini's experiments from discussions of science in her own home. In any event, when Shelley created *Frankenstein*, the idea of revivifying human body parts by electrical stimulation was already old.

Frankenstein is a charming and engaging book about the exploits of Victor Frankenstein, self-trained in natural philosophy. Shelley constructed it as an autobiographical memoir of Frankenstein, who in turn frames it as a cautionary tale about tragedies that can follow from scientific passion unchecked.

As a youth, Frankenstein becomes captivated by the possibility of discovering "the elixir of life," for it might enable him to "banish disease from the human frame, and render man invulnerable to any but a violent death!" [111] Frankenstein's undisciplined readings in natural philosophy acquaint him with the "more obvious laws of electricity." A turning point comes when, instructed by an unnamed natural philosopher, Frankenstein learns about his theory "on the subject of electricity and galvanism, which was at once new and astonishing to me." Suddenly he sees that they turned the muddled old theories he had been reading into nonsense.

Frankenstein embarks upon a university education, still impelled to learn "the causes of life." He studies chemistry, physiology, and anatomy, but in these subjects he finds no answers. So in cemeteries and charnel houses, the methodical scientist examines bodies in various stages of decay, beholding in minute examinations "the corruption of death." After this seemingly endless toil, Frankenstein admits that "I succeeded in discovering the cause of generation and life; nay, more, I became myself capable of bestowing animation upon lifeless matter."

People acquainted only with movie versions of *Frankenstein* might be surprised to learn that, in the book, Victor reveals neither the secret of life nor the technology used to animate his monster. He withholds this crucial information in a benevolent spirit, wishing to spare readers from the same "destruction and infallible misery" that befell him. At the same time, Shelley enticed the reader to infer that electricity, perhaps galvanism in particular, is the secret ingredient. Indeed, when describing his first success, Frankenstein says that he "collected the instruments of life around me, that I might infuse a spark of being into the lifeless thing that lay at my feet." An eye of the creature begins to open, "it breathed hard, and a convulsive motion agitated its limbs." And so echoed in *Frankenstein* the experiments and electrical apparatus of Galvani, Volta, and Aldini. This connection Shelley finally made explicit in the preface to the 1831 edition of *Frankenstein*: "Perhaps a corpse would be re-animated; galvanism had given a token of such things; perhaps the component parts of a creature might be manufactured, brought together, and endued with vital warmth." [112]

Frankenstein captivated readers, not merely because it told in graceful prose a macabre, spellbinding tale, but because it challenged the ideology that science, pursued unfettered, could bestow only benefits on people and on society. Scientists themselves simply did not entertain the possibility that their products—new knowledge and technologies—might be put to nefarious uses. It was left to a young woman, perhaps reacting to Erasmus Darwin's imaginative projections, to raise this issue—an issue that resonates today more strongly than ever.[113]

Electrical experimenters of the eighteenth century were especially enthusiastic about the potential of their technology to alleviate human suffering. A half century before Volta's invention of the electrochemical battery, the physiological findings of Nollet, Jallabert, and others had already formed a basis for using electricity in medicine. In these applications, electrical technology rarely did any harm and sometimes brought patients comfort and relief. We now turn to medical electricity viewed in relation to other medical therapies commonly employed in the late Enlightenment.

7.   First, Do No Harm

Nestled on a knoll near the majestic Potomac River in Virginia, the Mount Vernon estate was home to George and Martha Washington. From their porch the couple could gaze across the river to the Maryland shore, watch squirrels cavort, or spy bald eagles soar. In spring they saw trees leaf out, the first vegetables inch skyward, and carefully tended gardens dazzle strollers with beautiful blooms. A time of renewal and rebirth, spring spread optimism. Summer sweltered and plants prospered, luxuriated, overwhelmed the land. Fall's coming brought a blazing mosaic of yellows, oranges, and reds. But in late fall and winter there was only drabness, for all of nature was painted in shades of gray and brown. The trees' vulnerable nakedness hinted that not all living things would survive to see another spring.

As was his habit, one day in December 1799 George Washington took a horseback tour of the Mount Vernon estate, overseeing projects and sometimes pitching in. By early afternoon, the retired president noted later in his diary, "it began to snow, soon after to hail, and then turned to a settled cold rain."[1] Undaunted by the bad weather and protected by an overcoat, Washington continued his rounds, not returning to the comfort of his warm home until after 3 P.M.

The next morning Washington awoke with a painful sore throat.[2] As the day wore on, the discomfort worsened and he became quite hoarse yet did not seek medical attention. That night Washington was afflicted with a fever and had difficulty breathing. By next morning he could speak only with much effort, and so at last Dr. James Craik of nearby Alexandria was sent for. In the meantime Washington directed his overseer, Rawlins, to bleed him, and—despite Martha's objections—it was done. Home remedies, including an emetic and bathing the feet in warm water, were tried, but to no avail. Dr. Craik came quickly and began ministrations: he applied a mixture of powdered blister beetles—known as Spanish fly—to the patient's throat, ordered him to gargle with sage tea and vinegar, and bled him once more. Choking on the gargle, the patient did not improve; and, after yet another bleeding, two more physicians were summoned.

When Drs. Gustavus Richard Brown and Elisha Dick arrived in the late afternoon, they at once administered a purgative and, not surprisingly, prescribed more bleeding. Washington, his breathing ever more labored and unable to eat or drink, convinced that this illness would be his last, finalized his will. In the meantime, the physicians applied to his feet and legs a poultice of wheat bran. Later that evening, his inflamed and swollen larynx completely closed, Washington expired without complaint.

⊣||⊢

With the benefit of hindsight, later critics contend that America's senior statesman was killed not by the disease but by his doctors.[3] The course of treatment, centered on repeated bleedings and purgings, produced severe dehydration that, at the very least, hastened his death. Although Washington's treatment was accepted practice at the time, recommended in leading medical texts, he might have been saved by a tracheotomy. His doctors did discuss this possibility; however, the senior-most physicians on his team rejected the operation, arguing that tracheotomy was still an experimental procedure that lacked the blessing of medical authorities.[4] And who among them would have been willing to risk opprobrium by slitting the great man's throat? Regardless of their efficacy (by modern standards), new therapies gained ground slowly among established physicians in the face of deeply entrenched medical beliefs and practices.

During the eighteenth century, bloodletting was often the physician's first intervention because medical theory maintained that contaminated blood had to be removed from the body. As we know today, bleeding could cause infection from a dirty surgical instrument and dehydrate the patient to a lesser or, in Washington's case, greater extent (he was relieved of more than five pints of blood). Ironically, dehydration was believed by Enlightenment physicians to promote wellness. Along with bleeding, then, physicians induced fluid loss by giving strong purgatives and emetics along with the seemingly obligatory enema, which the doctor or his assistant administered with a sizable brass syringe. In addition to being dehydrated, patients were sometimes slowly poisoned by prescription medicines containing toxic substances, including compounds of lead, mercury, and arsenic. Small wonder that calling for a physician was for many people a last resort, done in desperation after every patent medicine and home remedy had failed.[5] Home remedies and patent medicines were a mixed bag, but most caused no serious injury. Undeniably, however, physicians did kill. Such was the state of medical practice in the Enlightenment.[6]

Usually, experiments with alternative treatments were carried out first on the fringes of established medicine, often by people who were not physicians at all. As one physician at the time admitted, "Very few of the valuable discoveries in medicine have been made by physicians."[7] The eighteenth century was in fact a time of simmering dissent from medical orthodoxy. Sometimes practitioners on the fringe brought to light promising therapies, but their acceptance or co-optation by medical authorities was a slow process. It was in this environment—a cacophony of conflicting medical concepts and treatments—that electrical treatment began as a fringe therapy, flourished modestly because of patient demand, and even achieved a modicum of acceptance by the establishment. Surprisingly, eighteenth-century electrotherapists initiated technological trajectories that ultimately led to some of today's electrical treatments. In this chapter I survey the place of electromedical practices and technologies during the late Enlightenment.[8]

⊣||⊢

That electricity had effects on living organisms, even humans, did not go unnoticed outside the small community of electrobiologists. If electricity could hasten the motion of bodily fluids, some reasoned, then might it not have therapeutic effects, especially on diseases caused by "obstructions" that hindered normal flows of blood, menstrual fluid, digestive juices, or "Nervous Juice"?[9] More generally, it was believed that a force capable of killing healthy animals might, when applied judiciously, heal sick people.[10] And, as in animals, perhaps electricity could cause a stopped human heart to resume beating.

Almost immediately after the first investigations in electrobiology during the mid-1740s, a community consisting of electrotherapists and their patients began to form. A motley crew, the people who made clinical use of electricity included electrobiologists, electrophysicists, instrument makers, physicians, surgeons, apothecaries, ministers, and people of countless other occupations. Some practitioners justified their treatments explicitly on the basis of theory and physiological findings, but others did not; some merely added electrotherapeutics to their treatment repertoires, and others specialized in electrifying patients; some became esteemed healers, others were called quacks. In any event, the patients came: rich and poor, men and women, young and old—many having exhausted all conventional treatments.

One of the earliest and still the most compelling account of an electrical cure came from Jean Jallabert, an electrobiologist and Swiss professor. Having shown that electricity could stimulate motion in paralyzed muscles, Jallabert was eager to learn if he could actually ease paralysis. He did not have to search far and wide for a paralytic, for one was brought to him the day after Christmas in 1747. The man, named Noguès, was fifty-two years old; he had suffered an accident fifteen years earlier that left his right arm immobile, completely devoid of feeling, the muscles badly atrophied. And so, under the supervision of a surgeon, Monsieur Guiot, Jallabert and Noguès began a lengthy course of electrical therapy.[11] Regrettably, Jallabert furnished few details on exactly how he delivered electricity to Noguès's lifeless limb. We do know that, following Nollet and others, Jallabert employed a globe machine, rubbed by hand, presumably by an assistant. Charges were accumulated in a Leyden jar, which delivered shocks, in some manner painstakingly applied to individual muscles in Noguès's arm and hand.

The first change Noguès noticed was the return of some sensation. Then came the voluntary movement, with great difficulty, of fingers. Repeated treatments brought back the arm's muscle tone and natural color. Two months into therapy, Noguès was able to place a hat upon his head; he was so overjoyed at this accomplishment that tears welled in his eyes. Soon the muscles in his arm grew visibly larger and firmer, and he was able to lift a moderate weight. Despite one mishap—an overcharged Leyden jar exploded, causing the patient some pain—he returned faithfully for his appointments. However, bad weather, impeding travel, terminated the treatments, and Jallabert's moving account ends.

The successful treatment of paralyzed limbs by Jallabert and others was widely publicized in scientific journals and in magazines; some even reported the cure of an immobile tongue.[12] Newspapers in the American colonies trumpeted great cures that had been achieved in Europe, raising the hopes of paralysis victims and their families. Desperate to try anything, the suffering sometimes besieged people possessing electrical machines, including Benjamin Franklin. Although lacking medical experience, the Philadelphian did treat several paralytics at his home.[13]

Probably unaware of the earlier accounts by Jallabert and others, Franklin chose a different course of electrotherapy, drawing strong sparks from the afflicted areas and then applying to them large shocks from two enormous Leyden jars. These quick treatments were not continued for many days. Even so, several patients enjoyed greater mobility of paralyzed limbs—but later relapsed. Circumspect in his claims, Franklin acknowledged that temporary improvements might have resulted from a patient's journey or elevated spirits. Perhaps, Franklin reflected, a different regimen of shocks might have produced lasting cures as others had apparently achieved. (Despite Franklin's rather limited foray into electrotherapy, during the nineteenth century treatments with electrostatic technology were often called "Franklinization.")

As electrotherapists gained experience and published their findings, a consensus emerged on the best treatment for paralysis. It was crucial, as Jallabert had shown, that a large number of small shocks be applied directly to stimulate the affected muscles. What is more, this treatment had to be continued almost daily for many weeks, perhaps months. Employing this regimen, electrotherapists achieved many successes. Yet because paralysis has many causes—stroke, accidents, and psychological trauma—shock treatments sometimes failed.

In addition to treating paralysis, many electrotherapists touted their work to relieve myriad other maladies. The more avid promoters compiled long lists of conditions responsive—at least once in a while—to electrotherapy, including fevers, deafness, blindness, headache, toothache, ulcers and sores, rheumatism, epilepsy, sciatica, cessation of menses, tapeworms, kidney stones, hemorrhoids, even severe sore throat.[14] Diseases with names long lost were also listed, such as King's Evil, chlorosis, dropsy, and St. Anthony's fire, as were fits, hysteria, heart palpitations, and feet—cold or "violently disorder'd."[15] In the course of dealing during many decades with these common ailments, electrotherapists throughout Europe and the American colonies arrived at relatively standardized practices and corresponding technologies for administering the electric elixir.

Afflictions such as hysteria and melancholy, which are today regarded as psychosomatic, were treated by means of the "electric bath."[16] In an electric bath, insulated patients held a metal wire connected to a machine's prime conductor that electrified or "electrized" them (Fig. 17). To ensure isolation from ground, they were made to stand on a resin cake or an insulating stool, sit in a chair with glass legs, or lie on a bed with glass feet. Such treatments could last hours, and so probably patients preferred not to remain standing. During an electric bath, patients, often with hair standing on end, might feel some warmth and be perspiring more freely than usual, but the treatment was painless. In fact, if the connection to the machine had been made first, before charge had built up on the prime conductor, they did not even receive a mild shock.

In addition to an electrical machine, administration of the electric bath

Figure 17.   A woman receiving the electric bath. Busts of Nollet and *right,*
Franklin decorate the treatment room. (Adapted from Barneveld 1785, pl. 1.
Courtesy of the Bakken Library, Minneapolis.)

required insulating furniture. The most common variety was a stool, having
four glass legs, on which patients could stand for brief treatments. Not surprisingly, some instrument makers sold standard insulating stools; Nairne
and Blunt, for example, offered a selection ranging in price from 12 to
18 shillings.[17] Beyond the stool, most electric furniture would have been
constructed by, or for, each electrotherapist, and the varieties were endless.
In addition to carefully crafted electrical furniture, practitioners made use
of jury-rigged arrangements. In one configuration, Pierre Bertholon placed
an ordinary chair upon a large low table measuring about 3 feet by 6 feet;
the table legs rested in large glasses or cups.[18] And one practitioner we encounter shortly used an electric throne. Electrical furniture was also sometimes employed in other kinds of electrotherapy.

Another variety of electric bath, this one more literal, was devised by Georg Schmidt in 1784. Seated in a half-barrel of electrified water, the patient allegedly obtained relief from hemorrhoids.[19] It is doubtful that Schmidt's electric *Sitzbad* was ever copied or commercialized.

A second mode of standard electrotherapy employed accessories, usually connected to an electrical machine, which could draw sparks from or apply an "electric wind" to an ailing part of the body. To treat a case of deafness, for example, the electrotherapist brought a pointed conductor, held by a glass handle, close to the patient's ear (Fig. 18, upper left). Some electrotherapists created accessories for treating specific limbs and organs. Conductors for eyes, especially, had wooden points, whose reduced conductivity relative to metal ones was thought to be safer for this most sensitive organ (Fig. 18, upper right).[20] Using a wooden point for three or four minutes a day, practitioners treated eye inflammation.[21]

An especially popular accessory for drawing sparks was a glass tube containing a metal wire. At the end of the tube that touched the patient, the wire came up just short of the opening. At the other end of the tube, the wire protruded through a cork where it could be attached by chain to the machine (Fig. 18, lower left). The operator grasped the glass tube, which provided insulation, and placed it where needed. A curved version of this accessory was said to be handy for treating conditions of the mouth, such as toothache and tumors (Fig. 18, lower right).[22]

The standard treatment for most localized afflictions began with the drawing of sparks. In the event that this brought no relief, the electrotherapist might turn to more drastic applications (see below).[23] Had electrical treatment been administered to George Washington, his doctors no doubt would have repeatedly drawn sparks from the throat. Spared bleedings and all the rest, an electrically treated Washington might have lived a lot longer.

The drawing of sparks was a therapy commonly applied to the "cessation" of menses. To receive treatment, a fully clothed woman sat on an insulated chair. A chain placed at the small of her back connected her sacrum to a charged prime conductor. In front of the patient stood a grounded stand holding a long pointed conductor that was aimed at, but did not touch, her abdomen (Fig. 19). Virtually every electrotherapist insisted that this treatment often worked, and even staid medical textbooks sometimes mentioned it as an option—but more on that below.[24]

Some treatments required that a "Current of the Electric Fluid" be passed through a part of the body using two conductors; in most cases, actual shocks were delivered, often from the discharge of a small Leyden jar.

Figure 18. Medical accessories for drawing sparks. (Adapted from Mauduyt 1784, pl. 2, figs. 11–14. Courtesy of the Bakken Library, Minneapolis.)

Figure 19. A woman being treated for "cessation of menses." (Adapted from Mauduyt 1784, pl. 2, fig. 16. Courtesy of the Bakken Library, Minneapolis.)

Needless to say, the patient felt each discharge as a weird sensation or even as pain. An operator could reduce the unpleasantness by placing a small metal disk or piece of foil directly on the patient's skin.[25] In contrast to most conventional therapies, the discomfort of shocks did not persist beyond the moment.

The appliances perfected for shock treatments were sometimes quite imaginative. Perhaps the most common was a pair of metal conductors, called *directors*, which had glass insulating handles. At the business end of directors were small copper or brass balls that could be placed, for example, on either side of a knee joint or near the elbow and wrist to electrify a forearm (Fig. 20). Typically, directors were connected to an electrical machine

Figure 20.   Electrotherapist using directors to apply current to a child's arm. (Adapted from Adams 1799, frontispiece. Courtesy of the Bakken Library, Minneapolis.)

or Leyden jar by chains or bendable wires, which provided flexibility for precisely positioning the metal balls (Fig. 21, left).[26] All such directors had one disadvantage: the electrotherapist or his assistant had to hold them throughout treatments sometimes lasting more than an hour. One solution was to employ jointed metal conductors that stayed in place without continuing human involvement. These extended directors could be hung on insulated stands and connected by chains to the machine or, as in Nairne's version, actually plugged into the machine's two prime conductors (Fig. 21, right).[27]

Still other technologies were created for shocking specific parts of the body. The ingenious Bertholon described a device for treating toothache.[28] It consisted of a tiny wooden box, open at opposite ends, with the shaft of a bird's feather inserted into one side. A wire ran inside the shaft—which furnished insulation—and protruded through the box's opposite side (Fig. 22, middle, right). The box was carefully positioned in the mouth so that the wire touched the aching molar; the lips held fast the feather's shaft. To make the second connection, a metal stud was placed on the face, secured by a silk band tied atop the head (Fig. 22, top, right). Because the mouth is very sensitive to electricity, this treatment may have been more uncomfortable

Figure 21. Flexible directors for administering shocks. (Adapted from Nairne 1783, pl. 3, figs. 2 [right] and 6 [left]. Courtesy of the Bakken Library, Minneapolis.)

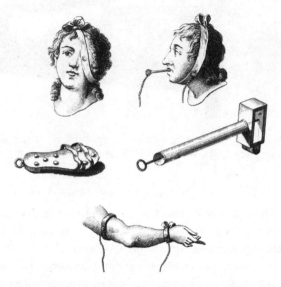

Figure 22. Accessories for administering shocks. (Adapted from Bertholon 1786, pl. 6, figs. 43 [bottom], 44 [top, left], 45 [middle, right], 46 [top, right], and 47 [middle, left]. Courtesy of the Bakken Library, Minneapolis.)

than the ailment. For foot problems, Bertholon devised an electric sandal with metal studs that contacted the heel and ball of the foot (Fig. 22, middle, left).[29]

The electric bandage, perhaps the most elegant and versatile appliance of all, was a silk ribbon with holes along its length that could receive a brass button.[30] In one application, two bandages were tied in place on either side of the area to be shocked (Fig. 22, bottom). Eyehole screws on the buttons permitted the bandages to be electrified through chains. An electric bandage could even be used to shock an eye (Fig. 22, top, left). And it was made in several sizes to fit arms and legs either slender or stout. According to Bertholon, the electric bandage could apply "the electric spark to any part of the body one wants, and with more certainty than any other method."[31]

Mindful that a shock too strong could at the very least diminish a patient's enthusiasm for repeating the treatment, electrotherapists tried to prevent overdosing. Timothy Lane, an apothecary who practiced electromedicine, invented the most commonly used device, a "discharging" electrometer that soon carried his name. He reported his invention to Benjamin Franklin, who communicated the letter to the Royal Society, and it was published in the *Philosophical Transactions.*[32] The Lane electrometer was simply two metal balls, one fixed, the other adjustable, which formed a variable spark gap—the greater the distance the spark had to jump, the stronger the shock would supposedly be. Electrotherapists, who doubtless received shocks all the time, surely knew that the Lane electrometer was inaccurate, its settings only erratically related to a shock's severity.[33] I suspect that electrotherapists used the device merely to reassure patients that they were taking precautions against overdoses.

Bertholon, noting that "a shock too great may be harmful," perhaps the voice of unhappy experience, offered several straightforward means for regulating the dose.[34] The first was to count the turns made by the electrical machine, which dictated the total charge delivered to a Leyden jar; a kind of cyclometer could be used to keep track of the revolutions. The second was to use Leyden jars of different capacities, commonly indicated by their volume or square inches of metal foil.[35] Another way to lessen— but not eliminate—the chances of an overdose was to use only a relatively small electrical machine, be it globe, cylinder, or plate. Ideally, the electrotherapist would acquire a machine designed for medical uses.

—|||—

Observing the growing interest in electrotherapy, instrument makers began to produce electrical equipment tailored to medical applications. Around 1766 John Read of Knightsbridge, near London, brought to market a compact cylinder machine that came equipped with a Leyden jar and Lane electrometer (Fig. 23).[36] The most popular medical machine, whose robust sales began by the early 1780s and continued into the nineteenth century, was manufactured by Nairne and Blunt (Fig. 24); even Bertholon used one.[37]

With its cylinder of modest size, the Nairne machine boasted two prime conductors, one negative, the other positive, which perched on glass pillars (Fig. 24). Holes drilled in the prime conductors allowed accessories to be

Figure 23.   Read's medical machine with built-in Leyden jar and, *in foreground on pedestal,* Lane electrometer. (Adapted from Lane 1767, tb. 20.)

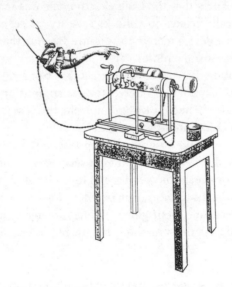

Figure 24.   Nairne's medical machine. (Adapted from Nairne 1783, pl. 5, fig. 1. Courtesy of the Bakken Library, Minneapolis.)

plugged in, such as extended directors and a Lane electrometer. Inside one prime conductor snuggled a Leyden jar, its ball protruding, offering charge. The compact machine was easily moved about and could be clamped to a small table for use. Two varieties were made, with cylinders 5 and 7 inches in diameter.

Nairne's medical machine, which was also advertised to be appropriate for "Philosophical Uses," was sold in several packages—accessories included. The most basic package was described as "A Patent Medico-Electrical Machine, Cylinder about five Inches Diameter, with two Conductors, Electrometer, two Sets of Flexible Tubes and Joints, two insulated Handles; a Pair of Directors for giving gentle Vibrations or Shocks, through any Part of the Body; three Pairs of different sized Metal Balls; two Brass and two Wood Conical Points, Chains, Clamp, and Box of Amalgam—in a coloured Deal [wooden-plank] Box." The price was £5 5s. Deluxe packages, one including the larger machine, an insulating stool, and a "luminous discharging rod," could be had for £12 12s.[38]

Although medical machines such as Nairne's were beyond the reach of the laborer's purse, urbanites of the middle class could have afforded one. Indeed, that these machines enjoyed healthy sales for many decades suggests that electrotherapists were not the only purchasers; apparently, some people were buying them for self-medication.

Another option for those desiring electrotherapy at home was to rent a machine and accessories. The ever-enterprising Nairne also made his medical packages available on a rental basis. Machines could be had for 10s. 6d. per week, provided that the renter placed a deposit on the package equal to its full value, which would be refunded at its return—less delivery charges and any damages assessed.[39]

Somewhat as George Washington's physicians did, electrotherapists made house calls, attested by the many case histories presented in electrotherapy books. To improve the portability of machines and accessories used for house calls, a few instrument makers sold compact equipment in carrying cases.

—||⊢

Despite a torrent of reports about electrical cures, most physicians of the eighteenth century failed to embrace electrotherapy. Why were Enlightenment physicians so wedded to traditional practices? One reason is that they had trouble determining which new therapies actually worked. At that time, medical knowledge was accrued by anecdotes called "case histories," not through rigorously controlled experiments.

Surprisingly, the bulk of folk remedies, apothecary concoctions, physicians' prescriptions, and the alcoholic or narcotic swill peddled by traveling medicine men appeared at least sometimes to cure. And in fact, as we know today, most acute illnesses—bacterial and viral infections, headaches, burns, sprains and muscular pain, cuts and abrasions, diarrhea—heal by themselves. Thus, almost any treatment might attract claims of curative powers. The illogic that leads to such conclusions is so beguiling that it still works on us: because the administration of a medicine preceded a cure, it must have been responsible for it. That is why pharmacopoeia grew very large, with many medicines recommended for treating any ailment.

The problem is that experimenting on living organisms one at a time, especially ill humans, cannot reliably link cause and effect. The same treatment, applied to the same condition, sometimes fails, at other times succeeds.[40] Complex organisms have complex response repertoires, and so the effects of a treatment must be sorted out by sophisticated experimental designs using large samples. Only in the mid twentieth century did the double-blind design, where neither doctor nor patient knows who is getting medicine and who placebo, become the gold standard for judging the efficacy of medicines. Well into the twentieth century, medical books still prescribed many eighteenth-century remedies, from purgatives to mustard plasters.

A second reason is that, at a time when many alternative medical theories and therapies competed for attention, most established physicians might have been loathe to depart from traditional practice for fear of tarnishing their reputation or being accused of quackery.

Third, most physicians, knowing little or nothing about electricity, were probably unwilling to adopt a novel and seemingly cumbersome technology, requiring new apparatus and training, in place of remedies that apparently worked, especially for acute illnesses. In the absence of clear, dramatic evidence of electrical cures' effectiveness, traditional treatments prevailed. The proponents of electrical technology, including a few distinguished physicians, came to believe that its greatest benefit was in helping patients who suffered from chronic conditions like paralysis and nervous disorders.[41] Precisely these long-suffering people were willing to try any new therapy that promised relief. If most physicians refused to deliver electrotherapies, other practitioners would.

—|ı|ⵏ—

The availability of electrical machines suitable for medical use in most instrument shops—for sale *or* rent—opened opportunities for electrical entrepreneurs. Many people having scant acquaintance with electricity began to practice electrotherapy, often alongside their previous professions. According to Dr. James Graham, whom we meet again shortly, unqualified practitioners were everywhere in London of the late 1770s: "I tremble with apprehension for my fellow creatures, when I see in almost every street in this great metropolis a barber—a surgeon—a tooth-drawer—an apothecary, or a common mechanic turned electrical operator."[42] The widespread availability of electrotherapy suggests that it was becoming accessible to an appreciable segment of London society. Even so, the vast majority of Londoners, not to mention the inhabitants of small towns and villages, might have had trouble coughing up enough coin for long-term shock therapy.

Inspired by Franklin's work and convinced of the manifold benefits of electromedicine when judiciously applied alongside assorted folk remedies, Reverend John Wesley, founder of Methodism, sought to make medicine—including electrotherapy—more widely available to the poor.[43] That a British clergyman would be concerned to bring benefits to the poor was a significant break from orthodoxy. As the historian William B. Willcox put it, the Anglican church "was of and for the propertied classes . . . and did not dirty its vestments in the poorhouse and the slum."[44] But Methodism taught that all who believed in Christ could be redeemed, even the lowliest beggar. Not surprisingly, Wesley was also an outspoken and early abolitionist.

In tending to the human body as well as the soul, Wesley wrote two medical books. The first, published in 1748, was a compendium of home remedies called *Primitive Physic*. For each disease or condition he supplied several remedies, sometimes including electricity. Baldness, for example, could be cured by rubbing the area, "morning and evening, with onions, till it is red; and . . . afterwards with honey." If that remedy failed, then the scalp could be washed with a boxwood decoction or electrified once a day. To the aged, he supplied nostrums for renewing strength: take tar-water twice a day, or drink a nettle decoction, or "be electrified daily," or "chew cinnamon daily and swallow your spittle." At the end of the book, Wesley also listed forty-nine diseases that could be cured by "ELECTRIFYING, in a proper manner."[45] The *Primitive Physic*'s dozens of editions in the eighteenth century suggest that innumerable families regarded it as a valuable medical resource. Evidently they exhausted its remedies before calling for a physician.

From Wesley himself, the poorer members of his congregations could obtain gratis the medicines prescribed in *Primitive Physic*. He established in Bristol a dispensary that served the poor for many decades; later he founded other dispensaries in London and Newcastle.

Wesley's second medical book—*The desideratum: or, Electricity Made Plain and Useful*—was devoted to electrotherapy. Like most such books, it was filled with case histories touting cures.[46] These accounts often mention the age, gender, and occupation of the patient and so furnish insights into who had access to electrical therapy. Apparently, men and women from all walks of life—gentleman, schoolmaster, plasterer; spinster, charwoman (in the language of the time)—as well as children of the rich and poor, enjoyed the benefits of electrical medicine.

Reverend Wesley himself obtained electrical machines, the earliest in 1756, and offered treatment during a specific hour daily. Through this practice, Wesley claimed to have brought relief to many hundreds, perhaps thousands, of people.[47] Other electrotherapists, including Bertholon in France, set up shop in hospitals. And by the end of the century "electrical dispensaries" had appeared in a number of British hospitals, usually operated by surgeons and apothecaries, seldom by physicians. However, a few books written by physicians did finally acknowledge the benefits of electrotherapy in some cases.[48]

⊣⏐⏐⊢

People on the higher rungs of London's social ladder could avail themselves of electrotherapy from a physician, Dr. James Graham. During a lengthy stay in the American colonies, Graham studied electricity in lectures at the University of Pennsylvania and resided for some years in New York City.[49] He returned to England and began practicing shop in Newcastle. After a controversy there with academic physicians had played out in newspapers, Graham retreated to London and opened a lucrative practice. His residence, which also housed treatment rooms, was situated on a terrace above the Thames River, not far from Westminster Bridge, within sight of the great abbey and Houses of Parliament—a distinguished neighborhood indeed.[50]

In his house, named the Temple of Health, people of wealth could get from Graham therapies that many mainstream physicians eschewed or condemned, including electrical treatment. His large and elaborate equipment was calculated to be awe-inspiring—and it was—capable of remarkable visual performance while merely giving an electric bath. Although instilling confidence in his patients, Graham's innovative technologies gave other physicians a somewhat less favorable impression. To them (and to many later observers) Graham was an infamous quack—but more on that later. First let us turn to Graham's dazzling electrical technologies.

Upon entering the Temple of Health, a visitor first passed through a long room containing a bench where patients of the "poorer sort" waited their turn for free treatment. Conspicuously lining the walls of this waiting room were "trophies" left by prior patients relieved of their afflictions, including crutches, walking sticks, eyeglasses, and ear trumpets (paying patients had their own entrance to Graham's establishment and saw none of these).[51] In a laboratory below, Graham fashioned potions with his own hands, endeavoring to ensure their purity, he insisted. This floor and the one above held treatment rooms whose furnishings—including electrical apparatus—were, to say the least, extraordinary.

The first treatment room, a spacious area 30 feet long, contained four Ionic pillars more than 10 feet tall, decorated with colorful lamps and candelabra.[52] Light from the lamps augmented the natural illumination streaming in from the chamber's stained-glass windows. In an arched recess facing the door was another pillar on which squatted an enormous Leyden jar, 21 inches tall and more than 1 foot in diameter. This was no ordinary Leyden jar, for on its exterior were figures and ornaments rendered in polished foils of tin, copper, silver, and gold. These colorful designs exhibited the "divine brilliancy of the electric or celestial fire . . . [and struck] with surprise, astonishment and delight, the eye and the heart, of every beholder."[53] The jar was purely for display, never used to deliver shocks. And there was much more.

Into a zebrawood frame was built an electric machine with a large cylinder of "snowy white polished glass."[54] An enormous prime conductor hovered above showy glass tubes and spirals, antique busts, and a "throne" on which patients sat—sometimes six or eight at a time. The prime conductor, fully 11 feet long and over 1 foot in diameter, was capped on both ends with large metal globes. Above the middle of the prime conductor were attached three more metal globes, in a triangle formation, topped by still another globe. Two glass pillars supported a 6-foot-long metal dragon, its tail resting on one end of the prime conductor.

Not far from the prime conductor was a metal shelf containing various medicines, both liquids and gases, in glass vessels. When electrified or subjected to magnetism, these potions were said to enter patients and work medical miracles. Many decades earlier, several experimenters claimed that, under electrical influence, odoriferous vapors passed through glass. And why not? Electricity was, after all, the subtle, etheric fluid that had so many surprising effects. However, several experimenters, including Benjamin Franklin, disproved the claim in the early 1750s.[55] Transferring fluids from glass vessels into patients was one bit of magic that not even electricity could perform. In the late 1770s only Graham's patients, and perhaps Graham himself, believed in the process.

Graham also employed more conventional equipment, including a massive air pump, to deliver medications "in the cure of diseases, especially in delicate, nervous and irritable constitutions."[56] With these machines, "aromatic and balsamic gums, herbs, seeds, flowers, chemical essences . . . are converted into mild balmy vapour for fumigating the lungs."[57] Hearing them described, did a prospective patient feel better already?

In lavishly furnished alcoves where both doctor and patient could sit in comfortable consultation, Graham magisterially prescribed further medicines. Prescriptions were sent to the floor below through a massive glass tube, decorated in brass, which terminated in the apothecary quarters. Once filled, the prescription was brought up by an assistant through a trapdoor in the floor.

A room down the hall was less plushly appointed.[58] Even so, a large and "noble" electrical machine impressed, its glistening glass cylinder harnessed in a mahogany frame, resting on a table covered with green cloth secured by gilt nails. The prime conductor, almost 5 feet long and about 8 inches in diameter, was supported by four glass columns. The electric fluid was conveyed by brass chains and rods to a bench raised on green glass insulators. As many as a dozen patients at a time could be treated on the bench, whose middle contained a "very powerful MAGNETIC SEAT." Doubtless, the patients treated gratis were herded into this second room.

Upstairs another treatment room was home to an electrical machine in the shape of a cross, its prime conductor covered in gold and silver foils.[59] Seated before this machine, on a chair raised upon an insulated platform, a patient could receive—through the entire body or diseased parts—"electricity, aetherial essences, vivifying air, and the magnetic effluvium." Graham claimed that this apparatus could cure, among other conditions, breast cancer and venereal disease. Although somewhat spartan by the standards of other rooms, the decor of this space still signaled elegance. Its carved doors were covered with creatures, the ceiling was painted with a "beautiful figure of health in full bloom," and opposite the central window was a royal print of the kingdom of heaven bearing an inscription in French.[60]

And then there was the pièce de résistance—the Great Apollo Apartment.[61] A cavernous space with a 15-foot ceiling, it extended 20 by 30 feet. Like the other treatment rooms, the Great Apollo Apartment had been designed for stunning visual performance. According to Graham, "words can convey no adequate idea of the astonishment and awful sublimity which seizes the mind of every spectator."[62]

What seized the spectator first was the existence of the temple itself, named for the Greek and Roman god who handled healing and light, prophecy and poetry, music and manly beauty.[63] The temple was roofed by a metal-covered dome, about 7 feet in diameter, supported by six glass columns. Divided into compartments, the dome's interior was decorated with vases, gilt flowers, and paintings. Although the dome itself was a prime conductor, its capacity was augmented greatly by three metal globes, more than 2 feet in diameter, placed above it in a triangle formation. The globes, capped by still another globe, held various medicines and were joined by brass pipes. More brass hardware connected the globes to metal fixtures under the dome, including a "magnetic crown," which could be raised and lowered. The fixtures sprayed upon the patient "nourishing dews, vivifying attractive or repellent effluvia and influences" propelled by air, magnetism, or electricity.[64]

During treatment in the dark, a patient seated at the Apollo temple's altar would be illuminated by the charged particles of the "nourishing dews" and undergo a "seeming beatification." Spectators would also see, on the dome, an evacuated "glass vessel in which the elementary fire is seen to play about like the most vivid and most beautiful *aurora borealis.*"[65]

In addition to the temple, the Great Apollo Apartment hosted a richly ornamented electrical pavilion, as well as "a magnificent electrical and magnetic throne," where patients could be bathed in essences—electric, magnetic, musical, and medicinal—in great splendor.[66] Music was supplied by an organ, but we do not know who played it or which composers were featured. A Bible was also conspicuously in view.

The Apollo apartment's extravagant decor, which included floor-to-ceiling stained glass windows, was dominated by the themes of fertility and marital bliss. One painting, for example, exhibited "a beautiful figure of fecundity or fruitfulness. She is a matron of pleasant countenance, resting on a couch; with one hand she is caressing two children, with the other she holds a cornucopia with fruits and flowers; and by one side, at her foot, is a rabbit."[67] Graham intended that these not-so-subtle symbols signal to patients that his treatments would promote "vitality" and thus procreation and conception.

To supply electricity for the varied equipment in the Apollo apartment, Graham had the service of two huge cylinder machines. These had been built a few years earlier for experiments carried out under the sponsorship of George III, on the proper shape of lightning conductors (see chapter 9). Charge was conveyed from the machines to the prime conductors by brass rods, wrapped with cords of white and blue silk, and a dragon (perhaps the mate of the other dragon)—6 feet long and covered with fine gold, whose crimson tongue and eyes blazed with "liquid fire."[68]

Two insulated chairs placed near the machines were used for directly administering an electric bath. Patients sat on a magnetic seat, covered with glass, their bare feet resting on rolls of sulfur. According to Graham, "there is no fever, rheumatism, cramp, spasm, or convulsion" that would not yield to this treatment, often in less than a minute.[69]

Somewhere in the Temple of Health there was also the Celestial Bed.[70] Troubled by the infertility many couples experienced, he offered a special electrical remedy for all who could afford the fee of £100. In this "very highly electrified" canopy bed, with magnetic and musical accompaniments, couples could "bask in a genial, invigorating tide of the celestial fire," where, he insisted, they could not fail to produce "strong, healthful, and most beautiful offspring."[71] To advertise the Celestial Bed, Graham gave public lectures on its manifold benefits. Enigmatic posters promised "A Naked Exhibition of ASSES, stripped of their ERMINE" and doubtless drew in scores of curious, well-heeled gentlemen.[72] And after Graham's lecture, curiosity may have led a few couples to try this most unusual electrical sleeping appliance.

Clearly, Graham's treatment rooms and electrical devices were a unique and colorful part of eighteenth-century electromedicine, in stunning contrast to the bare-bones machines and conductors used by other electrotherapists. No one had gone further and at such great expense to impress his patients. Graham not only boasted about the cost of his electrical installations—"several thousand pounds"—but also touted their superiority compared to all other apparatus he had seen in the world, which were "mean, awkward, and contemptible."[73] Indeed, they were, for they lacked cruciform conductors and electric thrones—not to mention Celestial Beds and fire-breathing dragons.

—|ı|—

Dr. James Graham is a notorious figure in the annals of British medicine. Roundly condemned in his own time, he is still regarded in many quarters as the paragon of quackery.[74] Quackery, however, is a judgment rendered by a particular social group—practitioners of conventional medicine. This group wields power by policing its boundaries, labeling as a quack anyone who strays beyond its norms of medical practice.[75] Hence those accused of quackery have not necessarily committed grave offenses against patients but may have merely run afoul of the establishment's prescribed procedures. One generation's quacks occasionally become the next generation's esteemed healers.

Graham's technologies and therapies were far from mainstream medicine. Moreover, he was an irritant to the establishment because he had the temerity to publish his views, questioning medical authority and condemning the methods of respected physicians. His evaluation of eye and ear therapies was typical. Mincing no words, Graham insisted that the principles underlying established practice were "fundamentally and in their nature wrong." Not surprisingly, he claimed that therapies built upon these principles were unhelpful, even harmful: "My blood creeps with horror when I reflect on the absurdity and barbarity of the methods generally employed for the relief or cure of their disorders, and on the shameful robberies and depradations of most of those men who have stiled themselves oculists or aurists."[76] Physicians did not delight in Graham's accusations that they were corrupt as well as incompetent.

Graham was also a serious threat because his relatively benign treatment methods, some of them perhaps even pleasurable, attracted patients in droves. He claimed to have helped countless thousands of sufferers, sometimes more than one hundred in a day, for a long list of ailments, including "putrid ulcerations of the throat."[77] Although a flamboyant self-promoter who exaggerated his successes, Graham probably never harmed anyone; after all, his patients were not cut into, bled or blistered, or dehydrated. And so they kept coming.

We need not look far for modern counterparts to Graham's practice, for aromatherapists, herbalists, diet doctors, and plastic surgeons come quickly to mind. Diet doctors and plastic surgeons harvest wealth in exchange for a socially desirable appearance. Rather than good looks, Graham promised his patients good health and vitality. And, if we can judge by testimonials, more than a few people believed that he had given it to them.

Why did Graham's ministrations cause many patients to improve? One explanation for his successes, and those of other electrotherapists, particularly in the treatment of nervous conditions, is the placebo effect. Through mechanisms still poorly understood, therapies that apparently lack a physiological basis for action can nonetheless have favorable outcomes. How strong is the placebo effect? In double-blind experiments, where placebos have been used as controls, commonly 20–70 percent of the patients receiving placebos report some improvement.[78]

Several factors seem to promote a favorable response to placebos.[79] First, the doctor must have a confident, authoritative presence, convincing in the claim that a specific medicine or therapy will bring relief; a charismatic personality and medical attire are also helpful. If the physician performs in a way that gives his words weight, the therapy is more likely to have a positive outcome. Judging by his written words, I conclude that Graham possessed a persuasive personality—ah, if only we knew what costume he wore.

Second, the place where the diagnosis occurs or where the prescription or therapy is provided should be "special"—that is, artifacts surrounding the patient must distinguish the medical setting from all others, advertising the healer's arcane knowledge and skill. Graham's Temple of Health was, if nothing else, a distinctive place. The patient was bombarded with a multisensory experience along with a lush pastiche of royal, religious, magical, and scientific icons. The treatment rooms created an aura of awe and emphasized Graham's role as an intermediary between the mundane here-and-now and the unfathomable etheric powers beyond. Even today, medical places have unique trappings, such as anatomical charts on bland-colored walls, scales and blood-pressure sleeves, examination tables covered in sterile white paper, and cabinets displaying an assortment of supplies and appliances. This assemblage of artifacts, found together nowhere else, serves to calibrate the patient's expectations and so contributes to wellness.

And third, the experience of undergoing a procedure—that is, intimate contact with medical artifacts, from swallowing pills to receiving injections to being cut open by surgical instruments—enhances positive responses. Electrotherapies were intrinsically intriguing and memorable, for the patient became joined to an electrical apparatus, experiencing unusual sensations—occasionally even pain. These procedures themselves promoted the belief that they were beneficial.

For the properly encultured patient, the factors of doctor, medical place, and procedure do, in many cases, ameliorate the condition. The successes achieved by electrotherapists are no more surprising than the cures chalked up by physicians who blistered, dehydrated, and poisoned their patients. People expecting to get better were primed for improvement—almost regardless of treatment. To exploit the placebo effect, the healer—whether shaman, physician, or quack—must choreograph a highly ritualized performance involving appropriate props, often on a special stage; Graham simply understood stagecraft far better than most.

-||-

Aside from a few practitioners like James Graham, Enlightenment electrotherapists were a circumspect and cautious lot, unlikely to make outrageous promises, much less endanger their patients. Sometimes they recommended against electrotherapy, as in advanced cancers or contagious diseases. Practitioners were also insistent, perhaps too insistent, that electrical treatments not be used on pregnant women. For example, while noting that electrotherapy was preferred for treating cessation of menses, the physician William Buchan warned, "Great attention and knowledge is required, in order to distinguish the arrest of the menses from a state of pregnancy. In the former, the application of electricity . . . is very beneficial; whereas, in the latter, it may be attended with very disagreeable effects."[80] He did not specify these effects, but a reasonable inference is that they included abortion of the embryo or fetus.

I wonder if electrical technology was actually used in the Enlightenment to induce abortions. At a time when giving birth to a bastard was still strongly stigmatized, an unmarried woman with child had a difficult dilemma, as did any woman with an unwanted pregnancy.[81] What were her options? After carrying the fetus to term, she could resort to infanticide, a practice common throughout the world. Dr. William Hunter reported that in eighteenth-century England women often killed their bastards at birth.[82] An alternative to infanticide was abortion, for which there were numerous folk technologies, many of which required the ingestion of specific plants.[83] Recourse could also be had to professional abortionists. In England, one practitioner brazenly boasted in the *Morning Post* that a "Lady" in need could retire to his home where "every vestige of pregnancy is obliterated."[84] He did not, however, mention her likelihood of surviving the procedure.

There is every reason to believe that abortions could have been induced electrically. After all, electricity applied to the uterus might initiate contractions or, if given as a severe shock, kill the embryo or fetus. Because abortion was already practiced by other means, women might try out new techniques that seemed safer or less unpleasant, thus aligning electrical abortion with an existing array of technologies—chemical and mechanical—for ending pregnancy.

And how might women, even poor ones, have obtained access to electrical abortions? Because electrical machines were available for rental, at least in England, a wealthy woman had the opportunity to perform the procedure in the privacy of her own residence. It is also possible that some electrotherapists specialized in giving abortions. More likely, a woman who missed a period, suspecting she was pregnant, could visit an electrotherapist or electrical dispensary, complaining of cessation of menses, and receive electrical treatment. Because there were no reliable tests then for the first months of pregnancy, even a wary practitioner could not challenge the woman's self-diagnosis of irregular menses. Moreover, as John Riddle has demonstrated in his recent book, *Eve's Herbs*, "cessation of menses" was a centuries-old euphemism or code for early-stage pregnancy.[85] Given the accessibility of electrotherapy, along with the almost universal claim (in the many books on electrotherapy from the eighteenth century) that electricity was a preferred treatment for cessation of menses, we may conclude that electrical technology was employed for abortions.

—|ı|⊢

One day in 1775 a child, Sophia Greenhill of Middlesex, England, tumbled from the window of her house. She suffered a depressed skull fracture; her breathing stopped immediately and she fell into unconsciousness. The seemingly lifeless child was transported to Middlesex Hospital, where the attending surgeons and an apothecary detected no pulse and gravely admitted that they could do nothing.[86]

Beginning in the late Enlightenment, accident victims such as Sophia Greenhill had a slim chance to avoid the quick pronouncement of death. However, new technologies and social arrangements were being put into place that made it possible sometimes to revive the "apparently dead."[87] Community organizations dedicated to resuscitating accident victims during the late eighteenth century ushered in emergency medicine as we know it. Almost from the outset, this significant social movement employed electrical apparatus in its repertoire of resuscitation technologies.[88]

In the eighteenth century the societies whose mission was to revive people who had drowned, been hanged or suffocated, had fallen or drunk to excess were "humane societies." Unlike the organizations today that warehouse stray dogs and cats, in row after row of dismal cages, awaiting adoption, the humane society movement arose to prevent unnecessary *human* deaths. It began in the Netherlands and was quickly copied.[89] Generally, a group of benevolent people of means formed a local chapter and publicized its activities. Chapters spread throughout the British Isles, including Scotland and Ireland, in the 1770s and 1780s. London's Royal Humane Society, endorsed by George III, was founded in 1774.[90] Americans also established humane societies, mainly in the larger cities, from Boston to Baltimore. The Massachusetts Humane Society, for example, held its first meeting in 1786 at Bunch of Grapes Tavern, on King Street and Mackerel Lane in Boston. Among its significant early works were lobbying Congress for the construction of the first Cape Cod lighthouse (at Truro), setting up huts on the beach to aid shipwrecked sailors, and promoting the use of lifeboats.[91]

Appreciating that a benevolent spirit could be stirred to action by material incentives, London's Royal Humane Society offered a reward of two guineas to people who attempted a resuscitation and two more if the attempt succeeded. Another guinea went to the proprietor of any place who would accept a body and furnish accommodations while a rescue was in progress. These substantial inducements were underwritten by donations and bequests of society members in what was, perhaps, a conspicuous display of enlightened charity.[92] Any claimed rescue had to be authenticated by society members.

In the London area, resuscitations were ideally performed by registered but unpaid "medical assistants," who could be called on as the occasion arose; only a very few were actually physicians. Assistants kept in their possession resuscitation apparatus, which they brought to the scene when summoned.[93]

By dint of experience, humane societies arrived at standard treatments for reviving people suffering "suspended animation."[94] The protocols were published in broadsides that could be posted widely.[95] In a case of drowning, for example, the first task was to replace the victim's wet clothing with warm blankets and occasionally to place the person in a bed between two much warmer bodies. Next, to remove water from the lungs and introduce air, assistants employed several means of "artificial respiration."[96] Lungs could be inflated by bellows, syringe, or even mouth-to-mouth resuscitation. In cases where breathing was mechanically obstructed, a tracheotomy might be performed.[97] (Perhaps if the Humane Society of Baltimore rather than ordinary physicians had attended George Washington, he might have survived to see another spring.)

When normal breathing and heart action did not resume, assistants resorted to stimulative measures. In addition to smelling salts or the tobacco-smoke enema, electricity was thought to be "admirably calculated to rouse the dormant powers" of the body.[98] James Curry, a strong advocate of electrical resuscitation, believed that electricity was "one of the most powerful stimuli yet known," which could excite contractions after other stimuli failed to produce any effect.[99] That electrical stimulation might be used to revive accident victims had been proposed as early as 1747.[100] However, sustained efforts with shock treatment did not take place until the 1780s. One member of London's Royal Humane Society recommended electrical shock "in all cases of suspended animation," and the electrical experimenter William Henly weighed in with support for the idea.[101]

In the Copenhagen society's published instructions for resuscitation, three of the thirty steps involved administration of electrical shocks: first to the heart, then to the spinal column and neck.[102] Detailed advice on electrical stimulation was furnished by Dr. Anthony Fothergill in his 1796 resuscitation manual. Once the lungs had been inflated, he recommended, "let the heart be excited by a gentle electric shock, passed obliquely from the right side of the chest through the left, in the direct course of the heart, and pulmonary vessels."[103] The lungs were then emptied, reinflated, and another shock administered. If this treatment did not restart the heart, the placement of the directors could be altered. Cautions were issued about ensuring good insulation and keeping the victim thoroughly dry.[104] By the end of the century, most humane societies had endorsed the use of electrical stimulation. It had clearly become one resuscitation technology among many employed in the first incarnation of organized emergency medicine. In the early nineteenth century, galvanism was added to the repertoire of stimulative technologies.

But what happened to the child, Sophia Greenhill? Pronounced dead, her body was turned over to Mr. Squires, who immediately administered electric shocks through the thorax. Shortly he felt a faint pulse, then, at last, the stirrings of very labored breathing: she had been reanimated. Although the young Sophia remained in somewhat of a stupor for several days— after all, she did have a fractured skull—"her health was restored."[105]

Other accounts of successful resuscitations leave no doubt that the technique of electrical stimulation of the heart was effective in the eighteenth century as it is today.[106] We may suppose, however, that this procedure was not employed as often as it could have been, owing in part to a lack of assistants skilled in using the technology.[107]

No complicated apparatus was needed for performing electrical resuscitations. Dr. Fothergill simply recommended "A SMALL ELECTRICAL MACHINE, with an electrometer annexed, and a coated jar of about 26 [square] inches . . . together with a pair of discharging rods properly insulated."[108] These specifications were easily met, for example, by the Nairne medical machine with its handheld directors and built-in Leyden jar. However, a surgeon, Mr. Fell, member of the Lancashire Humane Society, furnished in 1792 plans for a very compact machine designed specially for rescues; that is, it could supply only shocks.[109] The machine was indeed diminutive, employing a glass cylinder 4 inches in diameter and 6 inches long, which sat upon two glass supports only 9 inches tall. Mr. Fell's machine also saved space by eliminating the prime conductor; charge was accumulated in a small Leyden jar. And annexed closely to the machine was a Lane electrometer. With its tiny directors, which looked like glass-handled nails on chains, the entire apparatus could fit in a box with a capacity of about a cubic foot. A close examination of Mr. Fell's design—from crank to cylinder to supports—reveals that it was essentially a stripped-down, miniaturized version of Nairne's medical machine.

Also in 1792 *The Gentleman's Magazine* reported that the instrument maker Savigny sold, for purposes of rescue, a portable machine designed by

Charles Kite, an early advocate of electrical resuscitation.[110] It is likely that some humane societies, and perhaps assistants, took advantage of the opportunity to purchase compact electrical machines. It is uncertain, however, whether they became an item of standard equipment.

The humane societies compiled an admirable record of rescues. Especially noteworthy was the success rate of the Royal Humane Society. In the London-Westminster area as of 1794, the society's medical assistants and apparatus had saved 1,865 lives in 2,572 attempts.[111] The heart-warming accomplishments of the humane societies were publicized in magazines, yet reports that the grim reaper had been cheated troubled some readers.

Revival of the apparently dead raised perplexing questions about the nature of life and the propriety of man's meddling in God's plan. Religious objections were voiced to these human interventions, but the humane societies were adept at defusing them.[112] For one thing, they secured the help of the clergy, involving them in their annual meetings. Commonly, a prominent minister gave an address that roundly endorsed the society's activities and justified them on religious grounds, sometimes quoting Scripture; many of these speeches were published in society proceedings and sometimes separately as pamphlets.[113]

Second, to buttress the claim that resuscitation was not blasphemy, influential physicians and clerics active in humane societies propounded certain theories on the nature of life and death, and on the relationship between body and soul. For example, the esteemed physician Anthony Fothergill argued that human life—and that of all animals—was merely a consequence of the proper action of vital organs connected by the motion of bodily fluids.[114] Man had every right to keep the body machine running, even to revive it after a sudden stoppage—like restarting a balky clock by giving motion to its pendulum.[115] After all, such interventions did not affect the disposition of the "rational soul," unique to man in the animal kingdom, which only God could bestow or withdraw. Actual *human* death, then, was occasioned by the rupture of body and soul: in Fothergill's words, "the Soul quits its residence, and the Body, that exquisite piece of mechanism, becomes a motionless, inanimate corpse" and hastens to decompose.[116] These kinds of endorsements and rationales, creative blends of Christian dogma and physiological principles, ensured that people who dared object to resuscitations on religious grounds could be dismissed as superstitious, ignorant, or worse.

The experiments of electrobiologists and, especially, the good works of humane societies underscored a progressive blurring in the distinction between life and death. The infallible signs of death, cessation of heart and lung action, were in fact fallible.[117] If individuals marked by these signs could sometimes go on living, then how many people, lacking access to resuscitation technology, had revived spontaneously—in the grave? This question tormented many minds in the late eighteenth century, especially after Dr. William Hawes, an officer of the Royal Humane Society, published in 1777 a pamphlet entitled "Address on premature Death and premature Interment."[118] George Washington may have been familiar with Hawes's ideas, for he specified that his mortal remains not be entombed until several days after his passing.

The fear of being buried alive led to changes, some of them institutional, in the treatment of corpses. No longer was there a rush to inter; indeed, some countries passed laws that required a three-day wait before burial. And mortuaries were established to care for bodies during the interim.[119]

Another legacy of this fascinating conundrum is found today in the literary works of Edgar Allan Poe, who in the early nineteenth century treated the subject of premature burial with much verve. Stories such as *The Cask of Amontillado* evoke precisely the unspeakable horror of being entombed alive.

-||-

Compared to other communities that adopted electrical technology during the eighteenth century, the activities of electrotherapists penetrated deeply into Enlightenment societies, touching—and sometimes improving—countless lives. Accounts of case histories, electrical dispensaries, and the open-door practices of many benevolent electrotherapists leave no doubt that many men and women, cross-cutting all social classes, had intimate contact with electrical technology. Perhaps a broader segment of Enlightenment societies was literally electrified than enjoyed the antics of electrical lecturers. Yet the existence of these two communities indicates that, already in the Enlightenment, electrical technology had a significant place in society beyond the activities of scientists.

Let us now turn to the community of earth scientists, who employed electrical technology for investigating terrestrial and atmospheric processes. In the next chapter we at last encounter Franklin and his kite.

8.   An Electrical World

From the very beginnings of sustained electrical experimentation in the early eighteenth century, investigators noted a marked similarity between lightning and the sparks created by friction. For example, when Hauksbee rotated a glass vessel against a dry woolen cloth, he "observ'd the Light to break from the agitated Glass, in a very odd Form, resembling that of *Lightning.*"[1] After the invention of the Leyden jar, explicit comparisons of this sort became more common because the jar's discharge not only looked like lightning but also sounded like thunder. As Benjamin Martin put it in 1746, "these flashings and snappings succeed each other sometimes very fast, and represent, as it were in Miniature, an *artificial Thundering and Lightning.*"[2] A few years later Franklin too hypothesized that lightning was electrical; more important, he also proposed a definitive test. Soon the hypothesis was verified, as scientists in several countries drew electricity from the clouds.

The confirmation that lightning is an electrical discharge was perhaps the most dramatic and far-reaching finding of eighteenth-century science. Above all, it showed that human-made microcosms might mimic cosmic phenomena. With an electrical machine and Leyden jar, for example, the scientist created, in miniature, lightning in the laboratory.[3] The ability to model nature's most awesome forces had obvious implications for enhancing elite Enlightenment ideology; it also emboldened investigators to study myriad environmental phenomena using electrical theory and models. Even the generally circumspect Martinus van Marum enthused in 1776 that "it is already fairly certain that many of the chief operations of nature depend either entirely or partly on the operation of Electrical Matter."[4]

In generalizing from their electrical models, these scientists often erred. Because they could seemingly mimic clouds, the aurora borealis (also known as "northern lights"), and volcanoes, they saw the agency of electricity lurk-

161

ing everywhere: the earth had become an electrical world. With the benefit of hindsight, today we know that many similarities between laboratory models and environmental processes were superficial or misleading. Nonetheless, during the latter half of the eighteenth century and well into the nineteenth, electrical explanations enjoyed some popularity. And, in light of modern understandings, these investigators did get a few things right.

Because atmospheric electricity was believed to play a role in everything from hail to earthquakes, researchers devoted the lion's share of their effort to monitoring the electrical state of the air. To facilitate these studies, they fashioned a host of collecting and measuring technologies, some of which are still around today.

The large and heterogeneous community that employed electrical technology to study terrestrial and atmospheric processes can be termed *earth scientists*.[5] This chapter furnishes an account of their major research activities and discusses how electrical technology underwent changes in their hands.

⊣i∣⊢

Benjamin Franklin, founder and most influential member of the earthscience community, implicated electricity in several environmental processes. In 1749 he drew up a list of a dozen similarities between lightning and electricity, concluding, like others before him, that "the fire of electricity and that of lightning be the same."[6] But Franklin surpassed his predecessors by supplying a detailed theory to explain rainfall and lightning in electrical terms.

In Franklin's theory, electricity was involved, not only in lightning, but also in the formation of clouds and in the mechanism of rainfall. Franklin proposed that thunderclouds form more readily over the sea because the constant churning of salt and water generates a charge on water particles. As seawater evaporates and rises, it forms clouds of identically charged particles that are kept apart by repulsion. Above land, however, evaporating water does not become charged to nearly the same extent. Thus, Franklin argued, most thunderstorms on land occur when clouds formed at sea come into contact with uncharged or oppositely charged clouds or mountains. Under these conditions, the water particles lose their charges, are compressed, and so fall to earth as rain. According to this theory, thunderstorms should rarely occur far out to sea, a prediction supported by "some old sea-captains" whom Franklin consulted.[7]

Seeking to show that a marine cloud could cause lightning and rain when it met a terrestrial cloud, Franklin performed a simple experiment on an electrical model. Construction of his artificial clouds began with two round pieces of pasteboard, about 2 inches in diameter. On one side of each pasteboard disk he attached seven silk threads, 18 inches long. From each thread, Franklin suspended a pea to represent an air particle. He transformed the small clusters of peas into clouds merely by dipping them in water. One set, representing the sea-formed cloud, was then gently electrified. When Franklin brought the "clouds" close to each other, attractions and repulsions caused the peas to move briskly, and they dropped their aqueous load.[8] Thus, he said, it rains.

Franklin's homely model would hardly sway skeptics, since other processes—like wind—could likewise agitate the particles in clouds and cause them to fall as rain. A more realistic experiment would be needed to demonstrate that thunderclouds are electrified, and Franklin furnished that too. Or rather, in 1750 he set forth a plan for the crucial experiment.[9] The plan was based on "the power of points," the ability of a pointed object to acquire or give up charge more easily than a blunt one. Although this effect had long been known to electrophysicists, Franklin attached great theoretical and practical significance to it, believing that, with a pointed object raised high, an investigator might draw electricity from a cloud.

Begin, wrote Franklin, by mounting atop a "high tower or steeple" a sentry box large enough to contain a man and an insulating stool.[10] Next, take an iron bar, perhaps 20 or 30 feet long, and place the lower end on the stool. Bend the bar slightly, so that it can pass through the sentry box's door; the upper part of the bar, aimed skyward, must end in a sharp point. An experimenter standing on the stool when thunderclouds are passing above should be able to detect, through sparks, when both he and the bar become electrified. For investigators somewhat faint of heart, who preferred to be left out of the circuit, Franklin also suggested grasping an insulated handle to draw sparks from the bar through a length of grounded wire. Fortunately, no one attempted the experiment precisely as Franklin originally envisioned it. Had lightning struck the bar while an experimenter's body was in contact with it, he might have received a large, perhaps fatal, shock—insulating stool notwithstanding.

Franklin's small book on electricity, with its plan for the sentry-box experiment, attracted the interest of the great French naturalist the comte de Buffon, a longtime adversary of Nollet. Sensing an opportunity to smite his enemy, Buffon enlisted an old friend and polyglot, Thomas-François Dalibard, to translate Franklin's work into French.[11] Because Dalibard, a botanist, was unfamiliar with electricity, he sought the help of a science lecturer, one Monsieur de Lor, and together they repeated Franklin's experiments. This preparation enabled Dalibard to craft a credible translation, which was published in March 1752. For this edition, Dalibard wrote an introduction that chronicled previous electrical research and extolled Franklin's many virtues as the true follower of the experimental method. Not by accident, the introduction made no mention of Nollet.

In the meantime, Dalibard had demonstrated some of Franklin's electrical novelties before the French academy, and word of them rapidly reached the king. Louis XV asked for a private showing and was duly entertained by Buffon, Dalibard, and de Lor with a display of the Philadelphian's wizardry. Their encroachment on Nollet's territory—he had been the royal lecturer—was not forgiven. As we have already seen, Nollet pounced on the innocent Franklin's one-electricity theory and his explanation of the Leyden jar.

Encouraged by the king's enthusiastic reception of Franklin's work, Buffon, Dalibard, and de Lor decided to take up the proposed project for proving the identity of lightning and electricity.[12] In the village of Marly-la-Ville, not far from Paris, Dalibard erected an iron rod 40 feet tall and placed it on an insulator. Before Dalibard returned to Paris, he entrusted a local man, Monsieur Coiffier, to carry out the experiment when thunderclouds appeared. On May 10, after some thunder rumblings, Coiffier approached the rod with an insulated brass wire. Sparks flew between them, just as Franklin had predicted. This obscure villager had been the first to snatch lightning from the clouds, but of course Dalibard and Franklin were the ones who gathered accolades for Coiffier's courage. The portentous event at Marly, picked up by the press and publicized throughout the Western world, became known throughout Europe as the "Philadelphia experiment." Not surprisingly, Dalibard and Franklin became close friends.

The many people privy to the happenings at Marly marveled at what the men of science could accomplish. Lightning rods joined electrical machines and Leyden jars in materializing the belief that human beings could, through study and experimental technology, make progress in understanding—perhaps controlling—God's world. More than a scientific apparatus, the lightning rod—thrust defiantly skyward—was also a new kind of link, literal and symbolic, between earthbound humans and the cosmos. With a technology of his own creation, a man had decisively broken through traditional physical and conceptual boundaries. Apparently, there was no limit to what humans could learn and do; Western societies would never be the same.

—|ı|⊢

Captivated by the Philadelphia experiment and eager to repeat it, researchers, seemingly everywhere at once, obtained charge from iron rods raised aloft.[13] Even Nollet carried out the experiment but downplayed its significance and crankily claimed that he had been first to propose that lightning was electrical.[14] A surprising discovery made by his countryman Louis-Guillaume Le Monnier during these heady months was that even on a cloudless day the atmosphere contains electricity, albeit weak.[15] Thus, investigators would have to probe the atmosphere under all weather conditions if they were to understand the role of this electricity in environmental processes.

Georg Wilhelm Richmann (1711–1753) was one of the first to undertake systematic studies of atmospheric electricity.[16] A salaried member of the Imperial Academy of Sciences at St. Petersburg, he had already made fundamental contributions to thermometry theory and believed, long before many others, that heat resulted from the motion of minute particles. Regrettably, his experiments on atmospheric electricity would have a less happy outcome. Like others, Richmann built apparatus based on the Marly demonstration. Upon the roof of his house he had placed an iron rod, to which he attached a brass chain leading to a prime conductor in the laboratory below. He devised a simple but sensitive electrometer: the electricity's strength was indicated by how far a linen thread was repulsed from the prime conductor. Employing this ungrounded apparatus, which also included a Leyden jar, Richmann had often obtained useful information. But the 6th day of August 1753 was different.[17]

On that day Richmann was attending a public meeting of the academy when, just before noon, he heard thunder in the distance. Immediately he made haste to his home laboratory to secure yet another measurement of atmospheric electricity. He was accompanied by an engraver, Sokolow, who was eager to see the electrical apparatus firsthand so that he could render it more faithfully in copper-plate drawings for Richmann's forthcoming book. Sokolow was destined to become the only witness to that day's bizarre events.

While Richmann leaned forward toward his apparatus, so as to get a better look at the feeble charge indicated on the electrometer, a mighty bolt of lightning struck nearby. It was, according to people on the street, an odd occurrence and was followed almost immediately by calm skies. But that was no consolation to Richmann, for a ball of blue fire had jumped the 12 inches from the prime conductor to the scientist's bowed head. The blow knocked him back, and he ended up seated on a wooden chest, slumped against a wall, the Leyden jar shattered. Sokolow was badly shaken but, after his senses returned, he dashed outside, informing everyone on the street that Richmann's house had been hit by lightning. Alerted by the commotion, Mrs. Richmann entered the laboratory and found her husband unconscious, seemingly lifeless. Soon a group gathered in the laboratory and ministered to the fallen philosopher. They bled him twice, but no blood was forthcoming; they rubbed him violently and breathed air into his lungs, but he did not stir. Plainly, the academician Richmann had been struck dead.

Other students of electricity, hoping to avoid Richmann's fate, took an uncommon interest in the manner of his death. The details of this freak accident were published in several versions, from which readers could draw the appropriate lesson: it could have been prevented had Richmann's equipment included a connection to ground.[18] In reprinting an account of the accident in *The Pennsylvania Gazette*, Franklin could not resist appending a claim: "The new Doctrine of Lightning is, however, confirm'd."[19]

This tragedy nonetheless presented an opportunity for observing the effects of lightning on the human body. Not only were areas of discoloration on Richmann's skin noted, but the next day he was dissected and the state of his internal organs described at length. The external examination and autopsy revealed that Richmann had been killed instantly, and that the electricity had entered his head and exited from one foot—confirmed by a shredded shoe. Finally, the fragmented mortal remains of this scientific martyr were placed in a coffin, and he was laid to rest. In a later eulogy, Priestley the minister noted that not every scientist was fortunate enough to die so gloriously.[20]

Electrophysiology was not the only beneficiary of Richmann's untimely demise. A young German scientist, Franz Ulrich Theodosius Aepinus, was appointed to take Richmann's place at the Imperial Academy of St. Petersburg. There he would publish an important theoretical treatise that took steps toward quantifying both electricity and magnetism.[21] But the work was largely ignored in his lifetime. Perhaps, as in the later case of Charles Augustin Coulomb and his torsion balance, Enlightenment experimenters did not wish to wade through mathematics that obscured phenomena they believed were already understood.

Disseminators, including Ebenezer Kinnersley, also profited from the Richmann saga by flavoring their lectures with its macabre details. In an advertisement in the *Pennsylvania Gazette*, on March 26, 1754, Kinnersley promised that he would relate the circumstances of Richmann's accident and explain its cause, which strongly confirmed Franklin's doctrine on the electrical basis of lightning.[22]

Americans who know little about Franklin's science do have an inkling that he did something significant with a kite. And throughout American history melodramatic reference has often been made to the man who "snatched lightning from the heavens"; even in Franklin's time his exploit captivated the public. Remarkably, Franklin never published a detailed account of the experiment, describing it dispassionately in a few hundred words in a letter to Peter Collinson that was published in the Royal Society's *Philosophical Transactions* in 1752. This note also appeared in Franklin's newspaper on October 19, 1752, and was reprinted in his book on electricity.[23] By this time the experiment at Marly had already taken place and garnered international acclaim, so Franklin's kite escapade actually contributed no new principles; but it did furnish another technology for studying atmospheric electricity that enabled researchers to draw charge from greater heights than iron rods.[24]

A controversy swirls around precisely when Franklin flew his kite, but its resolution is of little scientific or technological import.[25] To make a long and inconclusive story short, in Priestley's 1767 account of the episode, doubtless based on recollections furnished by Franklin himself, the kite flight is said to have taken place in June 1752. At that early date, Franklin had not yet heard the good news from France about his triumph at Marly.[26] If he had flown his kite as early as June, some wondered, why did Franklin not take immediate credit for showing that lightning and electricity are one? This question, of course, proceeds from the dubious assumption that the intense competition for credit and priority permeating modern science was already present in the eighteenth century. Although some Enlightenment scientists, such as Nollet, fervently sought recognition for their discoveries, the generally modest Franklin might have been quite content with having proposed how to do the crucial experiment.[27]

In any event, Franklin's publications on the kite flight, prepared after he learned about Marly, acknowledged that others were first to show the long-suspected identity of lightning and electricity. Thus, Franklin chose to frame the kite experiment as presenting a more convenient technology for achieving the same result. Kites were already familiar to Franklin, for he had flown them as a child. One day in his youth, floating on his back, he allowed his kite to pull him across a sizable pond. Of course the ever-provident Franklin had arranged for another boy to carry his clothes to the distant shore.[28]

For the electrical experiment Franklin made his own kite, using wooden arms of cedar covered by a large silk handkerchief. A silk kite was able "to bear the wet and wind of a thunder-gust without tearing," presumably better than a child's kite of paper. To collect charge, he affixed a pointed wire, more than a foot long, to the kite's tip. The celebrated key was tied to the twine string with a silk ribbon. A Currier and Ives print depicts Franklin bravely standing outside, holding fast to his kite, during a ferocious thunderstorm (see also Fig. 1). In fact, he was perched just inside a door or window, not fully exposed to lightning; nonetheless, the kite flight appeared to be dangerous because it took place in the rain with thunderclouds menacing overhead. Many believed that had lightning actually struck the kite, it would have fried Franklin.[29]

Fortunately, the experiment proceeded as planned. After rain moistened the string, it conducted enough charge to be easily detected when Franklin's knuckle approached the key. From the key he then charged a Leyden jar, which enabled him to reproduce common electrical effects. In this way, Franklin "completely demonstrated" the identity of lightning and electricity.

Although Franklin was neither first to argue the electrical basis of lightning nor first to carry out a crucial experiment, as the years passed and precise dates were forgotten, Franklin—whose theoretical ideas were profound and influential—accrued ever more credit for these accomplishments. As we have seen with the invention of the Leyden jar, when several people independently make an important scientific discovery, the most famous one usually gets the lion's share of the credit. But it should also be clear that had Franklin or Dalibard been preoccupied with other matters, unable to conduct the crucial experiment, doubtless others would have done it within a matter of months.

The experiments at Marly and at Philadelphia furnished earth scientists with the rudiments of two technologies for collecting atmospheric electricity: the lightning rod (or conductor) and the kite. Let us see how these and other technologies were modified and deployed.

After Franklin's feat, kites became a tool of earth science in a veritable kite craze. And soon investigators showed that the amount of atmospheric electricity varies directly with the elevation of the collector.[30] Those who followed Franklin quickly learned that a child's kite was in fact a toy, often unable to tolerate high winds and the rigors of rain. Moreover, an off-the-shelf kite was ill suited for reliably conducting electricity downward. Seeking to improve these performance characteristics, many investigators made their own electrical kites.

Not everyone was as willing as Franklin to undertake research during a thunderstorm. Experiments in dry weather, however, required that the kite string be far more conductive than twine. Thus, an important modification was to embellish the string with a metal wire. In France, Jacques de Romas wrapped a hemp cord with a fine copper wire so that it resembled a violin string.[31] To carry this heavy cord, he constructed a huge kite of oiled paper. With his kite flying several hundred feet in the air, in June 1753 de Romas obtained a spark using a discharger to link a brass chain to a grounded iron rod; in one demonstration before two hundred observers, de Romas drew a spark 30 cm long.[32] Most investigators used kites of more modest proportions, simply joining a fine wire to other strands of the string.

Noting that even oiled paper was too weak to withstand wind and rain, Pierre Bertholon perfected a kite with a waxed taffeta body. To compensate for the heavy taffeta, he replaced the wooden supports with lighter ones of baleen and fastened them together with a copper joint. Copper points also tipped the ends of the baleen cross. His kite in flight must have been an inspiring sight, for it sported a tail and ears of silk ribbon.

Bertholon's kite design, and those of many others, also exhibited a greater concern than Franklin's for the investigator's safety. Bertholon's solution was to build an insulated, double-reel system for raising and lowering the kite.[33] The wooden reel for winding the string was fixed upon a glass insulator; a metal wire in the string could then form a spark gap with a grounded conductor. In the event the kite was struck by lightning, the electricity would jump the gap and be carried harmlessly to ground. A second wooden reel, the one actually turned by hand, was connected mechanically to the first and permitted the investigator to be at some distance— and completely insulated—from the metal wire whose other end ascended with the kite. A similar system, with the investigator safely ensconced in a structure, was illustrated by John Cuthbertson (Fig. 25). Also interested in protecting himself, Tiberius Cavallo (whom we meet again shortly) used a grounded connection and stood on an insulated stool. Cavallo flew his kite hundreds of times, suffering at most a few "slight" shocks to his arms.[34]

Another technology, rather more exotic, was adapted for collecting atmospheric electricity at elevation. Let us now examine its background.

-|ı|-

In the fall of 1783 Benjamin Franklin remained in France after having negotiated the Treaty of Paris with England. Not yet engulfed in revolution, this France—of urbane and courtly culture and vast social inequality, so far from American aspirations—still offered something of interest to the septuagenarian scientist. For it was there that the Montgolfier brothers, Joseph and Étienne, were experimenting with hot-air balloons.[35] Franklin had witnessed an early balloon flight in Paris on August 27, just before the

Figure 25.   Apparatus for the safe study of atmospheric electricity. (Adapted from Cuthbertson 1786, pl. 8. Courtesy of the Bakken Library, Minneapolis.)

treaty was signed. A few weeks later a highly ornamented balloon carrying a duck, cock, and sheep ascended before witnesses, including Louis XVI. More impressively, it brought the animals back to earth alive.

On November 21, on the grounds of the château de la Muette, near Paris, another Montgolfier balloon—74 feet tall—stood majestically, awaiting a cannon to signal the release of its tethers. Instead of farm animals, this balloon's gondola, resembling an enormous bread basket, carried two aeronauts.[36] One was Pilâtre de Rozier, a scientist and founder of an important scientific journal; the other was the marquis d'Arlandes. These highly courageous men represented well the mutual dependence of grandiose scientific technology and wealthy patrons.

At 1:54 P.M. the cannon sounded, the tethers were released, and the world's first manned aircraft headed to the heavens, its ascent witnessed by thousands of nearby villagers and Parisians, many of whom had purchased tickets. A more public scientific experiment could not have been imagined. After ascending a few hundred feet, d'Arlandes saluted the enthralled masses with a waving handkerchief, and the crowd replied with applause. The balloon continued climbing, but people on the ground could no longer distinguish its occupants. On this flight Rozier and d'Arlandes were hardly passive passengers; to stay aloft they had to stoke with straw the furnace that furnished the hot air to lift the 1,700-lb. machine. Driven by wind over the Seine, they passed low over Paris and continued southward. Appreciating the success of their flight, the men stopped feeding the fire and made a gentle descent in a field beyond the new boulevard, around 20 minutes and 5 miles from their starting point.

In addition to members of royalty, distinguished Parisians, and ordinary citizens, the witnesses that day included Benjamin Franklin, who observed the event from the terrace of his Paris residence. Along with other dignitaries he signed the official affidavit testifying to the Montgolfiers' feat. In the aftermath of their triumph, there was much debate over the merits of manned ballooning. In answering a skeptic, who had inquired what good a balloon might be, Franklin glibly responded, "What good was a new-born baby?" Franklin believed that the balloon flight was not only a "beautiful spectacle" but also an important proof of principle. He wondered if someday balloons might not become a common conveyance, perhaps relieving travelers of the punishment meted out by horse-drawn carriages lurching and bobbing on cobble roads. Optimistically he predicted that, as devastating offensive weapons, manned balloons might convince "sovereigns of the folly of wars." [37]

In the meantime, balloons did find *scientific* uses. Indeed, it took earth scientists precious little time to appreciate that balloons made an ideal tool for probing the atmosphere (and are still useful for such research today). After all, balloons—unlike kites—needed no wind to stay aloft. Bertholon was the first—probably in 1783—to employ a hot-air balloon to sample atmospheric electricity. He was quickly followed by Horace de Saussure of Geneva, who used an alcohol burner to supply hot air. [38] The balloons built by these men and other earth scientists were miniatures compared to the huge demonstration devices of the Montgolfier brothers; also, they "flew" tethered to the ground.

Following the Montgolfier brothers, who had shown that hydrogen had greater lifting power than hot air, Claude Veau Delaunay used this technology to raise an electrically outfitted balloon. [39] It was made of "gold-beater's skin" (a soft leather) coated with natural rubber to help it retain the hydrogen. At the top of the balloon Veau Delaunay placed a conductor, which communicated with the metal wire used to raise and lower the balloon. Suspended under the hydrogen bag was a small gondola made of water willow or silk; in the gondola was an electrometer that retained its charge long enough to be reeled in and read.

Although requiring skill, building this balloon was the easy part. Where did Veau Delaunay get the hydrogen to fill it? In the late eighteenth century researchers could not simply order a tank of hydrogen gas from the local apothecary or instrument maker. They usually had to make it in their own laboratories. Fortunately, hydrogen could be evolved by a number of well known and relatively simple chemical processes, such as immersing iron in sulfuric acid. Thus, to accompany the discussion of his balloon, Veau Delaunay described and illustrated a hydrogen generator.[40]

$$\dashv | | \vdash$$

To place collecting devices ever higher, on kites and balloons, was one obvious way to capture the atmosphere's sometimes feeble electricity. Another strategy was to adopt earthbound technology—variants of the lightning rod or conductor—sometimes in concert with more sensitive electrometers. With the memory of Richmann's misfortune still fresh, however, investigators took precautions to avoid a similar fate.

Not surprisingly, Franklin himself employed this technology, but with a new wrinkle. In September 1752 he mounted a tall iron rod on his chimney. From its base he ran a wire, the diameter of a goose quill, through a glass insulator into the house and attached its lower end to a metal pump—a solid connection to ground. Franklin cut the wire where it was exposed on a staircase, opposite his bedroom, and then separated the ends horizontally by about 6 inches. He then affixed a small metal bell to each end; the clapper was simply a brass ball hanging by a silk thread. When sufficient atmospheric electricity was present in the rod, the bells sounded, and Franklin could then rush to his apparatus and charge a Leyden jar. This equipment enabled him "to draw the lightning down" where he could conveniently carry out experiments.[41]

Franklin was especially interested in determining whether atmospheric electricity had a positive or negative charge; he suspected that it was uniformly negative. Although the first trials seemed to confirm this conjecture, Franklin later encountered a contrary instance. He concluded that atmospheric electricity could be either positive or negative, though more often it was the latter. These unexpected findings, also supported by Ebenezer Kinnersley's observations, were profoundly puzzling.[42] Although Franklin altered his theory to accommodate them, the modifications were far from convincing, as even he acknowledged. Surprisingly, to this very day there is no complete explanation for the charges that clouds acquire.

Perhaps the most elaborate earthbound system for collecting atmospheric charge was built early on by the Italian Giambatista Beccaria (1716–1781), long before the advent of sensitive electrometers.[43] Trained to teach, Beccaria at the age of twenty-one took the first of a succession of positions, ending up—in 1748—as a professor of physics at Turin. There he remained for the next twenty years, achieving no small measure of distinction for his teaching as well as for researches in electricity, astronomy, and geophysics. Incidentally, it was Beccaria's idea about "vindicating electricity" that led to Volta's invention of the electrophorus.

A member of several electrical communities, Beccaria was above all a fervent Franklinian, propounding throughout Europe the one-electricity gospel. In a letter addressed to Franklin at the beginning of his major book on electricity, Beccaria confessed that "I always shall be the most obsequious admirer of your eminent merit."[44] Not surprisingly, Franklin was warmly disposed toward Beccaria and even arranged for the English translation and publication of his magnum opus on electricity. After Beccaria sent him a copy of one work in 1762, Franklin penned a lengthy reply. Thanking him for the gift, from which he "received great information and pleasure," Franklin also acknowledged "the generous defence you undertook and executed with so much success, of my electrical opinions."[45]

Although in this letter Franklin reported no new experiments, he did describe his latest invention, a musical instrument. The Philadelphian thought it might be of interest to Beccaria because he lived in Italy, a "musical country." The Armonica had dozens of blown-glass hemispheres of varying size, which were mounted through their centers on a spindle that could rotate in a wooden frame. After wetting the glasses, set close together, with a sponge and water and pressing a foot pedal to start up the Armonica, the musician produced a tone by merely rubbing the edge of a glass with a moist finger. Franklin assured Beccaria that, when played skillfully, the Armonica's sounds were "incomparably sweet."[46] Apparently they were, for the Armonica enjoyed about a half-century of popularity in Europe; even Mozart and Beethoven wrote pieces for it.

I do not know if Beccaria made his own Armonica, but he did follow Franklin's lead in earth science, devising several technologies for studying atmospheric electricity. At the top of a hill in Mondavio, facing northward toward the Alps, Beccaria built a large antennalike apparatus. The collector consisted of an iron wire, more than 100 feet long, which stretched from a stack of chimneys to the top of a cherry tree; both ends were insulated. A second iron wire, attached to the middle of the first, ran below to his observatory—where he also lived—and entered through a crystal insulator fixed into a wooden windowpane. The strength of the charge was indicated by a pith-ball electrometer.[47] A cautious Beccaria made copious observations, but mostly in "serene" weather.

Bertholon built a similar permanent installation, on a smaller scale. All he needed was an iron rod, which terminated in a gilded copper needle that was set upon an insulator at the highest point of a building. An iron wire conveyed the charge into the laboratory; suspended from silk cords, the wire reached a copper conductor that hovered a few inches above a secure connection to ground. Like Franklin, Bertholon also used bells, fixed between the ends of the copper and grounded conductors, to announce when a sizable charge was coursing through the circuit. This sort of arrangement, which repaired the defects of Richmann's apparatus, was very common in the late Enlightenment. Bertholon also emphasized that, with this apparatus, he could use atmospheric electricity to carry out virtually any experiment requiring a source of charge; in effect, it could replace an electrical machine.[48]

For the earth scientists who were eager to sample atmospheric electricity in different locales, both airborne collectors and fixed installations presented obstacles and inconveniences. Not only was the apparatus expensive, but investigators could not easily make observations wherever and whenever they wished. A solution was found to these performance deficiencies in the invention of more sensitive, portable electrometers that could be taken anywhere and employed with little hassle.

Tiberius Cavallo (1749–1809) was one of the most inventive earth scientists, and his electrometer designs were widely adopted.[49] Born in Naples, the son of a physician, Cavallo moved from Italy to England as a young man, eager to learn about the business world. Before he could realize any business ambitions, he fell in with a different crowd—electrophysicists—and from then on the pursuit of science consumed his life.

Cavallo's interests were wide-ranging and included magnetism, music (he played the violin), pneumatic chemistry, and ballooning—in addition to the entire gamut of electrical researches. Perhaps his most memorable work is *A Complete Treatise of Electricity in Theory and Practice;* published in 1777, it went through several editions, growing to three volumes in 1795. Its clear exposition, well-organized presentation of principles, and ample illustrations make this book an exemplary reference on electrostatics from an eighteenth-century source. It cannot be said that Cavallo's re-

Figure 26. *Center and right*, Cavallo's portable electrometer and *left*, case. (Adapted from Cavallo 1780b, tb. 1, figs. 1–3.)

search contributed much new science to any electrical community, but he did develop new and important electrometers for studying atmospheric electricity.

One of Cavallo's early inventions was a miniature pith-ball electrometer, which he installed in a toothpick carrying case so tiny it easily disappeared into a pocket.[50] The electrometer itself—pith balls suspended from linen threads—was built in a glass tube; the latter also had a silk cord so that it could be hung easily. The exterior of the carrying case was fitted with pieces of amber and ivory. By rubbing one of these, and thus creating a charge of known polarity, Cavallo could conveniently assess the state of the atmospheric electricity monitored by his pocket electrometer. This design, however, had one major performance shortcoming: the linen threads were apt to become tangled.

Not surprisingly, the enterprising Cavallo brought to fruition another portable electrometer of considerable elegance that avoided tangles (Fig. 26).[51] Including its gleaming brass carrying case, the "improved" electrometer was only 3 inches tall. The heart of the device was a glass tube, capped by a hollow brass hemisphere. Inside the tube were two fine silver wires, suspended from an ivory support and connected electrically to the brass dome. On the bottom of each silver wire was a small cork ball. The glass tube was fastened to a brass base having screw threads. With the base and a screw-on brass cover as its carrying case, the electrometer could be moved about and deployed at will. Edward Nairne, for one, brought Cavallo's improved electrometer to market.[52]

$$-|\,|\,|-$$

The drive by earth scientists to build ever more sensitive electrometers culminated in one device so elegant and exquisitely sensitive that its design is still relevant today. The gold-leaf electrometer was the invention of Reverend Abraham Bennet (1750–1799), the curate of Wirksworth in Derbyshire. Profoundly interested in atmospheric phenomena, Bennet invented not only new apparatus but also theories. For example, he offered an appealing hypothesis to explain "falling stars"—our present-day meteors. Different zones of air, he suggested, can become oppositely charged, depending on local variations in evaporation, condensation, and temperature. As electrical equilibrium is restored by a discharge, fire may fly across the sky with "astonishing brilliance and rapidity."[53]

Like Franklin and countless others, Bennet also offered an explanation for the electricity in clouds, built on the belief that evaporating water acquires a small charge.[54] In the atmosphere, these charges can intensify and coalesce as the water vapor rises, eventually becoming electrified clouds. Differences in the polarity of clouds arise, for example, through wind and electrical induction. Throughout the presentation of his theory, Bennet referred to experiments that he and others had performed; every part of his theory was seemingly supported by results obtained from the manipulation of simple apparatus in the laboratory.

Bennet appreciated that certain manipulations, as well as the detection of atmospheric electricity on cloudless days, required an electrometer even more sensitive than Cavallo's. Knowing that gold is the best conductor of electricity and, in the form of leaf, is especially responsive to charges, Bennet incorporated this material into his new electrometer (Fig. 27, right).[55] He began with a glass cylinder, about 5 inches tall and 2 inches in diameter, which provided protection from wind. The cylinder was capped with a metal disk and rested on a metal base. Attached to the cap's underside were two strips of gold leaf, about 2 inches long and .2-inch wide. Joined only at the top, the dangling pieces of gold leaf separated when a charged object was presented to the metal cap. To prevent extreme movements of the gold leaf, Bennet glued pieces of varnished tinfoil to the inside walls of the cylinder where the gold leaf might strike them. Surprisingly, he found that this simple modification also increased the instrument's sensitivity.

With his ultrasensitive electrometer, Bennet made many observations on atmospheric electricity. In addition, he employed the electrometer in

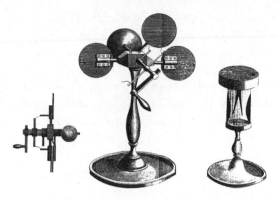

Figure 27. *Center and left*, Nicholson's doubler and *right*, Bennet's gold-leaf electrometer. (Adapted from Adams 1799, pl. 1, figs. 9 [left, center] and 10 [right]. Courtesy of the Bakken Library, Minneapolis.)

studies that examined the charge on steam produced by dropping various substances in water; he tested everything from hot metals to olive oil, recording the polarity of the resultant charge—if any. The new electrometer also allowed Bennet to investigate charge acquired by different powders forced through the nozzle of a bellows.[56] Clearly, the gold-leaf electrometer was a versatile instrument well suited for all sorts of earth-science experiments; it was, however, a rather fragile instrument, in no way portable. Despite the latter limitations, Bennet's invention was commercialized and adopted in many communities.[57] And gold-leaf electrometers can be found today among the demonstration devices used in introductory physics classes.

⊣|⊢

Sensitive though it was, the gold-leaf electrometer still did not meet the earth scientists' every need, for sometimes the atmospheric electricity was still too slight to detect. A new strategy was needed for registering such minuscule charges. And so earth scientists in the late 1780s devised devices for amplifying feeble charges, raising them above a sensitive electrometer's threshold of detection. As it turned out, an existing technology—Volta's electrophorus—was available.

In Volta's own version, which he called a "condenser of electricity," an atmospheric conductor is connected to the metal plate of an electrophorus and allowed to remain there for around ten minutes.[58] During this time the electrophorus—a capacitor—is being charged. When the connection is removed, the amplified charge induced on the metal plate can often be detected with an electrometer. Volta recommended using a very smooth, varnished marble plate in place of the usual resin disk.

Other investigators also employed an electrophorus for amplifying weak atmospheric charges, but their manipulations differed. Recall that, after rubbing, the electrophorus's disk stubbornly refuses to part with its charge. In fact, an investigator can repeatedly "remove" charge by means of the grounded metal cover. If the charge induced on the cover is transferred, again and again, to a Leyden jar, the latter eventually accumulates a charge much greater than the original. Although the perfect apparatus for amplifying minute charges, an electrophorus available from instrument makers tended to be large and bulky, as appropriate for experimenters as well as the disseminators who found its magical performance irresistible.

Even before Bennet's invention of the gold-leaf electrometer, Cavallo had made a more compact electrophorus, coupled to an electrometer, which became known as a "condensing electrometer." It was precisely this device that Volta himself employed in early experiments on contact electricity. Bennet modified the condensing electrometer and placed it atop his gold-leaf electrometer. He called the resultant apparatus an "electrical doubler," since after every transfer of the induced charge, the intensity on the metal cap approximately doubled, spreading the gold leaves farther apart.[59]

One inconvenience of the condensing electrometer was the need to transfer charge manually. Erasmus Darwin sought to make the device more user-friendly by employing a lever that moved the disks. William Nicholson modified Darwin's design by using disks disposed vertically that could be rotated with a crank (Fig. 27, center and left). Also, as the crank turned, contacts made and broke appropriate circuits. According to Bennet, who carried out extensive tests with Nicholson's doubler, this complex device could amplify, above the threshold of detection, even the low level of atmospheric electricity always present in a room.[60]

With various doublers and Bennet's gold-leaf electrometer, after the late 1780s earth scientists had a variety of apparatus for conveniently monitoring the atmosphere, however slight the charge. In addition to monitoring electricity under threatening and serene conditions, earth scientists also assessed the charge in fog, dew, and in rain itself. Rooftop collectors could be used to study the electrical state of fog, but new apparatus was needed for rain.[61]

That raindrops themselves could carry charge downward from clouds was proposed as early as 1741 by J. T. Desaguliers, chaplain to the prince of

Wales. Building on Stephen Gray's discovery that water could be electrified, Desaguliers performed a clever experiment with a copper fountain.[62] Immediately above the fountain, its spout pointing down, he placed a highly charged glass tube. Holding a thread on a stick, his assistant was able to detect a charge on the water streaming out of the fountain. With his copper cloud, Desaguliers made plausible the possibility that raindrops carry charge. Ahead of Franklin, he also entertained the idea that electrical processes might be involved in evaporation and cloud formation.[63]

In order to gather charge from actual raindrops, Tiberius Cavallo invented an "electrometer for the rain."[64] The starting point for assembling this combined collector-electrometer was a glass tube about 2.5 feet long, coated with sealing wax, which was attached firmly to a windowframe—one end inside the window, the other outside. To gather the charge on raindrops, he attached to the outside end of the tube several brass wires that formed an open mesh on a cane framework. From the brass-wire mesh he ran a conductor through the tube to a pith-ball electrometer inside the window. With this apparatus Cavallo confirmed that rainwater carries a charge, usually negative, which is sometimes sufficient to charge a small Leyden jar.

Although the new electrometers invented in the 1780s were sensitive, they still had to be watched closely, their readings dutifully recorded at brief intervals. Given that atmospheric electricity could fluctuate in a matter of minutes, an investigator might find himself gazing at an electrometer for hours on end, perhaps in the middle of the night, hoping to find a correlation between atmospheric electricity and the weather. The solution was to build a recording instrument. The Italian Marsilio Landriani, for one, invented a clockwork mechanism to register, on a resinous plate, variations in atmospheric electricity. The varying intensity of the charge could be made visible merely by sprinkling the plate with powder.[65]

Beccaria invented another device that mainly recorded the time and magnitude of lightning strikes. He called it a *ceraunograph* (a name still applied to lightning recorders), and it joined other apparatus in his "electric observatory."[66] The ceraunograph took as its starting point a spring-driven clock. In place of an hour hand, he affixed a paper disk to the clock motor, which rotated inside the ring-shaped clock dial. On each side of this disk, near the circumference, Beccaria placed a pointed iron wire; the wires were offset just slightly. One wire was connected to his rooftop conductor, and the other was grounded. When lightning struck nearby, a spark would fly between the two points, perforating the paper disk. If the strike was large, then the spark would be as well, leaving an even bigger hole in the paper. As long as the clock was still running when he read the instrument, Beccaria could easily calculate, from the stationary clock dial, when each strike had occurred. Unhappy with the performance of this model, Beccaria built a better one that operated on the same principle.[67] Although Bertholon and other atmospheric scientists were aware of Beccaria's invention, it does not appear to have been widely copied or commercialized at that early date.

Technology to collect, monitor, and sometimes even record atmospheric electricity joined older instruments for studying meteorological phenomena.[68] In the closing decades of the eighteenth century, atmospheric scientists were providing detailed information on rainfall, relative humidity, temperature, wind speed and direction, and air pressure in diverse places.

The academies and societies took an active role in collecting and publishing these data. The Royal Society of London and the Berlin academy, for example, had their own meteorological observatories. Periodically, the data obtained were collated and published by the societies.[69] No less a personage than Henry Cavendish reported results from the Royal Society's meteorological observatory. Such relatively systematic observations furnished a basis for empirically based theories of climate.

Although the observatories of the Royal Society and the Berlin academy occasionally reported on atmospheric electricity, gathering such data seems not to have had a high priority. Rather, most reports of atmospheric electricity were submitted to the societies by an assortment of earth scientists working in many places; lengthier data compilations were also published in books.[70] Despite somewhat haphazard reporting, observations on atmospheric electricity contributed to the construction of meteorological theories.

--||--

Perhaps taking a cue from Benjamin Franklin's "clouds," made from pasteboard and peas, other investigators built apparatus to model various atmospheric and geological phenomena. From the operation of these models, earth scientists often generalized to the large-scale processes they were mimicking.[71] Research in later centuries would support some of these theories, and we applaud their authors' efforts; but others did not fare as well. For example, electrical theories for earthquakes and volcanoes now seem absurd, and the laboratory analogs far-fetched if not ridiculous. Yet who can know which models and which theories posterity will condemn as silly science? To appreciate the context of the apparatus these early earth scientists built, we need to juxtapose their theories with those propounded by scientists less enamored with electricity.

Prior to about 1750, theories to account for the more familiar meteorological phenomena tended to invoke chemical mechanisms. For example, in early editions of his textbook Petrus van Musschenbroek asserted that the aurora borealis resulted from the burning of matter—exhalations deep inside the earth that had been liberated by strong temblors.[72] After the Philadelphia experiment, electrical explanations for terrestrial and celestial phenomena were offered in place of chemical ones. Electrical explanations for lightning and the aurora borealis rapidly gained adherents, perhaps because they could be so convincingly modeled in the laboratory. But the mechanisms of many phenomena remained, in the late eighteenth century, highly contestable: chemical, electrical, and other explanations all had their champions.

Not surprisingly, the aurora borealis *was* a favorite phenomenon for modeling. In an echo of Hauksbee's early experiments, Cavallo described a method for producing an aurora borealis artificially: simply attach a vacuum pump to a stopcock in a Florence flask and evacuate the flask. When rubbed, the outside of the flask will "appear luminous within, being full of a flashing light, which plainly resembles the Aurora Borealis."[73]

Bertholon offered a similar electrical rendering of the aurora borealis, presented in excruciating detail, which he supported with around half a dozen creative models.[74] One impressive apparatus was made from ten tubes of glass—some straight, others serpentine—that had been evacuated and sealed on both ends. Disposed like the rays of the sun, the tubes were attached to a central disk of copper. Also radiating from the copper disk were iron wires, situated next to each tube. According to Bertholon, when this apparatus was supported by an insulating stand and brought close to an operating electrical machine, all of the glass tubes glowed with a "beautiful electric light."[75] Another model consisted of just a single glass tube, also evacuated and sealed, but it was 7 feet long. When placed near an electrical machine, the tube's interior glowed white or reddish—a "magnificent spectacle."[76]

After experimenting with numerous models over the years, Bertholon at last arrived at a design he believed so perfect that witnesses to its visual performance could not doubt that the aurora borealis was electrical. It consisted of a glass receiver of the sort used with a vacuum pump. In the receiver he placed a metal, forklike affair whose ten or so tines ended in points aimed upward. The opening at the top of the receiver was sealed around a wire that plunged downward, making contact with the fork. When the latter was electrified, after the receiver had been evacuated, the points gave off "brilliant rays perfectly representing those of the Aurora Borealis."[77]

Although we believe today that the aurora borealis is an electrical phenomenon, eighteenth-century models only superficially resembled the underlying process. At least one eighteenth-century model of the aurora borealis was brought to market; doubtless it found the favor of disseminators.[78]

One of the most active and inventive earth scientists of the Enlightenment, Bertholon also addressed the cause of earthquakes and volcanoes, claiming that they had an electrical basis. Earthquakes were not the result of underground fires or low-pressure water vapor, as others had suggested.[79] Rather, an underground accumulation of electrical fluid caused them by creating a charge imbalance. The natural equilibrium was restored by a discharge of sufficient size to shake the ground. Noting that earthquakes and volcanoes often coincide in time and place, he argued that they are likely to have the same cause—an electrical imbalance between earth and atmosphere.[80]

To illustrate the efficacy of his explanation, and to test the devices he had invented for preventing earthquakes and volcanic eruptions (see chapter 11), Bertholon built a model village. On a magic table, he placed rows of tiny paper houses. To represent a volcano, he used a large vessel containing small, lightweight particles. When the electrical machine connected to the magic table was set in motion, houses moved and the volcano—through repulsion of the small particles—erupted in a manner that "perfectly resembles Vesuvius and Etna."[81] This must have been a very satisfying display, exhibiting the scientist's power to model harmlessly in miniature some of the earth's most terrifying phenomena.

St. Elmo's fire, the stream of light rising at night from a ship's mast, had long been observed by mariners. That this was an electrical phenomenon was swiftly accepted (and remains so today). But typically, Bertholon fashioned a laboratory model. Take, he suggested, an insulated piece of metal, pointed on one end, and hold it above an operating electrical machine. When the lights are low, the point will issue forth a "luminous plume."[82]

The energetic Bertholon also contrived electrical theories for the "aqueous meteors"—rain, fog, dew, hail, snow, and waterspouts, arguing that they all serve to circulate electrical fluid between the earth and the atmosphere.[83] Wind, he added, merely arises as a byproduct of the aqueous meteors reestablishing electrical equilibrium. Uncharacteristically, Bertholon described no special models to illustrate these ideas.

From time to time, other investigators did build models to represent the aqueous meteors. Claude Veau Delaunay, for example, employed pith balls in a glass receiver to mimic hail. When an external charge was supplied, the balls moved up and down.[84]

Given that electricity could produce circular motion, it should not be surprising that someone would suggest that electrical attraction is the force keeping planets in their orbits around the sun. This theory—bad physics even in its day—grew out of experiments by Stephen Gray and Granville Wheler on circular motions. Dr. Charles Mortimer, taking dictation from Gray as the latter lay on his deathbed, recorded that Gray's experiments with the motion of a light body around a charged electric might have resulted in "a new Sort of *Planetarium* never before thought of, and that from these Experiments might be established a certain Theory for accounting for the Motions of the Grand *Planetarium* of the Universe."[85] A few years later, after extending the experiments on circular motion, Wheler himself alerted astronomers to the possibility that "electric *Effluvia*" might have a role to play in explaining "the heavenly Phaenomena."[86] Only a few people, including Ebenezer Kinnersley, took these ideas seriously, so easily could they be refuted by Newtonian mechanics.[87]

—| | |—

With the conspicuous exception of the electrical theory of planetary motion, eighteenth-century earth scientists offered plausible theories for the involvement of electricity in geological and atmospheric processes.[88] In the nineteenth century, new research, for example in geology and chemistry, eliminated electricity as a cause of earthquakes and volcanoes and greatly diminished its role in the functioning of aqueous meteors.[89] But other findings—on the aurora borealis or St. Elmo's fire—have withstood further scientific scrutiny and stand firm today. Perhaps more important, eighteenth-century earth scientists fashioned technologies for collecting and measuring atmospheric electricity, from balloons to sensitive electrometers, which saw service well into the twentieth century.

The greatest legacy of Enlightenment earth science was an eminently practical technology to prevent lightning from damaging buildings and ships. Let us now turn to the community of property protectors that grew up around Benjamin Franklin's invention of the lightning conductor.

9. Property Protectors

Sudden, searingly bright, lightning inspires awe in most societies. Perhaps because lightning had no obvious mechanical cause, most people before Franklin's time ascribed its action to supernatural powers. Countless pantheons across the globe have lightning gods. Greek myths held the god Zeus responsible for sending rain and lightning to earth; sometimes he aimed his thunderbolts at misbehaving mortals. According to the Navajo in the American Southwest, lightning sent by offended gods could, in addition to killing sheep or destroying the Navajo's dwellings, cause disease. Whether regarded as the wrath of an angry god or an electrical discharge, lightning is a phenomenon feared equally by most peoples, ancient and modern. And with good reason.

As Richmann's misfortune reminds us, lightning is not something to trifle with. Yet his tragedy pales into insignificance compared to other lightning-caused disasters of the eighteenth century. In Brescia, Italy, lightning struck a powder magazine in 1769, setting off its entire store; the blast not only leveled the town but killed 3,000 people.[1] Often taller and set apart from other structures, powder magazines were among lightning's likely targets. Indeed, anything more conductive than air, especially if taller than its immediate surroundings, was somewhat vulnerable. Hence ships were at special risk, for wooden masts were a potent lightning attractor. When it struck a mast, the blow shattered the mast into small pieces; in the vernacular of the time, lightning "shivered the ship's timber"; sometimes a stricken vessel sank on the spot.

A lightning strike did not always exact a toll in human lives, but it often caused serious property damage. Some people took steps to protect their structures; the ancient Greeks, for example, worshiped Zeus and offered sacrifices to appease him, hoping thereby to save their structures from heavenly assault.[2] Needless to say, no measure was effective until Franklin invented the lightning conductor, sometimes called a lightning "rod." However, the rod itself is merely the uppermost portion (on the roof or at the tip of a mast); the entire conductor also includes a substantial and continuous metal connection to ground. According to I. Bernard Cohen, the lightning conductor was the first "practical" technology to emerge from science.[3]

People interested in safeguarding their property against lightning solicited the advice of Franklin and others having electrical expertise. In playing this role, such people became the first electrical engineers, hired to design in detail—and sometimes to build and install—lightning-protection systems, especially for important and costly structures.[4] In order to refine the design of lightning conductors, these engineers studied lightning-damaged buildings and sometimes built models to simulate lightning strikes in the laboratory. Although most lightning-protection systems were customized for specific structures, a few general-purpose conductors were also commercialized. Surprisingly, this simple yet effective technology, still employed in essentially the same form today, was not universally adopted and in some quarters met active resistance. And heated controversies erupted among electrical engineers over the best design for the lightning rod's tip.

Here I tell the story of how the diverse community I call *property protectors*—including the electrical engineers as well as the technology's users—came to adopt this "practical" invention. I also bring to light unique and intriguing electrical technologies that the engineers invented in the course of evaluating designs for lightning-protection systems.

—⊣|⊢—

Before Franklin's invention of the lightning conductor, people in the eighteenth century wishing to protect their property from thunderbolts had few technological choices. An obvious one, of course, was not to put up tall structures in exposed locations. Because of their generally modest heights, the dwellings of middle- and working-class people in cities or countryside had only a slight risk of being struck. On the other hand, churches and public buildings usually stood proud, brazenly inviting bolts from the sky. These structures had to be massive and tall to carry out their symbolic functions without diminishing the social power of the institutions they served. Likewise, elite homes were often large and obtrusive.

In return for impressive visual performance, the builders and stewards of these properties had to accept a greater likelihood of lightning strikes.

Churches, with their spires and steeples reaching toward the heavens, towering above the surrounding built environment (and even tall trees), were especially prone. One well-documented example is St. Mark's basilica in Venice. Although guarded by an angel, in the years 1388 to 1762 St. Mark's steeple had been struck and damaged nine times, putting the lie to the adage that lightning never strikes the same place twice. An equally telling case comes from Brittany. During one storm in April 1718, twenty-four churches received a thunderbolt.[5] Buildings constructed of robust masonry could be repaired after a strike, but many churches in the colonies and on the continent, especially in rural areas, were built of wood. Lightning's mischief could quickly reduce a parish church to smoldering embers and ashes.

In view of this constant threat, clergy and church elders over the centuries had arrived at a technological solution to ward off lightning. Believing that thunderstorms were caused by evil forces or a displeased deity, they invoked the ringing of bells as protection.[6] When a thunderstorm approached, bell ringers scurried to the belfry to do their duty. The pealing of bells, many of them specially consecrated to chase away storms, often went on for hours, occasionally lasting through the night—a terrifying concert.

The use of this ritual technology made perfect sense in the context of contemporary belief systems.[7] Even as churches continued to suffer the ravages of lightning, the practice of bell ringing flourished. Actually, this should not be surprising: anthropologists from the time of James G. Frazer, the author of *The Golden Bough*, note that religious belief systems supply persuasive rationales for the conspicuous, even frequent, failure of a ritual technology. For example, in explaining why lightning struck a church, parishioners could rationalize that the bell ringer was just a little tardy in getting started or had played the wrong tune. Alternatively, had the bells not been rung, the church would have suffered damage far greater. Musschenbroek, the distinguished Dutch Newtonian, offered an acoustic explanation for how bell ringing could disperse thunderclouds.[8] Moreover, because lightning strikes tend to be infrequent events—at least on a given building—it would have been difficult to determine whether a ritual technology like bell ringing actually worked.

Many ritual technologies are essentially innocuous and some—like the killing of witches or the ringing of bells during thunderstorms—do great harm; indeed, bell ringing was a dangerous calling. Lightning, in wending its way to ground through the better-conducting objects, often traced a course from the bells, down their ropes (sometimes damp from rain) and through the hapless bell ringer. Bell ringers died by the score. In one estimate by a German physicist, within twenty-three years "lightning had struck 324 bell towers and killed 103 imprudent bell ringers."[9] Even so, German magistrates continued to ordain that bells be rung upon a thunderstorm's approach. The death toll of bell ringers in other countries was probably comparable.

Clergy and church elders doubtless regretted the loss of life occasioned by the bell-ringing ritual; perhaps that is why some churches were receptive to experimenting with alternative technologies. Among the first structures to be protected by lightning conductors were houses of worship; even popes advocated its use. But, as we shall see below, this science-based technology often met resistance.

-||-

Franklin was an extraordinary natural scientist for the mid eighteenth century because he often concocted ways to put his discoveries to use in everyday life. After sundry experiments, including those employing his pasteboard-and-peas cloud, Franklin advised people on how to avoid harm when caught outdoors during a thunderstorm. Appreciating that a man had once died crouching under a leafy canopy, Franklin counseled others not to take shelter under a tree because it will "draw the electrical fire."[10] It is far safer to be in an open field because electricity will pass to ground through wet clothing in preference to one's body. This claim he supported with a simple experiment: a dry rat, he found, was more easily killed by the discharge of a Leyden jar than a wet one.[11] The number of people who followed Franklin's advice cannot be known, but perhaps it saved some lives.

On the basis of many experiments, Franklin had become convinced of the power of points: "points have a property, by which they *draw on* as well as *throw off* the electrical fluid, at greater distances than blunt bodies can."[12] This important effect, Franklin believed, might also have practical applications, perhaps in safeguarding structures from lightning. Before actually proposing his point-tipped lightning conductor, in 1749 he experimented on a model cloud, one far more impressive than the pasteboard-and-peas version. The new cloud was a big prime conductor, "made of several thin sheets of clothier's pasteboard, form'd into a tube, near ten feet long and a foot in diameter."[13] He covered the cloud entirely in gilt paper, then hung it from silk cords. This cloud had a large electrical capacity, much greater than his smaller, metal prime conductor. When fully charged the artificial cloud delivered a "pretty hard stroke" to his knuckle 2 inches away and left it aching.[14]

In the experiment proper, a grounded Franklin compared the ability of pointed and blunt objects, held in his hand, to acquire charge from the gilt cloud. With a blunt silversmith's punch made of iron, he was unable to draw charge from more than 3 inches away; at closer distances the discharge was sudden, with a "stroke and a crack."[15] In contrast, holding a needle toward the cloud, he could gradually and silently draw away its charge at a distance of 12 inches or more. Franklin also found that the greater the cloud's charge, the farther he could bleed it off with the needle.[16]

For Franklin, his room-size model perfectly captured the full-scale phenomenon of lightning, and he generalized accordingly. The silversmith's punch represented hills or high buildings, susceptible to a sudden strike when a cloud had enough charge to bridge the distance to their blunt prominences. In contrast, a grounded, pointed object—even when farther away—should slowly and silently relieve the cloud of its electricity.[17]

On the basis of the laboratory model, Franklin offered his revolutionary proposal in the form of a rhetorical question:

> may not the knowledge of this power of points be of use to mankind, in preserving houses, churches, ships &c. from the stroke of lightning, by directing us to fix on the highest parts of those edifices, upright rods of iron made sharp as a needle, and gilded to prevent rusting, and from the foot of those rods a wire down the outside of the building into the ground, or down round one of the shrouds of a ship, and down her side till it reaches the water?[18]

And, he added, by gently dissipating a thundercloud's charge, might we not be "secure . . . from that most sudden and terrible mischief?"[19]

For people who found themselves in homes without lightning conductors, Franklin offered additional advice. First he cautioned them not to remain near chimneys, mirrors, gilt pictures, or wainscoting. A good plan was to sit down in the middle of a room, putting their feet on a second chair. Even better, place a chair on several mattresses resting on the floor. Best of all, suspend a hammock from silk cords, equidistant from the walls.[20] The ultimate protection, of course, was the lightning conductor.

Franklin's proposal for the lightning conductor was tucked into a longer paper that he sent to Peter Collinson in July 1750. The cover letter to Collinson and the paper were published together in the earliest incarnations of Franklin's book in 1751–1754. During that period, he also announced his invention more publicly, both in *The Pennsylvania Gazette* and in *Poor Richard's Almanack*.[21] Curiously, in 1750 a brief abstract of his idea had already appeared anonymously in *The Gentleman's Magazine*.[22]

Although Franklin's invention received lots of good ink, there was hardly a stampede to install the new technology. The lightning conductor attracted controversy, not electricity, its pros and cons debated heatedly in both theological and scientific circles. Departing from his usual practice of eschewing scientific controversy, Franklin himself took part in the debates, addressing the criticisms and clarifying his earliest statements. The lightning conductor eventually enjoyed significant adoptions in the late eighteenth century, but it failed to go into general use. Surprisingly, that is still true today—and for similar reasons.

-||-

As he traveled around the colonies, propounding Franklinian science and technology, Ebenezer Kinnersley often encountered religious objections to lightning conductors.[23] American preachers had long cited natural disasters, such as ferocious thunderstorms, as evidence of God's displeasure with mankind. Scientists tinkering with God's messaging system thereby denigrated His providence and power. Kinnersley, a rather diplomatic minister, was able to defuse the religious arguments, and oversaw the installation of many lightning conductors.

When Dr. Lining, a correspondent of Franklin's, experimented with lightning conductors in South Carolina, he was accused of "atheistical presumption,"[24] a charge that was bandied about often during the lightning conductor's first decade. Additional insight into this charge is furnished in a 1758 diary entry by John Adams, who would become a frequent correspondent of Franklin's and the second president of the United States. Adams reported that an acquaintance, one Ben Veasey,

> began to prate upon the presumption of philosophy in erecting iron rods to draw the lightning from the clouds. His brains were in ferment, and he railed and foamed against those points and the presumption that erected them, in language taken partly from Scripture and partly from the disputes of tavern philosophy.[25]

Adams, of course, was a strong proponent of lightning conductors, believing that opposition to them stemmed from "Superstitions, Affections of Piety, and Jealousy of New Inventions."[26]

In 1760 the Junto—the society founded by Franklin in Philadelphia—deliberated over whether the use of lightning conductors was a presumption against God. The analysis, undertaken without benefit of Franklin's input (he was in England), is revealing. It concluded that thunder or lightning "is no more an Instrument of Divine Vengeance than any other of the Elements."[27] After all, the Almighty would not waste His wrath on objects so trivial as houses and trees. The Junto members affirmed, in good Franklinian fashion, that lightning was simply the restoration of electrical equilibrium between clouds and the earth. Installing lightning conductors, they reasoned, could be no more a presumption than trying to prevent or treat diseases. Moreover, failure to employ this prophylactic might amount to negligence if a bolt from the sky harmed a person's family or property. It is likely that the Junto's discussions mirrored those taking place elsewhere among educated—but still pious—people. Many would come to agree with the position articulated by Kinnersley: the installation of lightning conductors was not "inconsistent with any principles either of natural or revealed Religion."[28] Even so, the Boston lighthouse, erected in 1783, lacked a lightning conductor for five years because of religious objections.[29]

In Europe as well, religiously tinged opposition to lightning conductors flourished for many decades and occasionally became militant. In 1771 the naturalist Horace de Saussure installed a lightning conductor on his house in Geneva. His fearful neighbors were so displeased, he worried that they might riot. To assuage their anxiety he published a pamphlet explaining how lightning conductors work. Not only did the pamphlet have the desired calming effect, but some neighbors also had their own conductors installed.[30] In Italy Giuseppe Toaldo oversaw the placement of a lightning conductor on the ancient cathedral at Siena, which had been struck often. Many of Siena's residents were profoundly frightened by the new technology, but they did not threaten to riot. And when the conductor grounded a heavy lightning bolt on April 10, 1777, without damaging their beloved cathedral, the inhabitants' attitudes softened.[31]

Other property protectors were not so fortunate. In St. Omer, in France, Monsieur de Vissery de Bois-Valé had the temerity to outfit his house with Franklin's technology. Upset neighbors sued to have it removed and won, but de Vissery was persistent and appealed in court proceedings that endured from 1780–1784. His final plea, based on arguments about sound science and on the need for progress, was tendered by a young lawyer whose skillful performance at last reversed the earlier ruling. Not only had the case drawn in many eminent French scientists on behalf of lightning conductors, but it also helped establish the reputation of that ambitious lawyer Maximilien Robespierre.[32]

Public discussions of lightning conductors in France also took a more whimsical turn. If buildings could be protected from lightning, then why not people as well? In 1778 it was proposed that hats be equipped with lightning conductors. From a metal ring around the hat, a chain would dangle downward, dragging on the ground behind the wearer of this most fashionable contrivance of millinery artistry (Fig. 28).[33] Veau Delaunay even illustrated a telescoping, portable lightning rod—6 meters long when fully extended—that could be used, for example, by farmers in their fields.[34]

Figure 28. Hats with lightning conductors.
(Adapted from Figuier 1867, vol. 1, fig. 279. Courtesy of the Burndy Library, Dibner Institute, MIT.)

Surprisingly, some scientists raised doubts about the safety of lightning
conductors in general and the effectiveness of Franklin's point-tipped design in particular. The most curious feature of the scientific controversies
is that proponents on all sides based their arguments on, and furnished
additional evidence for, the very same experimental effect: the power of
points. But they drew vastly different implications about which lightning-
conductor designs—if any—would protect property best.

Almost predictably, one of the first—and most vociferous—critics was
Jean-Antoine Nollet. Remarkably, he maintained that Richmann's accident
proved the folly of Franklin's technology.[35] Detractors also fastened on the
possibility that, by attracting electricity from clouds, conductors would actually increase a building's chances of being hit.

Franklin responded in 1755 that his pointed lightning conductors had
both a preventive *and* a protective function.[36] The first reduced the charge

in a cloud, making it less threatening; the second conveyed the charge to ground in the event of a strike and so provided protection. Although graciously suggesting that he might have been misunderstood, Franklin could not fathom why Nollet and other detractors overlooked the protective function. In a 1768 letter published in later editions of his book, Franklin directly addressed Nollet's criticisms, quoting the French savant extensively.[37] By this time, however, Nollet's arguments against all lightning conductors had little credibility.

In the late 1760s Nollet stood alone among scientific men in claiming that Franklin's invention was "useless or dangerous."[38] The evidence was clear: well-documented cases, accumulated during the previous decade or so, had shown that lightning conductors worked. In his own house, for example, graced by a lightning conductor since the fall of 1752, Franklin occasionally saw a white stream of fire, "seemingly as large as my finger," pass between his alarm bells.[39] In the light provided by this dazzling display, Franklin insisted that he could see well enough to pick up a pin. A charge that large was almost surely an actual lightning bolt being carried harmlessly to ground. Thus, the bell-bracketed spark gap gave Franklin an extraordinary window into how a lightning conductor really works. That is why he could speak so confidently about its protective function.

In 1761 Kinnersley furnished definitive proof. After learning that a Philadelphia merchant, William West, believed his four-story house had been hit by lightning during a storm the previous summer, Kinnersley paid him a visit. He relentlessly quizzed West about the alleged lightning strike, interviewed witnesses, and climbed to the rooftop. Neighbors concurred that, on the day in question, the bright flash and explosion occurred simultaneously in the vicinity of West's house. What is more, one testified that he had seen streams of fire diffuse from the conductor's base several yards over the surface of the street. But there was more. West's clerk, sitting in the house, leaning against the brick wall where the conductor passed outside, had felt "a smart sensation, like an electrick shock."[40]

Not yet satisfied, Kinnersley inspected the rod on the roof, finding proof that not even Nollet could negate. The rod's brass tip, originally 10 inches long, showed unmistakable signs of melting and was now more than 2 inches shorter. Plainly, lightning had struck the conductor, yet Mr. West's house was unharmed. Kinnersley immediately congratulated Franklin, for the lightning conductor worked precisely as he predicted, and added: "May this method of security from the destructive violence of one of the most awful powers of nature, meet with such further success, as to induce every good and grateful heart to bless God for the important discovery."[41]

Franklin, God's agent for this discovery, continued to gather cases of buildings—with and without lightning conductors—that had received a bolt from the sky. No longer did he have to rely solely on theoretical arguments and laboratory models to defend his invention. Indeed, in his most vigorous response to Nollet's attacks, Franklin drew upon more than a dozen years of experience in analyzing the technology's operation. The pattern was clear: in conductor-equipped houses in North America, "not one so guarded has been materially hurt with lightning, and several have been evidently preserved," whereas unprotected buildings "have been struck and greatly damaged, demolished or burnt."[42] In Franklin's view, this hard evidence strongly outweighed the speculative arguments of Nollet or others.[43]

Franklin did have to worry about lightning-conductor advocates whose claims went too far. His friend Dalibard, for one, suggested that the entire city of Paris could be protected by a mere hundred or so lightning conductors disposed "in different quarters and in the highest places."[44] Somewhat later, August Witzmann argued that a network of huge, conductor-equipped kites could protect entire countries. He floated this proposal by the St. Petersburg academy, but it was shot down.[45]

A second objection to Franklin's invention raised by scientific adversaries was much more difficult to refute. It concerned the proper way to tip the rod, in a knob or point. Franklin's most able and energetic opponent on this seemingly small matter was Benjamin Wilson (1721–1788).[46]

Born into a large and well-to-do merchant's family in Leeds, Wilson had the means to cultivate an interest in art. But the comfortable life did not last past his teens, when the family descended into poverty. He moved to London, found employment as a clerk, and intermittently continued his studies in art. Wilson also read natural philosophy and became friends with William Watson, who was—in the 1740s—England's most accomplished electrophysicist. But the attraction of electricity was not enough to pull Wilson away from painting, the profession he would pursue throughout his life. Like many painters of that era, his bread and butter was portraiture, creating images of, and for, the wealthy. Wilson was successful at his craft and eventually found a patron—the duke of York. He even painted a very flattering portrait of Franklin.[47]

Painting paid the bills, but Wilson kept up a lively interest in electricity. Following Watson's lead, he performed many experiments and published his first findings in 1746. A Franklinian insofar as the one-fluid theory was concerned, Wilson also sought to show that Newton's "aether" was none other than electricity. During the 1750s he continued to publish on electrical research and coauthored a book with Benjamin Hoadly.[48] In recognition of his contributions to electrical science, Wilson was elected a fellow of the Royal Society, and in 1760 he received the society's prestigious Copley Medal (awarded as well to Franklin seven years earlier). Clearly, in the firmament of English science Wilson's star shone brightly; like Nollet, he was already a notable when he took on Franklin.

As a distinguished scientist with electrical expertise, Wilson was one of many men consulted in the design of lightning conductors for important buildings. On March 6, 1769, the dean of St. Paul's Cathedral applied to the Royal Society for advice on "the best and most effectual method of fixing electrical conductors" on the aging structure, already twice damaged by lightning.[49] A stellar committee was duly appointed, consisting of William Watson, Benjamin Franklin, Benjamin Wilson, John Canton, and Edward Delaval, which dispatched its duty without delay. Three months later, on June 8, the final report was read to the society and a copy transmitted to the dean of St. Paul's.

The cathedral, a magnificent edifice designed by Christopher Wren, was thoroughly inspected, inside and out. Committee members noted that the main roof, of lead sheet, was already secured against lightning, being grounded through metal drains that terminated at the sewer. However, two towers and the great cupola, capped by a stone lantern with a metal cross on top, were exceedingly vulnerable because they lacked continuous metallic links to ground. Not surprisingly, the committee proposed that these conductive gaps, in one case extending 48 feet, be bridged with iron bars or thick lead sheet, depending on location. Aware that the iron bars would gradually waste away through corrosion, the group also suggested that the bars be more than 1 inch thick. The committee's report, whose recommendations were fully implemented, remains a model of soundly reasoned engineering analysis and design, valid today.

Because the very tips of the towers and the cupola already had metal ornamentation, the St. Paul's case could not presage the conflict that soon erupted over points and knobs. After all, everyone recognized that the metal ornaments, when connected securely to ground, would themselves protect the cathedral. But soon another case in England, this one entangled in national security, would become a flash point of proper design for lightning conductors.

—|||—

The Purfleet armory was a series of buildings that included a boarding house, proof house, and five powder magazines loaded with barrels of gunpowder. The Board of Ordnance, believing that such stores deserved lightning protection, in 1772 requested help from the Royal Society.[50] A committee of electrical notables was again appointed, consisting of Henry Cavendish, William Watson, Benjamin Franklin, Benjamin Wilson, and J. Robertson. After inspecting the premises, the group noted that the long, rectangular buildings housing the powder barrels were primed for disaster. Along the full ridge of each roof ran a lead coping that, ominously, had no connection to ground; were lightning to strike, it would probably follow a path through the powder kegs inside. Mindful of the Brescia disaster, the committee sounded the alarm, urging a timely fix.

Doubtless influenced by Franklin, who drafted the report, the group proposed that tall iron rods, with 12-inch pointed copper tips, be mounted at both ends of the long buildings, connected to the metal ridges, and grounded securely in specially dug wells. Benjamin Wilson, however, openly dissented from the committee's consensus. He not only refused to sign the report, but he also authored an even longer report disputing the committee's conclusions on the matter of the rod's tip. It is a measure of Wilson's stature that the Royal Society published his dissent together with the committee's report.

Wilson's main argument against pointed rods was based on the same sort of experiments that Franklin had carried out, indicating that a point attracts charge more easily and at a greater distance than blunt rods. Wilson acknowledged that pointed rods had served science well by enabling investigators to collect atmospheric electricity. He maintained, however, that pointed rods were inappropriate in the practical context of designing lightning protection. His words were themselves blunt: "I have always considered pointed conductors as being *unsafe,* by their great readiness to *collect the lightning in too powerful a manner.*"[51] Thus, emplacing pointed rods that projected high above the roofs—the committee's recommendation— would invite lightning to strike the Purfleet armory.

To support his own recommendation that knob-tipped conductors be placed beneath the roof line, Wilson introduced the case of the Eddystone Light House, which sat highly exposed on a rock reef, 14 miles across treacherous seas from Plymouth. A marvel of engineering that had cost £40,000, the 90-foot-tall lighthouse, of 1,493 interlocking stone blocks, was designed and built in 1759 under the direction of the esteemed civil engineer John Smeaton. This was the first structure in England, perhaps in the world, that included a lightning conductor when erected; its clever design incorporated metal parts also serving other purposes. The lantern was framed with iron and topped with a copper ball that was grounded through a sequence of metal strips, kitchen sink, water pipe, and an iron chain bolted to the rock below the low-tide line.[52] More than slightly misrepresenting the situation, Wilson claimed that the lighthouse had been deliberately equipped with a blunted rod—in preference to a pointed one—and "has since received no injury from lightning."[53] This case of course showed only that lightning conductors do protect buildings, which by the 1770s was beyond dispute; Wilson could not claim that the lighthouse's copper ball had never received a strike.

Despite Wilson's lapse in logic, other members of the property-protector community seriously pondered his objections. William Henly, who also consulted on the design of lightning conductors, had ample reason to want the controversy resolved. Toward that end, in 1773 he conducted a series of laboratory experiments, which he hoped would at last lay the matter to rest. Apart from employing some unusual objects in his apparatus, including a gilded bullock's bladder, Henly's experiments yielded no new effects. They merely confirmed, yet again, that points were better at receiving charge than knobs and so were preferable for tipping lightning conductors. Although Henly believed that his work "completely decides the question," the matter was far from settled.[54]

The controversy entered a more highly charged phase after May 15, 1777, when the Purfleet armory was damaged by lightning. Actually, none of the powder magazines was hit; only the boarding house was harmed, and it was minor damage at that. In its original recommendations for this building, the committee had noted that a continuous series of metal fixtures, from roof to ground, already rendered the boarding house safe. They did, suggest, however, adding a pointed rod to the house's summit.

Not surprisingly, and again at the request of the Board of Ordnance, the Royal Society authorized an investigative committee, this one consisting of William Henly, Timothy Lane, Edward Nairne, J. Planta, and Benjamin Wilson. The good Philadelphian was absent this time because, by signing the Declaration of Independence the previous year, he had committed treason against the Crown: no longer in England, Franklin was a wanted man. Also of interest is the inclusion of Edward Nairne, an instrument maker, who had been selling and installing lightning conductors since at least 1772.[55] The composition of this committee apparently reflected a growing appreciation that the expertise required for this job was not electrophysical theory but experience in practical electrical engineering.

After inspecting the damage—a corner of the parapet wall had lost some stone- and brickwork, and a metal cramp around the edge was badly bent—the group concluded in a brief report that lightning had violently bridged a gap of 7 inches between metal parts of the upper parapet wall.[56] Graciously, it made no mention of the earlier committee's failure to observe the obvious lack of electrical continuity. The committee agreed on the simple remedy of filling the gap with metal plates; they would at last properly ground the roof.

His previous dissent brushed aside, an exercised Benjamin Wilson dissented again, refusing to sign the report and writing his own assessment of the Purfleet incident.[57] Ignoring the continuity gap, Wilson fixed blame on the pointed rod on the roof, claiming that lightning had not hit this building prior to the rod's installation. Wilson, his rhetoric ratcheted up, asserted in conclusion that the use of pointed rods was poor science that did not promote the welfare of society. But once again, he failed to persuade.

—|||—

An undaunted Benjamin Wilson, convinced he had truth on his side, sought to make his case anew with help from a higher authority. It was not God that Wilson petitioned, but the king. Wilson wanted to carry out an experiment on a scale large enough to realistically simulate the Purfleet accident. However, a project of this magnitude was beyond the personal means of this painter, so Wilson presented his proposal to George III, believing the king to be "always disposed to promote every pursuit which tends to the advancement of science and the good of the public."[58] Through the Board of Ordnance, the king granted Wilson the funds to proceed.

The plan was grandiose: simulate a large thundercloud moving over the Purfleet house, testing whether pointed or knobby rods provided better lightning protection. To implement this plan, Wilson constructed an apparatus equally grandiose (Fig. 29). The cloud was represented by a four-part prime conductor, around 155 feet long and over 1 foot in diameter, arrayed in the shape of a horseshoe.[59] The three main parts of the prime conductor were made from more than 100 drum rims joined by wooden slats and covered with cloth; the surface was rendered conductive by the application of 87 lbs. of tin foil. The prime conductor's fourth part, about 10 feet long and easily detachable for some experiments, was composed of brass drums. Needless to say, hanging this weighty prime conductor about 5 feet above the floor required stout silk cords. To augment the electrical atmosphere of the massive prime conductor, Wilson also suspended 6,900 feet of wire in the shape of a spiral.[60]

Charge was at first furnished to the "cloud" by two huge cylinder machines, driven in tandem by a single wheel. Unfortunately, this arrangement involved so much friction that the wheel did not turn easily. For most experiments, then, Wilson used only one machine.

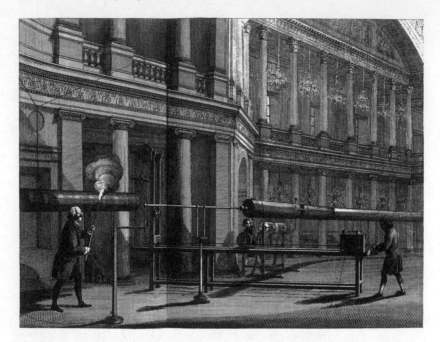

Figure 29.   Benjamin Wilson's model for studying the Purfleet incident. (Adapted from Wilson 1778c. Courtesy of the Bakken Library, Minneapolis.)

To represent the building, he built a scale model from wood, extending less than 2 feet in each dimension, with metal parts corresponding to their actual locations.[61] Realizing that it would be easier to move the model house than the cloud, Wilson put the house on wheels so that it could run on an elevated track; by this means he could easily position the house at various distances from the cloud and roll it at different rates.

Housing this apparatus was itself a formidable problem. Happily, owners of the Pantheon, a cavernous edifice in London used for shows and exhibitions, offered the "great room" as space for the research.[62] Wilson and an assistant, Mr. Wyatt, conducted countless experiments, varying factors such as the speed of the house's movement, the nature of the house's conductor (if any), and sundry configurations of the wire spiral.[63] To measure the quantity of electricity remaining in the "cloud" after each discharge, Wilson was "obliged at last to have recourse to the sense of feeling," taking the shock on the hand.[64]

Even allowing for variation in the human measuring instrument, Wilson's results were highly consistent: the power of points held fast. Pointed rods initiated discharges at greater distances than knob-tipped ones and also bled off more charge, whether the house moved or not. This was, according to Wilson, irrefutable proof of his contention that pointed rods, especially when raised above the highest part of a building, were *very* dangerous.

Among the witnesses to Wilson's various demonstrations at the Pantheon was the king himself. George III came away a believer and ordered that the pointed rod already installed at his Buckingham residence be replaced by a knobby rod.[65] When Franklin heard about the king's edict, while he was wooing the French to join forces with the revolutionaries, he commented that the matter was of little importance to him, but "If I had a wish about it, it would be that he had rejected" all lightning conductors.[66]

At last satisfied with his findings on November 12, 1777, Wilson reported a grand total of fifty numbered experiments to the Board of Ordnance and the king in a long paper that was published the next year by the Royal Society.[67] Immediately following Wilson's tedious recitation of results in the *Philosophical Transactions* was a brief report authored by yet another committee of the Royal Society, which made further recommendations on protecting the Purfleet armory. A gang of nine heavyweight electrical researchers, the committee included Priestley, Cavendish, Watson, and Charles Mahon; only Franklin the traitor was missing.[68] The report judged Wilson's entire series of experiments as "inconclusive."[69] Taken alone this judgment might suggest that Wilson's efforts had kept the issue alive, but that was far from its authors' intent. In fact the entire report was a ringing endorsement of Franklin's position, as it recommended adding more pointed rods to the Purfleet buildings. The committee's unwillingness even to discuss Wilson's experiments was itself a brutal rebuttal.[70] Now virtually alone in his views, even among experts in England, the painter and electrical experimenter had finally been silenced.

The prevailing opinion among scientists and engineers today is that the shape of a lightning conductor's tip makes not one iota of difference to its effectiveness.[71] Any lightning rod, properly installed and well grounded, will do the job; what matters most is electrical continuity. The lightning-related experiments of Franklin, Henly, Wilson, and others all suffered from the same defect: none *realistically* scaled down the electrical interactions of clouds, buildings, and lightning rods to a laboratory setting. In Wilson's case, relative to the size of his cloud, the model house was way too close or far too large—as were the rods. Had more realistic proportions obtained throughout the entire artificial system, Wilson would have seen no difference between the behavior of pointed and knobby rods.[72]

Although Wilson's experiments were imperfect, they set an important precedent. In the immense great room of the Pantheon, he had assembled

apparatus in a model system for studying pressing issues of electrical engineering. Wilson's project—an expensive, state-supported endeavor involving painstaking experiments and much new technology—did not achieve
its goal, but the general approach itself was sound. Better contrived model
systems would become important tools of electrical engineering in the nineteenth and twentieth centuries.[73]

The experiments of Benjamin Wilson also had a more immediate legacy, as the apparatus used in the Pantheon experiments were dispersed. Although he finally had an electrical system fit for a king, George III declined
to add it to his collection. The fate of the elephantine prime conductor remains unknown, but Dr. James Graham bought the cylinder machines for
his Temple of Health. How could the king part so readily with such seemingly significant—and unique—technology? By this time George III was
no longer a very active collector of scientific instruments, and a growing
war against the American colonies preoccupied him. Most likely the king
decreed that Wilson's apparatus be sold, as even his deceased mother's pos-
sessions had been (including her thimble).[74] George III apparently did not
indulge in nostalgia.

⊣∣∣⊢

While a few Englishmen were locked in esoteric engineering debates, small
numbers of people throughout Europe and the American colonies were
adding lightning rods with points to their buildings and ships. After all, reports continued to appear in scientific and popular literature about lightning conductors that had protected property brilliantly.

Regrettably, reliable information on the acquisition patterns of lightning conductors is hard to come by. And Franklin himself furnished only
tantalizing hints about the adoption of his invention. In a 1772 letter to de
Saussure, Franklin reported that in the colonies, "Numbers of them appear
on private Houses in every Street of the Principal Towns, besides those on
Churches, public Buildings, Magazines of Powder, and Gentlemen's Seats
in the country."[75] This letter perhaps exaggerated the prevalence of lightning conductors, but it did hint at an intriguing acquisition pattern that, we
shall soon see, also characterized European adoptions.

Many members of the property-protector community published books
that included descriptions of their lightning-conductor installations. Toaldo,
for example, compiled all his previous reports into a single volume whose
copious cases proclaimed the success of Franklin's technology.[76] Such books
give us insights into the works of specific individuals but not about general
acquisition patterns.

Insofar as installations on ships are concerned, we have scarcely more than anecdotes to rely on for the late eighteenth century. In a 1770 letter to Franklin, for example, one Captain J. L. Winn—who never set sail without a lightning conductor—reported with dismay that "very few vessels are furnished with them."[77] Another captain, the well-known explorer James Cook, circumnavigating the globe at the behest of the Royal Society, employed a lightning conductor on his ship *Endeavour*. On October 12, 1770, while anchored during a horrendous thunderstorm in Indonesia, Cook's ship was struck but survived without damage; a nearby Dutch ship, unprotected, was not as fortunate.[78] As late as 1831, W. Snow Harris chastised the Admiralty for failing to mandate the installation of lightning conductors on naval vessels, despite a decades-long litany of damaged British ships and dead sailors.[79]

Happily, one member of the property-protector community—Marsilio Landriani—did assemble some data on lightning conductors installed on structures across Europe. In 1784 Landriani published an advocacy book on lightning protection in which he not only documented the effectiveness of lightning conductors but also furnished guidance on their design. For example, he suggested that the best conductors were made from copper, but ones of lead, tin, and iron were also usable (copper, lead, and tin are metals that corrode slowly in the open air).[80] Iron conductors, he noted, could be made more rust-resistant by coating them with oil. Landriani also identified common defects in previous installations, including conductors that were too thin, had poor continuity, and were grounded imperfectly.[81] Although the use of multiple points on the tip of a rod was gaining some advocates, Landriani believed that a single point was more effective.[82] A large part of the book abstracted his correspondence with many other European members of the property-protector community, including de Saussure, Bertholon, Toaldo, Cuthbertson, Buffon, and Franz Achard.[83] In short, Landriani's book—published in Italian—was a comprehensive manual that furnished state-of-the-art information on the design and installation of lightning-protection systems. Landriani sent a copy of the book to Franklin, who "read it with great Pleasure."[84]

For present purposes, the last part of Landriani's book, an appendix, is of most interest because it listed all the European installations that had come to his attention. For each entry he recorded the kind of structure, its location, and usually the owner's name. Regrettably, the raw numbers of installations in this appendix—a total of 323 entries—cannot be taken at face value; doubtless numerous lightning conductors had escaped his correspondents' notice.

Beyond mere absences, the data set appears afflicted with significant national biases. For example, the largest number of entries—one hundred twenty—comes from Landriani's country, Italy, followed by France and Germany, both with a little more than sixty each. There is a mere sprinkling of installations reported in Poland, Switzerland, the Netherlands, Monaco, England, Denmark, and Belgium. I suspect that this pattern reflects only the extent of Landriani's communication network. England furnishes a case in point. It is listed as having only five installations, including two cathedrals and two palaces; there is no mention of the Purfleet armory, the Eddystone lighthouse, elite houses, or ships.[85] Likewise, Russian lightning conductors are missing. It is true that Richmann's death retarded the adoption of this technology in Russia, but by 1784 there were six installations in St. Petersburg alone in addition to those on powder magazines.[86]

My hunch is that these data merely record the activities of a handful of property protectors known personally to Landriani. In the Netherlands, for example, the reported installations almost certainly document the products of just one person—John Cuthbertson. On the basis of these data, we cannot offer strong conclusions about the lightning conductor's popularity in different places.

Yet if we accept Landriani's appendix as a good sample of the work done by *some* property protectors, we can discern patterning in the kinds of buildings protected as well as in the social positions of the purchasers. Let us look first at the type of structure. Not surprisingly, houses—by far the most abundant structure in almost all communities—constitute more than half the installations (54 percent). This is followed distantly by religious structures (14 percent), palaces and castles (8 percent), military structures (7 percent), and public buildings (6 percent). Schools and factories together make up only 3 percent.[87] The prevalence of military lightning conductors is vastly underestimated because, in some cases, Landriani's entries (each of which, in the absence of specifics, I counted only once) correspond explicitly to multiple installations, such as *all* powder magazines controlled by the grand duke of Tuscany.

Setting aside houses for the moment, we see that the structures of institutions, sometimes wealthy ones, were often protected, especially if they were vulnerable. The stewards of such properties, such as church and state functionaries, were apt to number in their ranks people familiar with electrical matters. Because many electrical experts were clergymen, they would have been strong advocates for outfitting religious buildings. And the grand duke of Tuscany was himself an electrical experimenter. In addition, persuasive instrument makers could have drummed up business through influential contacts. No doubt many castles, palaces, and military installations in wealthier nations, such as England, were protected, whereas poor parish churches in countless tiny hamlets continued to rely on bell ringing. I suggest that access to electrical expertise and sufficient wealth were two major factors favoring the acquisition of lightning conductors by the stewards of nonresidential structures.[88]

Costs of lightning conductors listed in the catalogs of instrument makers are misleading. Although off-the-shelf, generic conductors were modestly priced from £3 to almost £8, these were suitable only for some ships and houses; installation, metal fittings, and upgrades, such as more corrosion-resistant metals and a gilded point, were almost certainly extra.[89] Moreover, the custom designs and complicated installations required on nonresidential structures (and many houses and ships) would have been very pricey. A custom ship installation, for example, could cost £100.[90]

Houses also present an interesting pattern. Because Landriani listed the names of homeowners, I created two groups: houses owned by a titled person, such as duke or earl; and houses whose owners lacked titles. In any community the nobility would be just a minuscule minority of all residents, yet nearly half the houses in Landriani's list—84 of 174—were owned by titled persons. What is more, many other houses, some listed as "country homes," doubtless also belonged to the elite. Among the untitled persons who had bought lightning conductors were scholars and scientists, including Lichtenberg, de Saussure, Buffon, Fontana, Cavallo, and Voltaire. And we may infer as well that others on the list were wealthy merchants and shop owners.

Although it is possible that the less well-to-do installed their own lightning conductors, cost would have been a deterrent. And of course people who opposed the use of lightning conductors on their churches would not be likely to set one up at home. In the many eighteenth-century books on lightning protection I examined for this chapter, very few illustrated modest structures outfitted with Franklin's invention. In one case, the drawings may have represented wishful thinking on the author's part: the hope that his market could come to include peasants' houses and barns.[91]

The conclusion seems inescapable that lighting conductors for houses were mainly an elite consumer product. Apart from the protection that it provided, a lightning conductor had another performance characteristic that may have fostered its acquisition by wealthy and well-educated urbanites. A house's lightning rod had great visibility to all passersby. It marked its owner as a knowledgeable and prudent person. If one elite house in a community was fitted with a lightning conductor, other members of that social class—especially neighbors—might also have opted to buy them.[92] The lightning conductor at home, I suggest, was another materialization of Enlightenment ideals and of participation in elite culture.[93]

—|||—

The installation of lightning conductors, even in highly vulnerable churches, was surprisingly spotty. A lengthy article in *The Gentleman's Magazine* in 1787 chronicled the effects of recent thunderstorms throughout the British Isles and the continent.[94] Not only did people fail to take adequate precautions to protect themselves when outdoors, but churches and many homes also suffered lightning's mischief. In Grenoble, for example, the tower of St. Mary's church was demolished by a strike that also rent the walls and tore up a pavement, exposing the burial vaults below. A French ship struck near the Isle of Wight sank with the loss of all lives. Clearly, neither case had deployed lightning conductors. We cannot ascribe French rejections of Franklin's technology to Nollet's lasting influence because sparse use was the rule everywhere. In Bavaria, for example, in 1774 alone lightning destroyed at least thirteen unprotected churches.[95] In surveying the damage to unprotected structures, experts such as Nairne and Henly continued to document the path of lightning, presumably having been called in to recommend a tardy remedy.[96]

Even today, in lightning-prone regions, many buildings remain without protection. One general factor at work, in the past and present, is our inability to predict the occurrence—and assess the risks—of inherently rare events. In such cases, we are inclined to believe that the chances of having our own property struck by lightning *and* suffering significant damage are negligible enough to ignore. Today, many highly educated, middle-class people live in lightning-prone areas and lack conductors on their homes.

Risk assessments, however flawed or implicit, probably figured in decisions to acquire lightning conductors during the eighteenth century. One line of evidence in support of this conjecture is the acquisition patterns for nonresidential structures. To wit, the most uniformly protected type of building, in all nations for which some data are available, is the powder magazine. The dire consequences of a lightning strike on such a structure—loss of military supplies as well as human lives—were so well known after the Brescia disaster as to occasion little overt discussion. Such knowledge clearly factored into decisions, given that powder magazines were preferentially protected in comparison to all other nonresidential buildings.[97]

—||—

In seeking vestiges of eighteenth-century electrical technology within our modern world (outside museums, of course), we find that the lightning conductor stands supreme. Indeed, some emplaced during the eighteenth century may still be functioning today. This electrical technology is almost unique in that the original design has undergone no significant revision. So thoroughly did Franklin understand electricity, especially the effects of lightning on structures, that the lightning conductors installed today differ only in trivial details from his earliest specifications.

Other "practical" electrical inventions of the eighteenth century, though few and far between, had a more mixed fate (see chapter 11). Some inventions languished merely as ideas and proof-of-concept prototypes, but others were judged significant and brought to a point of technical adequacy and commercial success —but not until the nineteenth or twentieth centuries. Before turning to the community of visionary inventors, I now examine the significant role of electrical technology in chemistry.

10. A New Alchemy

The recognizably modern chemistry that emerged in the last decades of the eighteenth century was an amalgam of many ingredients, one of which was alchemy. Building on traditions of Islamic chemistry, alchemists—Isaac Newton was one—did more than seek ways to turn base metals into gold. Their explorations accumulated much empirical knowledge and many useful technologies. An even more important ingredient contributing to modern chemistry was craft traditions, especially in glass and ceramics, mining and metallurgy, munitions, dyeing and leather working, brewing, and medical preparations.[1] Specialists in these industries developed numerous technology-intensive processes for analyzing and synthesizing substances. By the middle of the eighteenth century, chemists were members of a well-established, international community skilled in studying chemical reactions and in transforming materials.[2]

So large was the chemist's toolkit that, in Antoine Lavoisier's classic textbook, which presented full-blown his "new chemistry," illustrations of the core apparatus, consisting of everything from glassware to scales to stills, filled 13 plates—a grand total of 159 figures. Lavoisier also cautioned students that learning the language and apparatus of chemistry required at least "three or four years of constant application."[3] Significantly, Lavoisier's compendium included one electrical device, a "gasometer," which enabled the combustion of confined gases—more on that later.

The advent of electrical machines and, especially, Leyden jars and batteries, gave chemists new tools for investigating substances and stimulating reactions. *New alchemists* is the tongue-in-cheek term I apply to the community that adopted electrical technology for chemical research. By defining "chemical research" broadly as the analysis and synthesis of substances and the study of their properties, I encompass activities that fall within today's disciplines of chemistry *and* materials science.[4] Members of this community included Franklin, Cavendish, Priestley, Volta, Beccaria, and Lavoisier himself; they not only invented new electrical technology but also made discoveries that helped foment in chemistry a theoretical revolution. Moreover, these investigators confected in their laboratories surprising technological effects that would become the basis of many commonplace electrical products. In this chapter, I highlight studies carried out by the new alchemists and discuss their electrical technologies.

—|ı|ⱶ

In March 1989 two chemists from the University of Utah, Stanley Pons and Martin Fleischmann, made an announcement that seemingly shook the foundations of physics: in an electrochemical cell resting on a table they had achieved nuclear fusion. According to accepted theory, however, only a monstrous, high-temperature device like a tokamak could bring hydrogen nuclei close enough together to fuse into helium. Pons and Fleischmann had claimed to accomplish what nuclear physicists believed was impossible.[5]

The Utah chemists' tabletop process, said to mimic the sun's source of energy, came to be called "cold fusion." In the ensuing months, hundreds of researchers around the world repeated the Pons-Fleischmann experiments, but the results were far from supportive. Their reputations at risk, Pons and Fleischmann insisted that other researchers were not following the original experimental procedures precisely. Meanwhile, a few theorists in chemistry and other disciplines, perhaps hoping to tarnish the luster of nuclear physics, struggled to explain the purported effect—an excess of heat not caused by any other known process. In the end, however, the most carefully controlled experiments failed to find excess heat, leaving intact theories of nuclear physics and dashing hopes for a limitless source of cheap energy. Pons and Fleischmann silently retreated into obscurity with no admission of error. Entirely discredited, the notion of cold fusion today denotes an infamous episode of sloppy science that chemists, especially, would prefer to forget.

Few people today are aware that the term "cold fusion" was coined in the middle of the eighteenth century by Benjamin Franklin. Like its modern counterpart, the first cold fusion described an apparently impossible process—the melting of metals without heat. The idea of cold fusion came to Franklin as he struggled to account for some of lightning's strange effects. In particular, it was said that lightning had once passed through a sword, partially melting it without scorching the leather scabbard. This could be explained, Franklin conjectured, by assuming that electricity "can insinuate itself between the particles of metal, and overcome the attraction by which they cohere." This enabled the solid metal to attain, however briefly, a heatless liquid state; "it must be a cold fusion."[6]

Not long afterward, Franklin performed experiments that appeared to give his cold-fusion hypothesis some support. He began with strips of plate glass, about the width of a finger, along with narrow strips of metal leaf, including gold, silver, and gilt copper. Between the pieces of glass he sandwiched a metal strip, allowing its ends to protrude. Next he bound the sandwich firmly with silk thread. When Franklin connected the metal strip's ends in "an electrical circuit" to a large Leyden jar, the metal leaf melted.[7] What is more, the metal became so tightly fused with the glass that not even the strongest acid could dislodge it. Most important, when Franklin touched the pieces of glass immediately after the discharge, he could never "perceive the least warmth in them."[8] He repeated the experiment, this time putting the glass-metal sandwich into a small bookbinder's press. In this configuration, the glass was less likely to shatter during the discharge, but otherwise the effects were identical. The phenomenon of cold fusion— confirmed by the touch of a finger—was for Franklin real.

The cold-fusion hypothesis, formulated around 1749, and supporting experiments were tucked into letters published in the earliest editions of Franklin's book. Deeply buried amidst his many experiments and seminal ideas, the claim of cold fusion nonetheless garnered attention. The combative abbé Nollet, then still in his prime, quickly took notice and attacked Franklin; cold fusion, he maintained, was impossible.[9] Perhaps the most thoughtful assessment of Franklin's hypothesis—congruent with modern understandings—was penned by Gowin Knight, an authority on magnetism and the principal librarian of the British Museum.[10] He suggested, for example, that a sword's thin edge could be melted instantly by lightning yet feel cool to the touch because the heat of fusion was immediately dissipated by the adjacent, large mass of cold metal. Perhaps out of respect for the great scientist, Knight made no mention of Franklin when arguing that cold fusion was merely "vulgar error."

Ironically it was Franklin's close friend and acolyte Ebenezer Kinnersley who, not long after the publication of Knight's letter, carried out a brilliant experiment that would decisively consign the first cold fusion to science's dustbin. Kinnersley devised an instrument, which he called an "electrical air thermometer," that could discern if heat was produced when electricity passed through, and melted, a piece of metal.[11]

The air thermometer (Fig. 30, right) exploited the venerable principle that air, when heated, expands. The device consisted of a glass tube, around

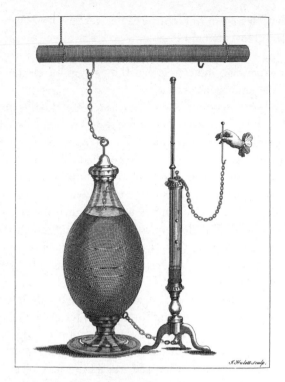

Figure 30. *Right*, Kinnersley's air thermometer and *left*, Leyden jar. (Adapted from Franklin 1769, pl. 5. [389], courtesy of the Bakken Library, Minneapolis.)

11 inches long, with an internal diameter of 1 inch. Sealed well on both ends with brass caps, the tube was set vertically upon a mahogany stand. Inside the tube Kinnersley could mount a small hooked wire for hanging various objects, such as fine wires; the top of the hook passed through the upper brass cap and became one conductor. The tube also held a ball-tipped wire, extending upward from the lower cap, which served as the second conductor in a circuit with a Leyden jar or battery. A small amount of liquid was placed in the tube, into which deeply dipped a tall, graduated tube of very small bore—much like today's glass thermometer. The inner tube extended through, and about 10 inches above, the upper cap. If heat were produced when the battery discharged through the test specimen, the air inside the larger tube would be rarefied, increasing the pressure and causing the liquid level in the indicator tube to rise.

Using a large Leyden jar (Fig. 30, left), Kinnersley tested assorted materials, including "a strip of wet writing paper, wet flaxen and woolen threads, blade of green grass, filament of green wood, fine silver thread, very small brass wire, and strip of gilt paper." [12] The results were uniform: in every case the air expanded, indicating that the passage of electricity through these materials had produced heat.

In order to leave no doubt, Kinnersley tested metal wires outside the air thermometer. With discharges from a thirty-five-jar battery, he found that the wires could be heated to incandescence; sometimes they even melted. Metals were not, therefore, electrically liquified by cold fusion. Rather, a large charge passing through a small-diameter wire met considerable "resistance," which produced heat, sometimes sufficient to bring about melting. [13] A larger wire, he added, offered less resistance and so conducted the charge without heating up. Kinnersley had laid the foundation for understanding resistance heating, the process at the heart of many present-day products, including toasters, electric stoves, soldering guns, ceramic kilns, and industrial furnaces. [14]

Kinnersley reported the electrical air thermometer along with his findings to Franklin. After considering Kinnersley's letter, which also touched on other scientific subjects, Franklin wrote a response and published both letters in later editions of his book. In contrast to Pons and Fleischmann, Franklin did not defend cold fusion and attack his critic's research methods. Rather, he regarded Kinnersley's experiment as "beautiful" and acknowledged—with no apparent pain—that cold fusion was "error." [15] Also, in a footnote to the earliest mention of cold fusion in his book, Franklin in effect retracted the idea; few concessions in science are more gracious.

A fascinating instrument, the "electrical air thermometer" has generated much confusion over the years because some people assumed, on the basis of the name, that it was supposed to measure air temperature. That feat, however, was well beyond the capabilities of Kinnersley's creation. [16] Nonetheless, it served admirably as a specialized device for vanquishing the first cold fusion. [17]

-||⊢

In the experiments with metal leaf and glass, which had bolstered his claim of cold fusion, Franklin also reported a new effect: different metals "stained" the glass different colors. Gold produced a "somewhat reddish" stain, while that of silver was "greenish." [18] In the course of extending Franklin's findings, others crafted new apparatus for testing specimens, which enabled them to explore myriad technological effects of resistance heating.

Giambatista Beccaria, an outspoken Franklinian, was an early and energetic new alchemist. Following Franklin's lead, he affirmed that different metals impress different colors into glass. He also showed that metal foils subjected to a discharge created "stains" of similar color on paper.[19] Most likely, these stains were metal oxides, but the term "oxide," which Lavoisier introduced around 1789, would not enter common usage until the early nineteenth century.[20] Instead, chemists referred to these compounds as "calces." Clearly, employing the most rudimentary apparatus—a source of sizable charge, a sandwich of glass or paper, and metal foils—an investigator could cause a metal to become its calx.[21]

But calces too could be transformed by resistance heating into still other substances. In demonstrating these further effects, Beccaria made use of a small glass tube, with internal metal conductors separated by a small gap.[22] He could pack the tube with powder, usually a metal compound, and deliver through the conductors a sizable charge; this electrical treatment caused the particles to melt and fuse, creating a glass. For example, powdered litharge (a lead oxide) became a glass having the color and transparency of amber. How delighted he must have been to find that verdigris (a green copper acetate) turned into a glass, red and glossy.

Many new alchemists introduced sundry powdered materials into similar glass tubes, subjected them to a large discharge, and studied the products. Doubtless this was great fun. Still, the results were not really new science—the reactions were already known. The only novelty was technological: the reactions had been fomented by direct resistance-heating of the powdered substances. What is more, none of these effects became the basis of a commercial process in the eighteenth or early nineteenth century. After all, abundant heat for these reactions could, in every case, be more readily supplied by traditional technologies such as coal-fired furnaces.

A perceptive observer, Beccaria noticed that some metals required large charges to calcine yet others did not. In seeking the cause of this variation, Beccaria suggested that it might be the metal's inherent electrical resistance.[23] To wit, more resistant metals, such as iron, calcine easily, whereas highly conductive silver and gold do so with greater difficulty. Experiments by Priestley had already yielded a ranking of metals by their electrical resistance: from most to least resistant were iron, copper, brass, silver, and gold.[24] Beccaria's hypothesis apparently had merit. That materials with higher electrical resistance also became hotter helped explain why calces, most of which conduct poorly, could be vitrified.

Along the way, Beccaria's experiments documented additional technological effects. One of the most surprising was that electrical treatment of cinnabar, a sulfide ore of mercury, brought forth the silvery metal itself.[25]

He also showed that mercury could be vaporized with a spark discharge and that particles of metals, such as iron or brass, fused at their points of contact.[26] These technological effects would have important applications in later centuries. The electrical fusion of metal pieces, for example, eventually became the basis of arc welding, widely used today.

Beccaria and other new alchemists also explored the ability of electricity to affect the magnetic properties of metals. Sailors had reported for decades that a nearby lightning strike could cause a ship's compass to go haywire. And electrical experimenters, perhaps beginning with Kinnersley, had noticed that a compass needle spun rapidly when brought close to a prime conductor.[27] Beccaria, after sending discharges through needles of several metals, formulated two important generalizations.[28] First, needles of iron—but not of gold, silver, or copper—could be magnetized by electrical discharges. Second, an iron needle so treated may acquire or lose magnetism, perhaps even suffer a reversal of polarity. In a tantalizing conjecture, Beccaria also suggested that magnetism was "produced by a universal systematical circulation of the electric element."[29]

Like Beccaria, many eighteenth-century investigators explored relationships between electricity and magnetism. Yet none "discovered" electromagnetism. The main reason is that they regarded electricity and magnetism as different phenomena that happened to exhibit some similarities. Thus, investigators focused on illuminating, experimentally, the similarities and differences, compiling in the process impressive lists and offering theories to account for some of the observed phenomena.[30] What is more, Charles Augustin Coulomb had pronounced the impossibility of magnetism and electricity interacting, and doubtless his views were given great weight, at least in France.[31] Beyond the compass-needle experiments, then, no one inquired whether or how one could be converted into the other. Such a question would not have made much sense until the early nineteenth century, after scientists had begun to sketch out the law of energy conservation. That law finally furnished a theoretical mandate for seeking technological means to convert any form of energy into any other.[32] Had anyone formulated this problem in the eighteenth century, he could have devised the apparatus to demonstrate electromagnetism; after all, none of the electromagnetic effects discovered decades later by Ørsted and Faraday—even Hertz—was beyond the reach of electrostatic technology.

—||⊢

Ascertaining the electrical properties of nonmetals, which would become a consuming concern of twentieth-century scientists and engineers, began in the Enlightenment. Since the time of Gray and du Fay, it had been noticed that some insulators, like glass, lose their ability to hold a charge when heated. This effect occurred because, as Kinnersley demonstrated, hot glass is conductive.[33] Some years later Beccaria showed, by heating a water-filled flask, that glass turns into a conductor at a temperature far above that of boiling water.[34]

Appreciating that perhaps other insulating materials, like oils, become conductive when hot, some investigators invented devices for testing liquids. One design employed a long glass tube, bent downward into a gentle arc. The investigator poured a quantity of liquid, such as wet pitch or various oils, into the tube, which was held above a source of heat. Next, conductors were inserted—from both ends of the tube—into the insulating liquid. After applying heat to the tube's bottom, the investigator connected the conductors to a Leyden jar and observed whether it discharged through the tested substance. Usually it did.[35] Regrettably, this device's design was flawed, since the hot glass itself, in the vicinity of the tested substance, was probably responsible for the conduction.

The passage of charge through various solids, liquids, and gases was clearly a kind of play activity—what happens if I try . . . ?—that racked up a roster of sometimes spectacular technological effects. Often grabbing materials close at hand, investigators learned, for example, that fruit and pieces of wood exploded, cotton could be set on fire, and alcohol burst into flames. And not surprisingly, gunpowder could be detonated with a spark. Franklin, for one, rammed dry powder into a small cartridge and inserted pointed wires at each end. With a battery of two huge Leyden jars—8 to 9 gallons each—he was able to set off the cartridge.[36] Such effects of course became staples in public lectures. More than that, these playful experiments contributed to an awareness that electrical discharges could cause the combustion of many nonconducting or poorly conducting materials.

Thus, electricity provoked both calcination and combustion, two processes that had preoccupied chemical theorists for centuries. The theory favored by most chemists during the mid eighteenth century, advanced by the German investigators J. J. Becher and G. E. Stahl, was that of "phlogiston." Instead of regarding these processes as the formation of a new compound by the *addition* of something (such as oxygen), the phlogiston theory posited that calcination and combustion involved the *loss* of an "inflammable principle" or the "food of fire."[37] This invisible substance, acquired or emitted by other forms of matter, was called phlogiston. Materials like charcoal, which burned almost completely, were said to be very rich in phlogiston, which they gave up during combustion. Likewise, when metals were heated in the process of creating a calx, phlogiston was also lost. A great many reactions were explained as the loss or gain of phlogiston.

We moderns find it difficult to fathom how so many smart scientists could have believed the phlogiston theory for so long; after all, it has "obvious" fatal flaws. For example, if phlogiston escapes during calcination, why do calces weigh more than the original metal? This question presumes an awareness that matter could be neither lost nor created in chemical reactions, a principle not fully formulated until the 1770s, when Lavoisier began to introduce his new chemistry.[38] Moreover, the question also presupposes an accurate measurement of all the products—solids, liquids, and gases—of reactions, but before Lavoisier it was seldom made. Even when anomalous weight gains were occasionally discerned, phlogiston theorists could explain them away (perhaps phlogiston has a negative weight). The phlogiston theory was immensely productive, underwriting much fruitful research; ironically, a few of phlogiston's adherents, including Priestley and Cavendish, made crucial discoveries that catalyzed Lavoisier's theoretical revolution. In particular, their experiments—some with electrical apparatus—laid the foundation for Lavoisier's recognition that calcination and combustion are both processes of oxidation.

Antoine Laurent Lavoisier (1743–1794), the son of a wealthy Paris merchant, was educated at the Collège Mazarin.[39] Although seemingly destined for a career in law, Lavoisier attended lectures and demonstrations in chemistry and soon became captivated by this demanding discipline. Fettered to his furnace, he spent unending hours learning laboratory skills and honing a knack for asking penetrating questions. Before age twenty-five, Lavoisier was already sending papers to the French academy on sundry scientific and practical subjects; he was elected a member while still in his midtwenties.

Interested as well in the affairs of state and in partaking of aristocratic privileges, Lavoisier as a young man joined the Ferme générale. This was a strange institution that functioned as an internal revenue service. But there was a big difference between the operation of the Ferme générale and a modern state agency of taxation. Members like Lavoisier paid a small fee to the French crown for the privilege of collecting taxes, which they kept for themselves. Needless to say, members of the Ferme générale—including Lavoisier—became high-profile targets during the Reign of Terror; the great chemist himself was guillotined in 1794.

In the meantime, Lavoisier rubbed elbows with other French elite, including notable scientists and industrialists. In the home of Jacques Paulze de Chastenolles, a fellow member of the Ferme générale, he became acquainted with many prominent Parisians. There he also met Jacques Paulze's fourteen-year-old daughter, Marie Anne Pierretti; their mutual attraction soon culminated in marriage. Her parents, pleased with the union, gave the newlyweds a lovely house. Held in the thrall of Lavoisier and chemistry, Marie Anne was more than eager to become the sorcerer's apprentice. Not only did she have a talent for drawing, but she studied science and learned several languages so that she could participate in Lavoisier's research. Theirs was, by all accounts, a happy marriage and fruitful collaboration.

Although a capable and meticulous experimenter, and inventor of much new chemical apparatus, Lavoisier also had a knack for folding other chemists' discoveries into his brilliant theoretical syntheses. And he placed a reliance on highly accurate balances, insisting on the need to weigh all substances before and after an experiment. Early on he propounded the view that matter is neither created nor destroyed in chemical reactions: "the quality and quantity of the elements remain precisely the same; and nothing takes place beyond changes and modifications in the combination of these elements."[40] Having emphasized the need to weigh reactants and products, Lavoisier never held the phlogiston theory in high regard; he would eventually put it out to pasture with the unwitting assistance of Priestley and Cavendish.

Lavoisier's own experiments in 1772 furnished tantalizing evidence that the phlogiston theory was vulnerable. He found that both sulfur and phosphorus, when burned in air, gained weight. Not only did he generalize this finding to all combustion processes, but Lavoisier also claimed that calces were simply metals augmented by some sort of air.[41] (At that time, all gases were called "airs.") This he showed by heating a lead calx with charcoal, which produced both metallic lead *and* a considerable amount of gas. But what was this extraordinary and elusive gas that combined with other elements to produce calces and combustion products? Joseph Priestley would soon supply the answer.[42]

While living in Leeds, Priestley occasionally visited a nearby brewery to study the gas evolved by fermentation. This gas, he showed, snuffed out burning wood chips; perhaps it was the "fixed air" that other chemists had already reported. In the process of making fixed air (carbon dioxide) at home, Priestley found that small amounts would dissolve in water, creating a mildly acidic solution indistinguishable from sparkling mineral water. He presented this discovery at a meeting of the Royal Society, and encouraged members to sample his new beverage. It passed the taste test and met with much acclaim; indeed, Priestley was awarded a medal for the invention.[43]

The study of fixed air whetted Priestley's appetite for investigating gases of all kinds, and in so doing he pioneered the field of pneumatic chemistry. But first he would have to develop new apparatus. When heating some table salt with vitriolic (sulfuric) acid, Priestley discovered a problem with conventional gas-collecting techniques. Usually, gases were bubbled through water and accumulated in an upside-down flask. However, the gas generated by his salt-sulfuric acid concoction was so soluble in water that he could scarcely get a sample. This problem he solved by collecting gases over mercury. The new gas, when dissolved in water, produced Priestley's "muriatic" acid, now known as hydrochloric acid.[44]

Armed with his new technology and assisted by John Warltire, Priestley heated countless substances and studied the gaseous products. In one experiment he obtained a gas from heating a mercury calx. Others had done this experiment countless times before, but their focus had been on the reappearance of the liquid mercury. Only Priestley paid any attention to the gas—oxygen—and sought to learn its properties. Predictably, he inserted a lighted candle into a container of the gas and, to his utter delight, witnessed the flame enlarge and burn with unprecedented brightness. Playing with other burning materials, Priestley was soon able to generalize that this gas, which he called "dephlogisticated air," fostered combustion. Perhaps, he conjectured, the new gas was a constituent of the air around us. Soon he would also show, in experiments with mice enclosed in glass jars, that dephlogisticated air supported life much longer than common air. After inhaling the new gas himself, Priestley predicted that it might have uses in medicine, especially as an aid to people with breathing problems.[45]

On a trip to Europe with his patron, the earl of Shelburne, Priestley met Lavoisier. The dissenting minister's spoken French was far from fluent, but with the aid of an intermediary, he was able to communicate to Lavoisier the gist of his experiments on dephlogisticated air. Still seeking the mysterious gas that combined with metals to form calces and also supported combustion, Lavoisier immediately appreciated the significance of the Englishman's discovery. The French chemist and aristocrat, almost overcome with excitement, hastened to his laboratory to repeat and extend Priestley's experiments. More than satisfied with the results, he reported them to the French academy as entirely his own.[46]

In August 1778 Lavoisier announced that the new gas, which a decade later he named oxygen, was responsible for combustion and calcination.[47] A lesser constituent of common air, oxygen was also respirable; indeed, many would call it "vital air" in recognition of its role in sustaining animal life. With these findings, which rested upon the work of Priestley, a firm believer in phlogiston, Lavoisier began to dispense with this imaginary substance. His fame growing in scientific circles, Lavoisier's laboratory received many distinguished visitors, including Benjamin Franklin.[48]

–|ı|⊢

While Lavoisier was preoccupied with oxygen and sketching out his new phlogiston-free chemistry, English investigators were using electrical technology to make a startling discovery about water. One of chemistry's longstanding questions concerned water's composition: was it a pure element, the view bequeathed by the Greeks and handed down by many generations of natural philosophers, including Musschenbroek and Nollet, or was water composed of other elements?[49] Experiments that synthesized water by burning hydrogen ("inflammable" air) would begin to lay the matter to rest and give Lavoisier more ammunition for his fight with the phlogiston phantom.

The ability to synthesize water depended on several key pieces of apparatus. First was the technology for preparing hydrogen. Fortunately, by 1770 Cavendish and others had shown that iron or zinc, when immersed in strong acid, would dissolve and form copious hydrogen bubbles.[50] This gas, which burned with a blue flame, could be collected with conventional techniques. The second indispensable technology, an apparatus for generating and collecting oxygen, had been invented by Priestley; and by the late 1770s there were many other processes for generating oxygen. So far, so good.

Although chemists now had available water's gaseous ingredients, merely bringing them together in a container was insufficient, they found: hydrogen would have to be burned in the presence of oxygen. As early as 1775, Priestley himself had ignited hydrogen-oxygen mixtures; he and other experimenters were surprised at the violence of the explosion.[51] These reactions were provocative and entertaining, but there was no way easily to capture and quantify the products. Accurate measurements would require that reactions take place in a sealed, leakproof vessel. But such a vessel presented an immediate obstacle: how could they ignite the mixture? Obviously, thrusting a lighted candle into a flask, which was adequate for demonstrations, was impossible here; ignition would have to be internal. Alessandro Volta's "electric pistol" led directly to new apparatus—hybrids of chemical and electrical technology—that met these performance requirements.

Volta had in these years taken an interest in the chemistry of gases, excited particularly by the findings of Priestley, with whom he corresponded. To facilitate these researches, and especially to burn hydrogen, Volta constructed a reaction vessel, made of metal or of glass, whose gaseous con-

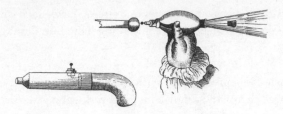

Figure 31.   Two versions of the electric pistol.
(Adapted from Seiferheld 1791c, figs. 34 [right]
and 36 [left]. Courtesy of the Burndy Library, Dibner Institute, MIT.)

tents were ignited by a spark jumping across internal conductors. In 1777–
1778 he reported his invention to Priestley and others.[52] The electric pistol—along with equipment to supply it with charge—appealed to many disseminators and quickly entered chemistry laboratories throughout Europe
(Fig. 31). Coupling two of Volta's inventions, one investigator even showed
that the electric pistol could be fired by the spark from an electrophorus.[53]

Although the electric pistol made it possible conveniently to ignite a confined gas, control over the reaction products was incomplete. This problem
was overcome by electrifying glass vessels that had been fitted with various pipes and stopcocks. In some versions, investigators actually passed the
wire conductors, often platinum, through the wall of the vessel while the
glass was still molten. When the glass hardened, the wire was firmly bonded.
The practice of sealing conductors in glass would make possible, more than
a century hence, the manufacture of lightbulbs and vacuum tubes.

Sometimes electrified reaction vessels were used not to ignite gases but
to study the light given off by the discharge passed through a given gas.
Priestley and later investigators carried out many studies of this kind,
which disclosed that different gases, when electrically excited, emit light
of characteristic colors.[54] These effects would become the basis of many
nineteenth- and twentieth-century technologies.

Given the widespread interest in gases at this time, chemists were bound
to play with electrified reaction vessels, igniting different gases in varying
combinations and studying the products. As William Nicholson noted in
his 1790 chemistry textbook, "It is frequently an interesting object to pass
the electric spark through different kinds of air, either alone or mixed together."[55] Clearly, this new technology served as a discovery machine, a
template for playfully producing new effects—scientific and technological;
the investigator merely had to vary the initial gases.[56] The finding that

oxygen and hydrogen combined to form water was thus a highly probable event. Predictably, several people made that discovery quickly; even Volta had observed dew on the inside of his pistol after firing.

When significant scientific discoveries are rendered nearly inevitable by shared questions, playfulness, and relevant apparatus, acrimonious claims over priority are apt to erupt from many quarters.[57] Among the several candidates advanced for this momentous discovery, Henry Cavendish's case is most compelling; his exhaustive experiments—and copious notes—leave no room for doubting what he did and when he did it.[58]

Stimulated by Priestley's work, Cavendish began to study gases in the 1770s. Not surprisingly, he framed his research questions in terms of phlogiston theory.[59] He was eager to learn, for example, the fate of the common air that disappears when something is burned (phlogisticated) as well as the composition of dephlogisticated air (oxygen). These researches led, eventually, to the crucial experiment, a determination of the nature of the "matter condensed" by the electrical ignition of hydrogen in the presence of oxygen.[60]

Cavendish's reaction vessel was a pear-shaped flask furnished with a brass stopcock and internal conductors.[61] After exhausting the flask, he filled it with a mixture of hydrogen and oxygen in the rough ratio of 2 to 1, which Priestley had calculated on the basis of his earlier experiments. Cavendish ignited the mixture, then repeated the entire process several times. Curiously, the resultant "liquor" tasted somewhat acidic. Believing that this liquid was water along with nitrous acid that had contaminated the oxygen, Cavendish ran the experiment again with various preparations of oxygen, even collecting the gas from leaves of plants (following the lead of Ingen-Housz and Priestley). In this way he showed that clean gases tended to yield nonacidic water. By attending to quantitative detail, he also concluded "that almost the whole of the inflammable and dephlogisticated air is converted into pure water." [62] And thus was expressed, without fanfare, one of the most monumental discoveries in the history of chemistry.

Cavendish carried out the experiments that pinned down the composition of water in the summer of 1781. Not until January 15, 1784, however, did the dilatory perfectionist present a paper on his results to the Royal Society; the paper was published that same year in the *Philosophical Transactions*. Although his biographers insist that Cavendish was usually cavalier about claiming credit, in this paper he took pains to point out that he had told Priestley about these experiments, and that during the summer of 1783 a friend had described them to Lavoisier. Both men, excited by Cavendish's work, had headed to their laboratories and confirmed his findings;

and both had reported their results before Cavendish. Perhaps that is why he bothered to assert priority.

The accomplished chemist Lavoisier appreciated the need to rule out known sources of contamination and so made a reaction vessel (with accessories) somewhat more complex than that of Cavendish.[63] His "gasometer" employed a 30-pint globular flask that rested on a metal stand with its aperture pointed upward. This opening was sealed by a brass plate, penetrated by two conductors that formed a spark gap inside the vessel. One of the conductors was in fact a pipe, which connected—above the brass plate—to three other pipes, all with stopcocks. Two of these pipes extended to nearby reservoirs of oxygen and hydrogen; the third was attached to a vacuum pump. To ensure the purity of the gases, he placed into the oxygen and hydrogen pipes some deliquescent salts to absorb residual moisture. By adding a very sensitive balance, Lavoisier was able to measure accurately the amounts of oxygen and hydrogen that yielded complete combustion. With this impressive apparatus, the skilled French chemist had confirmed Cavendish's seemingly outlandish claim that water, the stuff of rain and oceans and life itself, was a compound of two colorless gases. In his publication of 1783, Lavoisier furnished a lucid theoretical account of the reaction, which made no reference to phlogiston: water was simply oxidized hydrogen.

Familiar with Lavoisier's publication, Cavendish concluded his 1784 paper with a theoretical discussion. Appreciating the explanatory power of Lavoisier's account, he acknowledged that "it will be very difficult to determine by experiment which of these opinions is the truest."[64] Although that statement was true given the protean character of phlogiston theory, others nonetheless saw in Lavoisier's formulation an undeniable clarity that contrasted sharply with the older theory's cumbersome locutions.

Lavoisier's elegant interpretation of the crucial experiment hastened the adoption of his new chemistry, even by some British chemists. Thus phlogiston theory became, almost suddenly in comparison to the glacial pace of other theoretical revolutions in science, a quaint curiosity. In assessing the impact of the crucial experiment in 1798, *The Philosophical Magazine* exulted that it had "effected a complete revolution in the theory of chemistry."[65] Needless to say, this revolution owed much to electrical technology, which enabled the design of reaction vessels that gave the experimenter unprecedented control over the combustion of gases.[66]

Electrical reaction vessels were adopted by many chemists, who fashioned them in an endless variety of shapes and sizes. Like Lavoisier, most added flasks and plumbing fixtures to enable the evacuation of air and the entry and exit of reactants and reaction products. The drawings of many

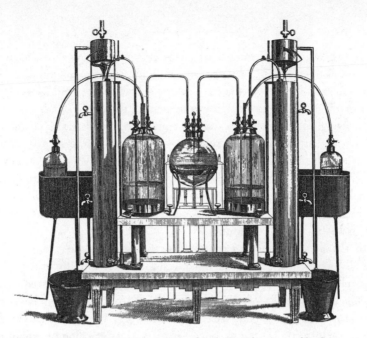

Figure 32.   Van Marum's apparatus for the combustion of hydrogen in oxygen. (Adapted from van Marum 1798, pl. 1. Courtesy of the Dibner Library, Smithsonian Institution.)

such apparatus in books (Fig. 32), including Lavoisier's, give few obvious hints of electrical ignition, for they include neither battery nor electrical machine. Usually, however, electrical ignition is betrayed by the conductors inside the reaction vessel itself, one curved and the other straight, the latter terminating just above the vessel in a ball (Fig. 32, center). Sometimes there is also a casual remark in the text about the use of a spark to set off the reaction. In these apparatus the electrical technology was clearly subordinated to the chemical. Friedrich Accum made few references to electricity in his 1803 chemistry textbook, but one comment tellingly indicates how thoroughly electrical technology had been assimilated into chemistry: "The effects of electricity are too well known . . . to need any description." [67]

Among the investigators using electrical reaction vessels in these years was Martinus van Marum. Ironically, the Teyler Museum's monster electrical machine that failed to produce any new physics found a niche in chemistry. In the late 1780s van Marum and his collaborators carried out many experiments employing this machine, which charged a huge battery of 135 Leyden jars. Although few if any new effects were produced, the massive charges did allow the investigators to clarify known phenomena such as the calcination of metals, the magnetization of iron, and the melting of wires. [68]

Most important, experiments on gases furnished important support for Lavoisier's theory on oxygen's role in oxidation and combustion. Convinced by these studies, van Marum became an early champion of Lavoisier's new chemistry, diffusing his doctrines throughout the Netherlands.

Although the huge Teyler machine gave great spark, its setup and operation—the latter sometimes requiring four men—were a hassle. Indeed, van Marum came to appreciate that the large machine was too cumbersome for everyday use in chemical research. So he designed a smaller machine, with heavily insulated rubbers and one 30-inch disk, which was not only easier to set in motion but also generated positive and negative electricity in equal—and large—amounts.[69]

In the hands of many experimenters, accessorized electrical reaction vessels yielded new effects and enhanced the understanding of reactions in pneumatic chemistry.[70] Significantly, a few people also wondered whether the electrical ignition of gases might have novel applications beyond the laboratory and lecture hall (see chapter 11). Volta, whose electric pistol had helped catalyze these studies, suggested that the explosion of inflammable gas might substitute in weapons for gunpowder.[71] Today hydrogen— burned in oxygen and ignited electrically—powers the space shuttle's mighty engines, as foreseen by Erasmus Darwin.

$$\dashv \mathsf{I} \mathsf{I} \vdash$$

Although Cavendish's crucial experiment in its many incarnations rapidly undermined phlogiston theory and raised the stock of Lavoisier's new chemistry, some chemists hesitated for a few years to accept the conclusion that water was a compound. If water was a compound, these skeptics wondered, why had no analysis ever convincingly revealed its constituents? After all, water had been spared no torture in laboratories, yet always it remained water. Although Lavoisier himself showed in the 1780s how to decompose water with red-hot charcoal, the interpretation of his experiments required many auxiliary assumptions that could be easily disputed.[72] However, skeptics were soon forced into silence or surrender by another crucial experiment, of consummate elegance, in which electricity played the major role. Not only did this experiment remove lingering doubts about the nature of water, but it also demonstrated electricity to be a tool of chemical analysis as well as synthesis.

The electrolysis of water took place in the Netherlands in 1789, carried out by collaborators of Martinus van Marum. Adriaan Paets van Troostwyk and Johann R. Deimann, with the assistance of John Cuthbertson, employed a surprisingly simple apparatus.[73] The key artifact was a tube,

12 inches long with a bore of .125-inch, oriented vertically; the tube's bottom was bent into an S-shape. A platinum wire was sealed into the straight end and extended downward 1.5 inches into the tube. The upper, straight segment of the tube was evacuated and filled with distilled water. In the bent portion a small quantity of ordinary air was left as a shock absorber to keep the tube from breaking during discharges. A second platinum wire was threaded into the tube, coming up just .625-inch short of the first. The bottom of the tube, still open, was then immersed in mercury.

A Cuthbertson electrical machine with two 31-inch plates was employed to continuously charge a relatively small Leyden jar connected to the metal wires of the apparatus. Careful adjustment of the distance between the jar's ball and the prime conductor produced visible bubbles of gas at the platinum conductors, in periodic discharges. (In an apparatus of this sort, hydrogen and oxygen are produced at both conductors because the Leyden jar's discharge alternates rapidly between positive and negative.) The gas gradually accumulating in the upper end of the tube pushed the water level downward. As the gas neared the end of the upper conductor, a discharge ignited the gaseous mixture, causing a tiny explosion and reforming the water. After the rise in water level, the evolution of the gases resumed, and the cycle was repeated. And thus was demonstrated the electrolysis of water.[74]

Although these results were widely accepted, another new alchemist, George Pearson, reported in 1797 that he and others had had difficulty repeating the Dutch experiment. Finally, with Cuthbertson's personal guidance (he had returned to England), Pearson succeeded. But Pearson was displeased with the design of the original device and so built his own. He used a large-bore tube, which remained straight, and extended the upper conductor nearly to the tube's base. With this apparatus he could accumulate much more gas.[75]

A few years later William Wollaston invented a device of radically different design that enabled hydrogen and oxygen to be evolved and collected separately. The key to this invention was Wollaston's appreciation that an electrical machine itself produces direct (nonalternating) current. Thus, if a machine's positive and negative conductors are immersed in water, without the intermediary of a Leyden jar, the investigator can collect hydrogen bubbles from one and oxygen bubbles from the other.[76]

‑||‑

Because the electrolysis of water required nothing more than a lengthy series of nonviolent discharges, it is likely that investigators had previously stumbled upon this effect in the course of other experiments. After all, many researchers had passed charges through water. Perhaps they had failed to notice bubbles forming on the conductors or, more likely, they were unprepared to assign such bubbles any significance. At least one investigator, Beccaria, did report the bubbles, but he drew no wider implications from the effect. It took Lavoisier's new chemistry to establish a theoretical context for appreciating the importance of this obscure phenomenon.

But there was another problem: creating measurable or usable products from the electrolysis of water—or any solution—was highly time-consuming. For example, Pearson reported that it required many hundreds, sometimes thousands, of discharges to generate the merest amount of the hydrogen-oxygen mixture.[77] The root of this problem—and here I must drop into present-day terminology—is that electrostatic generators produce precious little current. Today's battery in a large portable stereo can supply more current than most electrostatic generators—and is capable of electrolyzing water. It would have been possible to augment the current of an eighteenth-century electrical machine by using huge networks of capacitors as voltage dividers (as Cavendish did in attempts to replicate the torpedo's electrical properties), but no one did. Before anyone bothered to make such an attempt, another electrical technology came along whose performance characteristics were well suited for electrolysis. Volta's electrochemical battery, as he himself appreciated, produced electricity of low voltage and high current.

Amazingly, an experiment using a voltaic battery for the electrolysis of water took place even before Volta's report on his pile and crown of cups was published in the *Philosophical Transactions*. Anthony Carlisle and William Nicholson, shown Volta's manuscript by Joseph Banks, then the president of the Royal Society, immediately perceived its possibilities. After assembling a voltaic battery, these men succeeded quickly in electrolyzing water. Other experimenters also appreciated the potential of Volta's invention for chemical analysis and soon began building rather large batteries. A glass jar containing two conductors, usually platinum, and the liquid to be electrified made up the core apparatus—an electrolytic cell—for experiments in electrolysis.

The electrolytic cell, even more than the electrified reaction vessel, became a discovery machine for chemistry. Water was just the first of countless liquids playfully subjected to the high currents of voltaic batteries. Over several years at the Royal Institution in London, Humphry Davy sustained a program of electrolysis that yielded many surprises, including new chemical elements such as the metals sodium and potassium.[78] Eventually these kinds of experiments—along with the accumulated principles of electrostatics—formed a basis for explaining chemical bonds in terms of the attraction of oppositely charged particles (ions); such understandings remain fundamental.[79] The electrolytic cell also pointed the way to explaining the operation of Volta's batteries in electrochemical terms, displacing Volta's own contact-theory of electricity. Clearly, the electrolytic cell was the paramount contribution of eighteenth-century electrical technology to chemistry and even today remains at the core of countless industrial processes. Even so, contrary to the claims of Pons and Fleischmann, electrolytic cells cannot produce cold fusion; electrical alchemy has its limits.

—|ı|⊢

Volta's invention of the electrochemical battery traditionally serves as both an end and a beginning for electrical technology. In this view, which pervades almost all nineteenth- and twentieth-century histories of electricity, voltaic technology at last provided researchers in many communities with an ample source of current. In the hands of electrophysicists, the electrochemical battery facilitated the key inventions of the electromagnet, magneto-electric machine (or generator), induction coil (a transformer), and electromagnetic motor. Undeniably, Volta's invention was a momentous beginning, for the descendants of these technologies are central to life in our modern electrical world.

At the same time, the account continues, a shortage of technological possibilities and the onslaught of the voltaic battery and the exciting new inventions that followed in its wake caused electrostatic technology's eclipse. Historically, this view is simply wrong. Not only did electrostatic technology continue to play important roles in various scientific communities throughout the nineteenth and twentieth centuries, but it also entered the lives of ordinary people in many guises, well beyond public displays and medical practice.[80] Indeed, the foundations of many important twentieth-century products, including xerography, telegraphy, and the internal-combustion engine, lie in electrical technology of the eighteenth century. To support this seemingly outlandish claim, I now turn to a number of electrical products devised by the Enlightenment community of visionary inventors.

11.   Visionary Inventors

Benjamin Franklin was in many ways the Thomas Edison of the eighteenth century. Like Edison, Franklin was a prolific inventor who created products—the lightning conductor, a wood stove, and bifocals—that he hoped might improve people's lives. Like Edison, Franklin lived to witness the widespread adoption of many of his creations. Like Edison, Franklin won acclaim for his inventions and became famous at home and across the Atlantic. As autodidacts from modest circumstances who achieved greatness, both men gave substance to the American dream.[1]

Beyond these important similarities lies a world of difference in the activities of these two remarkable men. Franklin earned a living from his printing and publishing businesses; Edison's earnings came through inventing. Franklin, believing that inventions should be given freely for the common good, never applied for a patent; Edison accrued more than one thousand—still the American record.[2] Franklin invented relatively simple products; Edison invented complex products as well as entire technological systems. These differences have little to do with ability or personality and everything to do with dramatic changes in society that transpired in the nineteenth century.

One far-reaching change was the transformation of patents into property—indeed, into commodities. During the eighteenth century, the few Americans who bothered to obtain patents received little more than a paper token of their ingenuity. The following century witnessed the emergence of what the historian Carolyn Cooper calls a "patent management system" involving, among others, patent examiners, agents, and attorneys.[3] By participating in this system, inventors could treat their patents as property subject to rental, lease, or sale. Patent management systems in America and in Europe coevolved with the explosive growth of modern corporations rooted in the public sale of ownership shares; the expansion of invention-hungry, large-scale technological systems, from steam- and water-driven factories to telegraphs and railroads; the mechanization of agriculture; and the creation of insatiable markets for new consumer products.

In this altered societal context, an inventor not only had many fertile fields to cultivate but also the institutional means to profit from patents. If entrepreneurs and capitalists deemed an invention useful, the inventor could employ the patent management system to derive income from his patent(s) and move on to new challenges. Such unlimited opportunities for making a career of invention, which Edison fully exploited, were largely unavailable to inventive people of the Enlightenment like Franklin (who died in 1790).[4]

Thus, the independent inventor, a self-employed person whose primary income comes from managing patents for new products and processes, lacked a viable role in eighteenth-century societies. This is one factor that helps us understand why so few individuals dedicated any effort to, much less built a career around, creating "practical" electrostatic technologies and systems, devices that could be used in homes, businesses, and factories along with the infrastructure to power them.[5] And yet certain people, supported by colleges and churches and sundry occupations, did have fascinating and sometimes important product ideas that highlighted potential applications of electrostatic technology. In this chapter I survey major efforts of these *visionary inventors*, focusing on products envisioned to be useful, at least in principle, in nonscientific activities.[6]

—|ı|⊢

Some possibilities for turning electrical effects into artifacts of everyday life were appreciated as early as the 1740s, when electrical phenomena were first becoming known to a public beyond natural philosophers. Prior to inventing the lightning conductor, but after he had begun electrical experiments, Franklin in 1749 acknowledged his chagrin at having produced "nothing . . . of use to mankind."[7] With summer approaching—a poor time for electrical experiments because of the high humidity—Franklin humorously proposed that he and his collaborators partake of an all-electric picnic dinner on the banks of the nearby Schuylkill River. The main course would be a turkey, killed by electric shock and roasted on a spit turned by his motor. The fire would be kindled by the discharge of a Leyden jar. Attendees would drink to the "healths of all the famous electricians in *England, Holland, France,* and *Germany* . . . in *electrified bumpers,* under the discharge of guns from the *electrical battery.*"[8] The electrified bumper was a tumbler of thin glass containing wine and electrified like a Leyden jar. When brought to the lips of a person closely shaved, it delivered a smart shock. Humor aside, Franklin's fanciful picnic nonetheless embodied the earliest published vision of an electrical world.

Another perspicacious electrical visionary of that time was D. Stephenson, who worked for the Office of Ordnance in the Tower of London and collaborated on electrical research with the instrument maker John Neale. Published in *The Gentleman's Magazine*, Stephenson's ideas reached a large readership. In one letter of 1747, he furnished several suggestions, at least two of which eventually became important technologies—but not in the eighteenth century.[9]

Aware that an electric spark could ignite gunpowder, Stephenson proposed that mining might be carried out more safely if the gunpowder used for blasting were electrically detonated at a distance. He furnished no details of this invention other than noting the need for insulated wires or chains to link an electrical machine to the gunpowder. I have found no evidence that the electrical ignition of explosives ever reached fruition in the Enlightenment, but the first appropriate technologies based on voltaic batteries and electromagnetic generators were adopted in the mid nineteenth century.[10]

Stephenson's second seminal idea was that of using electricity to reduce chimney emissions. Because electricity could effect movement in a candle's smoke, Stephenson suggested that the interior of a chimney could be outfitted with wires or chains connected to an electrical machine. The electrified smoke could then be "redirected," but how and where he did not say. To power the electrical machines, he proposed use of "the force of water, wind, a man, or horse, or by a weight."[11] Indeed, the same machine could serve several houses. When developed in the twentieth century, electrostatic precipitators would become an important technology for reducing industrial air pollution.

—|ı|⊢

Without doubt, the paradigm of "practical" electrical inventions is the lightbulb. Surprisingly, the electric lamps that Thomas Edison first commercialized in 1880 were based on a simple effect, first reported by Kinnersley and others: when sufficient electricity is conducted through a thin thread or wire, it heats up and radiates visible light. Today we have additional kinds of electric lights, such as fluorescent and neon, based on different effects. In a fluorescent light, an evacuated glass tube is coated on the inside with a phosphorescent material, which glows when excited by electrons streaming from an incandescent filament. Glass tubes filled with neon (or other gases) directly give off light in a characteristic color when electricity is passed through them; neon, for example, glows red. Surprisingly, the effects that underlie both fluorescent and neon lights were also first described in the eighteenth century.[12]

If three major ways that electricity can generate light (beyond mere sparking) were already known in the Enlightenment, why did no one develop and commercialize electric lighting? The answer to this question turns out to be rather general and accounts for the lack of activity on many fronts of electrical invention.

In most areas, electrical technologies would have had to compete against products already adopted, deeply entrenched in daily activities. Electric lighting in particular would have had to replace the candles and oil lamps already used for interior illumination. In principle it was easy to manufacture glass globes or tubes that could be used for electrical lighting by phosphorescence or by gas discharge; after all, Hauksbee had shown early on that just one globe furnished sufficient light to read by —assuming the text was in large print.[13] With further development, fully functional electrical lamps doubtless could have been produced.

Even if coated or gas-filled glass globes and tubes had been brought to market, there remained the problem of supplying them with electricity. The lack of a convenient, reliable, and inexpensive source of continuous current was a significant barrier to commercializing many electrical technologies in the eighteenth century, just as it was throughout most of the nineteenth. Surely no one would have seriously contemplated cranking an electrical machine all evening or periodically recharging Leyden jars in preference to lighting a candle or lamp. In addition, candles and oil lamps were cheap, accessible, and reliable; electrical lighting would have seemed—to instrument makers and virtually all others—an utterly impractical alternative in the absence of an electrical infrastructure that could furnish power without constant human attention.

Even so, instrument makers made lavish and expensive devices at the behest of wealthy customers to serve mainly symbolic functions. But the majority of large electrical machines bought by the elite were not especially innovative designs, nor were they used to create new effects. Rather, they were present in someone's cabinet for visitors to discuss and admire and covet—hardly a "practical" purchase. Why, then, did instrument makers not also sell them electric lighting systems for display? The answer, I suspect, is that wealthy people interested in scientific instruments bought deluxe versions of *existing* electrical technology; they did not underwrite the development of truly novel electrical things, much less new electrical systems. To make electric lighting workable would have required numerous and costly experiments to come up with reliable steam- or water-powered electrical machines along with many accessories. For members of the Enlightenment elite, financing the development of an entire electric lighting system might have represented a challenge too formidable. There were more convenient and less risky ways to acquire prestigious technologies.

Apparent practicality and economy notwithstanding, members of the elite did sometimes pay for the development and commercialization of new technologies, though this practice became important in the electrical realm only in the late nineteenth century.[14] A case in point: after his brilliant show-and-tell of electric lighting at Menlo Park, Edison's first customer was the transportation magnate Henry Villard, who contracted for the installation of an Edison system on his new steamship *Columbia*, despite what appeared to some as manifest diseconomies.[15] This installation in 1880 required the invention of numerous accessories from lamp sockets to switches. In the eighteenth century, however, electrical systems would have had to compete on their own merits in the absence of wealthy benefactors willing to subsidize their development merely to acquire a unique symbol of modernity.

Only in those few cases where the alternatives to electrical devices had an even less favorable mix of performance characteristics, as in medical therapies, could electrical inventions be brought to the point of being judged "practical," especially in activities of ordinary people. In short, the greatest immediate obstacle to developing and commercializing electrical technologies in the eighteenth century —*and long after*—was the lack of a system to supply electricity conveniently, reliably, and inexpensively. More important, there was an absence of appropriate social and economic incentives for inventors to solve this problem.

─┤│├─

One form of "electric" lamp needed only occasional recharging and apparently enjoyed some popularity during the late eighteenth century. In fact, it was not an electric lamp but a gas lamp that was ignited electrically by the spark from an electrophorus. Volta himself, inventor of the electrophorus, fashioned the first lamp of this kind, an outgrowth of his many experiments on gases.[16] In its simplest form, the electrophorus, as we recall, is a plate of resin that is easy to charge by rubbing it with a cat's pelt. Because it could hold a charge for weeks, the electrophorus was a convenient

Figure 33. Two versions of Volta's "electric" lamp.
(*Left*, adapted from Adams 1799, pl. 5, fig. 91;
right, Gütle 1791, tb. 5, fig. 13. Both courtesy
of the Bakken Library, Minneapolis.)

source of spark for igniting the lamp's gas, usually hydrogen. Volta suggested that it could also be used to light candles.

The lamp itself was made by many people in many versions. In the simplest design, a glass globe, resting on a sturdy base, stored hydrogen gas (Fig. 33, left). Above the globe was attached, by means of a pipe with a stopcock, another glass vessel containing water. Just below the stopcock was a small-diameter brass pipe, which terminated in a jet. On either side of the jet were brass electrodes supported by glass insulators that formed a spark gap. When the stopcock was opened, water flowed from the upper to the lower chamber, forcing the hydrogen to issue from the jet where it could be ignited by a spark. In complex examples of this lamp, one of which was built by Jan Ingen-Housz, a mechanical linkage simultaneously opened the stopcock to release the fuel and set off the spark from an electrophorus built into the base (Fig. 33, right).

Following Volta's advice, Ingen-Housz used his lamp for lighting candles; with 100 cubic inches of hydrogen, about 100 candles could be lit.[17] This seems like a very complex apparatus for merely lighting a candle, but consider the alternatives. Phosphorus matches were available at that time but also dangerous (prone to self-ignition) and expensive.[18] For the vast majority of people, who still created a flame by striking flint with steel, the electric lamp would have been an intriguing—if not always affordable— alternative.

Adams and Jones, London instrument makers, brought a simple version of Volta's lamp to market, which was advertised as "A new perpetual inflammable air-lamp, lighted by the electrophorus, a curious and useful apparatus"; it sold for £4 4s.[19] Similar devices, with and without electrophorus, appear often in electrical texts and experiment books. According to some sources, these lamps achieved a measure of popularity, especially in Germany, where they were sold by several instrument makers.[20]

Ingen-Housz also proposed the use of Leyden jars to ignite candles, believing that this was preferable to and quicker than employing flint and steel or calling a servant.[21] The jar's spark was passed through a piece of cotton and wrapped around a nail that had been rolled in powdered resin. The resin immediately caught fire and ignited the cotton, which in turn lighted the candle's wick. Ingen-Housz even carried around a Leyden jar in a pocket, good for twenty lights, which could be used anywhere. The tiny Leyden jars sold by instrument makers had several uses (recall, for example, Thomas Jefferson's pocket shocker), so we cannot assume that all were bought as lighters.

-||-

That electricity could produce mechanical effects was already known by Gilbert, who caused small particles to dance about under the influence of rubbed amber and whose versorium needle rotated in the presence of a charged object. By the middle of the eighteenth century, electrophysicists and disseminators had contrived many simple devices that, when connected to a source of charge, displayed rotary or reciprocating motion.[22] From carillon to horse race, these inventions exhibited technological effects brimming over with possibilities for building everyday electrical technology.

There were, in fact, at least three ways to produce rotary motion electrostatically. The first and simplest results in what I call a "pinwheel" motor, invented in the 1740s, which operates by generating an electric wind. Surprisingly, the pinwheel motor was occasionally incorporated into a more complex invention, an electric orrery. Although at least one electric orrery came to market, it was advertised along with electrical, not astronomical, apparatus.[23] Apparently, the instrument maker judged it to be noncompetitive against existing orreries. Its performance had serious shortcomings: after all, mechanical orreries could be very precise in illustrating the motions of many bodies in the solar system, but electric ones were drastically simplified. What is more, mechanical orreries were powered by self-contained mechanisms, such as springs; they did not need to be connected to Leyden jars or electrical machines. The electric orrery was simply another clever device for showing that electricity could produce rotary motion.

Jakob Langenbucher, inventor of many display devices, also employed a pinwheel motor in "the electric scale." [24] The motor, mounted horizontally, could wind a silk thread that was draped over a pulley. Suspended from the thread below the pulley was a pan like that used in an apothecary's balance. Langenbucher's scale could lift up to 7 ducats, but it was hardly a precision instrument; presumably it could only distinguish between weights above and below 7 ducats. Today, of course, almost all scales and balances are electronic, but they operate on different principles. [25]

A second kind of motor works like a turbine, applying the electric wind — the flow of air molecules streaming from a charged point — to turn the vanes of a rotor. Assembled in the shape of a windmill or water wheel, the vanes were made of stiff paper. The electric wind emanating from a point pushed against a vane until it turned the rotor on its shaft or pivot point, which also brought the next vane into the electric wind. [26]

Although no one in the eighteenth century suggested enlarging the miniature electric turbine, James Ferguson (1710–1776) did design provocative models. [27] Originally a painter of miniature portraits in his native Scotland, Ferguson found a higher calling in London as a scientific lecturer and designer of apparatus, especially globes and orreries. In his twilight years he took an interest in electricity and built his own display devices. Apparently a gifted sculptor in the media of paper and wire and wood, Ferguson installed his turbine motors in tiny models of existing machines: clock, orrery, grist mill, and water pump. Audiences responded favorably to their performances, and several instrument makers included a set of the ingenious devices in their catalogs. [28]

What makes these models especially intriguing is their incorporation of gears — paper, to be sure, but gears nonetheless — which more than hinted that electric motors could in principle operate machinery (Fig. 34). But there was that pesky problem of supplying electricity. It doubtless occurred to Ferguson and many others that the use of electric motors in this way was bedeviled by a daunting inefficiency: one had to use considerable mechanical power to generate electricity, which, by means of the motor, returned in the end only a feeble amount of mechanical power. Actually proposing to harness such a system to compete with existing prime movers, such as steam, water, animals, and wind, would have seemed ludicrous at the time, as it did well into the nineteenth century.

Only one electrostatic motor of that era produced, in the judgment of Joseph Priestley, "appreciable power," and that was Franklin's. According to Franklin, the wheel of his first motor rotated "with great rapidity twelve or fifteen rounds in a minute, and with such strength, as that the weight of

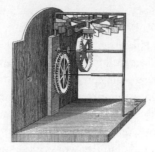

Figure 34.   Ferguson's model of an electric clock.
(Adapted from Ferguson 1775, pl. 2, figs. 1 [right]
and 2 [left]. Courtesy of the Bakken Library,
Minneapolis.)

one hundred *Spanish* dollars . . . did not seem in the least to retard its motion." [29] As already noted above, in connection with his picnic, Franklin
suggested that his motor could drive an electric spit for roasting a large
fowl. There is no evidence that Franklin actually roasted a bird in this manner, but the quintessentially pragmatic printer had thought of a use for his
invention beyond the display of effects.

Franklin's second motor, with built-in plate capacitor, offered opportunities to construct self-powered devices. Had anyone desired, the motor
could have driven automatons or even models of wheeled vehicles, but I
have found no record that it did. The closest anyone came to harnessing the
motor was Franklin himself who, with his friend and collaborator Ebenezer
Kinnersley, applied it to the ringing of small bells, which repeated a simple
tune.[30] No one, however, brought to market music boxes containing a Franklin motor and bells.

In recent decades electrostatic motors have made a comeback. Nanotech-
nology requires microscopic motors, which can actuate valves and other
mechanisms in micromachines, but electromagnetic motors are far too
complex and bulky to be miniaturized that far. In contrast, there is no lower
limit to the size of an electrostatic motor (an atom, after all, is a kind of
electrostatic motor) and so experimenters have returned to the oldest—
and in some ways, the simplest—of electric motor technologies. Made on
computer chips, these micromotors are likely to play a role as nanotechnol-
ogy develops in the decades ahead.

⊣∣∣⊢

The Franklin motor demonstrated, beyond a doubt, that electrostatic forces
could produce nontrivial mechanical effects. Had anyone in the eighteenth

century wished to build an electric orrery, powered by a Franklin motor that was every bit as complex and realistic as ones operated by springs, it would have been feasible; yet, to my knowledge, no one did. But the basic operating principle of the Franklin motor, creating motion by the exchange of charges and the consequent repulsion of like charges, was incorporated into an electric clock.

The clock was the creation of Georg Heinrich Seiferheld, a prolific inventor of intricate electrical devices who published its design in 1802. An invention of consummate complexity, the clock was built into a box no larger than a cubic foot that contained dozens of wooden and metal parts.[31] The mechanism included a weight and ratchet gear and advanced through a series of repulsions and attractions of ball-tipped conductors connected to push rods. On one charge from a Leyden jar, Seiferheld claimed, the clock could run for six hours. To announce a new hour, the clock chimed and a pointer rotated to the next number.

Because of the need to recharge the Leyden jar at six-hour intervals, one of which might have been in the middle of the night, we may suppose that Seiferheld's clock would not have appeared competitive with clocks powered by springs and falling weights. We may also presume that, lacking an escapement, it was not very accurate.

Perhaps some hobbyists built clocks following Seiferheld's plans, but I have found no evidence that the invention was commercialized. Yet in principle it might have become a successful elite product. By the end of the eighteenth century, clocks had been widely adopted, penetrating even the least wealthy social stratum.[32] Wealthier folk might have eagerly embraced a novel electric clock, which others could not have afforded. With an escapement mechanism and a sizable battery of Leyden jars, an electric clock could have run for weeks.[33] Perhaps no other invention of that time went so far in suggesting that electricity could operate everyday machines.

Another Seiferheld invention, "color tubes," was a toy having interesting design features.[34] The starting point of color tubes was a wooden container about the size and shape of a relatively flat cigar box. At one end, he placed on its edge a rectangular piece of glass. The glass was covered by paper except for six dime-size disks in a row just below the upper margin; each disk was painted a different color with transparent paint. At the other end of the box, he placed a removable wooden dowel. When the dowel was turned to one of six permissible positions, a corresponding disk was illuminated. To change the colored light, the operator removed, rotated, and reinserted the dowel.

The dowel was in fact a switch that could close any one of six circuits by making contact with metal strips leading to a vertical tube behind each colored disk. When a circuit was completed, sparks from the rear illuminated the correct disk. The device was powered by a Leyden jar through the intermediary of a conductor perched inelegantly atop the dowel and connected internally to metal strips below. Probably the color-tube toy had the capacity to entertain both children and adults, not unlike similar electrical playthings of the twentieth century.

The first time I glanced at one of Seiferheld's many calculating devices, I became quite excited, imagining that he had made the first analog electrical computer. But, my excitement waned after further study, for all were simple mechanical calculators that merely used an electrical display. Although still noteworthy, they were not *electrical* computers. A typical example was the multiplication wheel. The operator could multiply any two numbers between 2 and 10 by lining them up on two concentric, rotating disks; their product appeared in a little box illuminated by sparks.[35]

My favorite amusement in the Seiferheld assemblage is the "electric card game."[36] The operator handed the player a set of eight cards, all hearts, consisting of 7, 8, 9, 10, jack, queen, king, and ace. After the player chose a card, it was placed, face up, upon the center of a cardboard disk, about a foot in diameter. A second disk of cardboard, the same size, was lowered carefully on the first, thereby covering the selected card. Around the circumference of the upper disk was arrayed another set of the cards. When the operator pressed down on the upper disk, the hearts in one of the cards—not surprisingly, the same card chosen by the player—lighted up; that is, their edges gave off sparks.

The workings of this all-electric game are simple. The hearts on the cards attached to the upper disk had been cut out, coated with metal foil, sprinkled with cinnabar (a red pigment), and then replaced. The trick was possible because the actual card selected by the player had on its underside a conductor that closed a circuit connecting the Leyden jar to the hearts on the corresponding card of the upper disk. Each card from which the player could choose had a unique arrangement of conductors, heading from the center to a corner or the midpoint of an edge. One can only imagine the operator's squeals of delight as a player again and again failed to figure out what had happened. Seiferheld is careful to note, however, that players should be prevented from seeing the cards' undersides.

Regrettably, we know of no demonstrable continuity between Seiferheld's inventions and later electrical technologies.[37] In a different societal context, his works' noteworthy features might have initiated new technological trajectories.

First is the demonstration that electrostatics could be readily controlled in devices having complex circuitry. The switches Seiferheld invented for the "color tubes" and "electric card game" embody the basic features of electrical switches still used today. The potential controllability and complexity of electrostatic technology would also be attested by electrostatic telegraphs (see below).

Second, the circuitry of several inventions hinted at possibilities for quantity, perhaps automated, production. In the "color tubes," Seiferheld—following precedents set by lecturers—cut out thin strips of metal foil and glued them to an insulating substrate. These circuits remind me of nothing less than the earliest printed circuits of the 1950s, adopted to reduce the labor needed for assembling radios.[38] Although Seiferheld's circuits were not printed, it is easy to envision how they might have been produced in quantity using mechanical, not chemical, deposition of metal foils.

Third, and finally, Seiferheld's inventions as a whole demonstrated that electricity could be incorporated into everyday devices such as clocks and games. They could, for example, be used as motive power, as replacements for mechanical linkages, as remote controls, and as illuminated displays.

-|ı|ı·

One everyday technology provides us with an example of how an electrical invention of the eighteenth century contributed, eventually, to the development of an entire industry: the automobile. I suggest that the electric ignition of internal combustion engines, and perhaps the engines themselves, had their beginnings in eighteenth-century electrostatic technology.

As the "electric" lamp demonstrates, a spark could ignite inflammable gases as well as volatile, explosive liquids. I could easily claim that technologies of electric ignition arose during the Enlightenment. But instead I assert that there was actual technological continuity between early electrical technology and the internal combustion engine. The linkage between these technologies is Volta's electric pistol, a self-contained combustion chamber whose gaseous contents could be set off by a spark. Volta's invention inspired a host of copies, in glass and metal, which became important investigative tools in chemistry.

But the main use of an electric pistol at that time was to display, in lectures, "the power of exploding inflammable air."[39] People who witnessed demonstrations of these pistols came away with the appreciation that an electric spark could unleash a formidable force. After all, some electric pistols shot not only corks but also metal bullets.[40] That someone would apply this principle to the invention of an internal combustion engine should

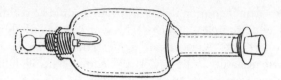

Figure 35. Electric pistol and spark plug. (Adapted from Cavallo 1781, pl. 2, fig. 19. Courtesy of the Burndy Library, Dibner Institute, MIT.)

not be surprising, for the latter required little more than the concept of a piston moving in a cylinder, which was already manifest in, for example, steam engines, air pumps, medical syringes, and even Ingen-Housz's electric pistol.[41]

Among the electric pistol's more interesting incarnations was a two-part device discussed by Tiberius Cavallo. Cavallo was not the inventor of this type of electric pistol, for he acknowledged that it was "commonly sold at the philosophical instrument shops," but I found it illustrated first in his work (Fig. 35).[42] Perhaps because this invention appeared in Cavallo's book on chemistry, historians of electricity and of the automobile have overlooked it.

Like many other electric pistols, Cavallo's consisted of a brass chamber, one end of which received a stopper that was expelled by the combustion of a hydrogen-oxygen mixture. Unlike the rest of the genre, however, this pistol had a screw-in conductor that was, in every important respect, a spark plug. It differs from the modern spark plug mainly in having a glass rather than a porcelain insulator.[43]

In the late eighteenth and nineteenth centuries, the electric pistol would have been known to anyone who had witnessed public lectures on chemistry or electricity or had studied these subjects in school. Moreover, the electric pistol, with and without spark plug, was not only available in instrument shops but was also a fixture in many chemical and electrical laboratories. Thus, the electric pistol and the effects it displayed were technological resources that any reasonably knowledgeable inventor of new engines would have been able to exploit. And one or two did.

James Watt, learning of early experiments with an engine that purportedly worked on heated air or gas, began to worry about competition to his lucrative steam-engine business. He encouraged Priestley, a fellow member of the Birmingham Lunar Society, to assess whether such engines really posed a threat. Collaborating on experiments with Matthew Boulton (Watt's partner), Priestley concluded in 1781 that Watt had nothing to fear, since any gas would be far more expensive to burn than coal.[44] Arguing at first that the explosive force of burning gases could not compare favorably with that of gunpowder, Ingen-Housz remarked "that little more than a pleasing amusement can be expected from the force of any inflammable air."[45] However, after experimenting with the impressive explosions of ether vapor, Ingen-Housz acknowledged that this gas might be useful for more than amusement.[46] Despite these mixed assessments, the seeds of new technology had been sown.

I suggest that the beginning of *sustained* efforts to build internal combustion engines during the 1790s was no accident, no inspiration de novo, for this was the decade in which discoveries of pneumatic chemistry, partly made possible through electrical combustion chambers, were trumpeted far and wide.

Inspired by the possibilities of harnessing the controlled explosion of gases, inventors came up with diverse designs for internal-combustion engines.[47] Some had one piston, others had two; some varied the pressure on both sides of the piston, others only on one. And they burned diverse gaseous fuels, including hydrogen, coal gas, alcohol, and turpentine. However, generalizing on the basis of more than three decades of experiments, in 1828 the physician Thomas P. Jones, the editor of the *Journal of the Franklin Institute*, had already rendered a negative judgment on the genre. Jones's conclusion, the kind that historians of technology are fond of quoting, was blunt: "we have strong doubts, whether under any ordinary circumstances, an engine operated upon by an explosive mixture of gas, or vapour, will ever be constructed, which will successfully compete with the steam engine."[48] Although Jones had actually seen a working model of an engine (made by one Samuel Brown), his judgment was based on the belief, eminently reasonable at the time, that the fuel for a gas engine would be much more costly than coal.

Although the electric pistol had alerted many inventors to the power of exploding gases, most early engines (ca. 1790 to 1850) employed lamps or candles for ignition. Perhaps inventors preferred these simple expedients when building models and prototypes or were unfamiliar with the nuances of managing electrostatic technology. Yet there were several significant exceptions. In 1801 the French engineer Philippe Lebon was awarded a patent for an engine that ran on coal gas ignited by an electric spark.[49] The electrical machine to supply spark would have been rotated by the motor's driveshaft. It is unclear whether Lebon had built a working prototype of the engine before his untimely assassination in 1804. Less than a decade later, Isaac de Rivaz, of Switzerland, employed electric ignition in a hydrogen-burning engine; in 1807 he was awarded French patent no. 731 for this invention.[50]

Curiously, the first internal combustion engine to be commercialized had electric ignition. Étienne Lenoir's simple but sturdy gas engine, put on the market in 1860, received its spark from an induction coil.[51] With the invention of the latter, the higher voltage needed to produce an adequate spark could be generated without an electrical machine; only a voltaic cell or battery was needed. Although a commercial success, Lenoir's engine was much more costly to operate than a steam engine, as had been predicted. Still, it did find a modest market among those who wanted an intermittent source of power that could be activated instantly.[52]

In the next half century, electric ignition fell in and out of favor many times, but its permanent place in automobile engines, which included spark plugs little different from those described by Cavallo in the eighteenth century, would be secured by 1910. This technological trajectory began, I suggest, with Volta's electric pistol.[53]

$$\dashv\,\vert\vert\,\vdash$$

In addition to designing the "electric" lamp, with his electrophorus Volta initiated an even more remarkable technological trajectory, that of electrostatic printing. The scientific roots of this technology were well known to many eighteenth-century electrical experimenters, as follows. When the knob of a Leyden jar touches the surface of an insulator, such as a plate of resin or glass, the induced charge remains localized, found only around the point of contact. If the knob touches the insulator in several places, all acquire similar charges.[54] These point charges fail to spread uniformly over the surface because the intervening areas are nonconductive.[55] The best insulators, such as the resin plate of an electrophorus, can retain localized charges for days, even weeks.

This effect seems unremarkable, and it is—scarcely worth the trouble to demonstrate. But in 1778 Georg Christoph Lichtenberg (1742–1799), a professor of mathematics at Göttingen University, reported an accidental discovery that revealed a far more interesting effect.[56] He had just constructed an enormous electrophorus whose resin plate reached a diameter of 2 meters. The room in which he built the electrophorus had not yet been cleaned, and so powder from smoothing and polishing the plate was everywhere. Postponing the tedious cleanup, an impatient Lichtenberg tried out his emperor-size electrophorus. In the midst of these shakedown runs, he noticed that air currents had deposited some of the ubiquitous resin powder on the metal charging plate. Intrigued, he lifted it out of the way, al-

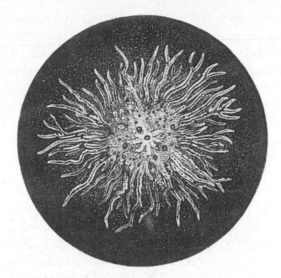

Figure 36. A Lichtenberg figure. (Adapted from Lichtenberg 1779, tb. 1. Courtesy of the Burndy Library, Dibner Institute, MIT.)

lowing the air currents to waft powder on the resin's surface. Instead of being uniformly distributed, as it had been on the metal, the resin arrayed itself "like small stars at certain points."[57] Lichtenberg next found that he could enhance the patterns by deliberately sprinkling on more powder. And when he passed the knob of a charged Leyden jar over the resin plate, he created configurations of powder so bizarre that they defy accurate verbal description.[58] These intriguing powder patterns, some vaguely resembling leaves or cells or nebulae, came to be called "Lichtenberg figures" (Fig. 36).[59]

After wiping a powder pattern from the plate, Lichtenberg discovered that it would reappear, sometimes even more spectacularly, when he applied fresh powder. This result was expectable given the propensity of an electrophorus to retain its charge despite many discharges. Studying these reemerging images but unwilling to draw them for his scientific report, Lichtenberg fastened upon a simple way to copy the figures. He took a sheet of black paper, painted it with glue, then gently pressed the glued side onto the powder patterns. When he lifted the paper, he found that the adhering powder had literally captured the figure. With fresh dustings of powder and new pieces of glued paper, Lichtenberg was able to take second and third impressions from the plate. This process of duplicating figures he referred to as a "new variety of printing."[60]

As a college professor, especially one afflicted with serious physical dis-abilities, Lichtenberg was disinclined to develop his printing process fur-ther, much less bring it to market. He did succeed in making more purpose-ful images, including letters of the alphabet, by carefully placing a charged chain on the resin plate. In one instance, he printed the letter *K* and gave it, under glass in a gilt frame, to his teacher Kästner, who was quite impressed.[61] It is doubtful, however, that such novelties would have inspired commer-cially minded people to regard electrostatic printing as a preferable way to reproduce handbills, newspapers, and books. It was, after all, a dreadfully complex process, for the printer would have had to set type anyway in or-der to induce a text image on the plate. And then there was the problem of fixing a colored powder to the paper without covering it entirely with glue. Surely electrostatic printing, if it could be brought technically to a satisfac-tory state, would be more expensive than conventional printing methods. Apparently, there was no obvious advantage to this process and many draw-backs, except in the special case of reproducing Lichtenberg figures or nov-elty items.

Even so, sometimes a complex and expensive technology simply per-forms in ways that no other technology can and so is adopted, usually to carry out a fairly specialized function. If electrostatic printing made pos-sible an entirely new industrial process that promised profit, then perhaps its shortcomings would not matter. After all, a manufacturer might have been able to gain a competitive advantage by using an electrical technology to produce a high value-added product. One famous industrialist, Josiah Wedgwood, collaborated with an electrical experimenter to perfect an elec-trostatic printing process for decorating pottery.

Josiah Wedgwood was a member of the Birmingham Lunar Society, along with James Watt, Erasmus Darwin, and Joseph Priestley. The Lunar Society embraced clergy, scientists, and factory owners, who gathered in a convivial social setting to participate in enlightenment. There they learned about the latest findings in science, the arts, and literature and found kin-dred spirits to collaborate on projects.

Coming from a long line of potters, Wedgwood was a relentless inventor and experimenter who drew upon his large social network to obtain infor-mation. And he constantly pestered friends to acquire, during their far-flung travels, samples of clays and minerals. With these materials Wedg-wood conducted thousands of experiments, creating new clay and glaze recipes. He also made an important contribution to scientific apparatus, the "Wedgwood thermometer," which could measure the high temperatures reached in pottery kilns.[62]

Always on the lookout for new techniques that might be useful in his factory, Wedgwood became intrigued by the possibility of employing electricity to deposit enamel glazes on pottery.[63] In the eighteenth century, colorful enamels were painted by hand, each design repeated endlessly on pot after pot by skilled artisans. Wedgwood wondered if this laborious and costly process could be replaced by some sort of electrostatic printing; after all, a glaze when first compounded is a powder, and powders were easily electrified.[64] In collaboration with Reverend Abraham Bennet, the inventor of the gold-leaf electrometer and voltage doubler, he embarked on experiments to find out.

Their studies on electrical deposition of enamel glaze were published in a remarkable book authored by Bennet. The book's third section reports dozens of new experiments, inspired by Lichtenberg, with electrified powders and gases—and a large electrophorus.[65] Bennet's apparatus, in addition to the electrophorus, included a bellows for spraying powders.

In a reasonably predictable finding, Bennet learned that some powders emerged from the bellows charged positively, others negatively. Taking it a step further, he began by inducing positive and negative charges in different regions of the resin plate. Then he ground together sulfur, which has a canary-yellow hue, and minium, a red oxide of lead, and placed the mixture in the bellows. When he sprayed the powder, the yellow and red powders separated, attracted to oppositely charged regions of the plate.[66] Few followed up Bennet's nifty demonstration, but today the electrostatic separation of powders is a technology important in industry.

An obvious limitation of Lichtenberg's printing method was the use of glued paper to capture the image. Bennet solved this problem by inventing a powdered ink. Made from shavings of Brazil wood, gum arabic, and alum, the concoction was boiled until all water had evaporated. The residue was ground into a fine but, Bennet acknowledges, quite disagreeable-smelling powder. After the powdered ink (a combined pigment and glue) was sprayed onto the plate, it was captured with damp paper, linen, calico, or even leather. Once dried, the image could be exhibited "to those who have not leisure or inclination to perform the experiments."[67] Elegantly solving the image-transfer problem, Bennet showed that Lichtenberg's printing technology held considerable potential for technical refinement.

And yet if electrostatic printing were ever to reproduce images other than Lichtenberg figures, a way would have to be found to draw fine lines on the surface of the insulating substance, be it a resin or porcelain plate. Bennet's solution was an "electrical pen," invented nearly a century before Edison's namesake.[68] The pen was made from a large thermometer tube, which was coated with silver on the inside and gold on the outside. The pen's point was a blunt needle fixed with sealing wax to one end of the tube and attached to the silver lining. The electric pen was, of course, a capacitor, which could be recharged periodically by touching its point to the knob of a Leyden jar. Bennet claimed that he could draw lines as fine as any made with a conventional pen.[69] Pleased with the results, he showed examples of his "electric picture" to Jan Ingen-Housz and the instrument maker George Adams, Jr.[70]

The first experiments on ceramics took place at Wedgwood's house. Unfortunately, the weather was uncooperative, and the images were "not so beautiful as might be expected."[71] In any event, Wedgwood apparently was not impressed with the commercial possibilities of the process. Bennet, however, continued experimenting on his own and obtained, he insisted, excellent results.[72] Unable to fire the ceramics at home, Bennet regrettably could not preserve a token of his progress.

Although the Bennet-Wedgwood experiments on ceramics were a technological dead end, electrostatic printing on paper eventually enjoyed immense commercial success. Building on the foundations laid by Lichtenberg and Bennet, in the middle of the twentieth century Chester F. Carlson perfected and brought to market a technology of electrostatic reproduction called xerography.[73]

–||–

Wedgwood's factories disgorged fancy and expensive ceramics that entered elite homes, especially in Europe and America. Those homes probably also held pricey musical instruments, such as a harpsichord and clavicord. During this time pianos, invented in 1700 by the Italian Cristofori, were undergoing development but did not reach an essentially modern form until the turn of the nineteenth century.[74] Another keyboard instrument known in the eighteenth century was the church organ for which J. S. Bach, who died in 1750, had written copiously. Although today we are familiar with all kinds of *electric* keyboard instruments, the first person to construct one lived in the eighteenth century.

Since the 1740s, lecturers had used the mechanical effects of electricity to produce sound. A few made elaborate electric carillons, using bells of varying sizes and thus tones, for playing simple tunes. R. P. Delaborde, a Jesuit, took electric music a significant step farther, creating "a new kind of harpsichord," whose practicality he sought to demonstrate.[75]

Use of the term "harpsichord" for his invention is a misnomer because, in a true harpsichord, depressing a key causes strings to be plucked. In De-

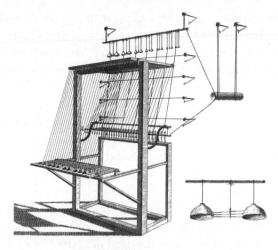

Figure 37.   Delaborde's electric harpsichord.
(Adapted from Delaborde 1761, pl. 2, figs. 3 [left]
and 4 [lower right]. Courtesy of the Bakken Library, Minneapolis.)

laborde's "electric harpsichord" sound emanated from ringing bells. Anticipating this criticism, Delaborde vigorously defended his choice of terms,
arguing that the instrument was far more perfect than an electric carillon.[76]
In fact, it used two brass bells instead of one for each note, with a clapper
oscillating between them. What did the electric harpsichord sound like?
The reader can only imagine, but Delaborde insisted that its tone resembled an organ's.

In Delaborde's harpsichord, an electric carillon writ large, the pair of
bells for each key hung, along with their clappers, from iron rods (Fig. 37).
A clapper could be activated by pressure on the corresponding key, which
completed the circuit to one bell in each pair, causing the clapper to move
briskly between them; however, there were no foot pedals. The instrument
was powered by a globe generator that charged a prime conductor and, in
turn, the iron bars. Under the best of conditions, the stored charge enabled
the musician to perform a long piece without having to crank the machine.

A performance provided, according to Delaborde, a sensory spectacle.
When the harpsichord was played in a dark room, not only did the music
captivate, but the listener's "eyes are agreeably surprised by the brilliant
sparks that attend the playing of each note."[77] Clearly, the modern light
show accompanying rock-and-roll performances was not unprecedented.
Despite its apparent versatility, the harpsichord was developed no further.

As a temperamental musical instrument, at the mercy of weather conditions, it probably would not have fared well competing against real harpsichords. As an expensive display device, it might have found a market among collectors and enthusiasts, but it was apparently not commercialized.

—|||—

From Priestley to Delaborde, Enlightenment clerics were an inventive lot. Pierre Bertholon had an especially fertile imagination and, apparently, technical skills as well. Having concluded that seed germination and plant growth are spurred by electricity, Bertholon set about building technology that could be used by farmers. To apply charge to plants in a garden or field, which could overcome supposed electrical deficiencies, Bertholon pursued two lines of technological development: one harnessed atmospheric electricity and the other employed an electrical machine.

For collecting atmospheric electricity, Bertholon built an "electro-vegetometer."[78] Atop a tall wooden mast but insulated from it, he affixed a conductor bristling with many points. Attached to the metal porcupine was a chain that descended to a horizontal conductor consisting of two iron rods. The rods were connected by a joint that permitted rotation and were supported on an insulated stand. One end of the horizontal conductor was attached by an insulated iron ring to the mast, which allowed movement in a complete circle. The business end, of course, was a phalanx of points for spraying charge. Because the horizontal conductor had considerable freedom of movement, the farmer could adjust it to approach individual plants.

Bertholon employed electro-vegetometers in his kitchen garden and was pleased with the results. Generously generalizing from his experience, he believed this invention could improve the yield and quality of produce from all plant species, everywhere. Bertholon also confidently predicted that the electro-vegetometer "will be adopted by all those who are instructed in the great principles of nature."[79] But despite testimonials on the invention's significance by Bertholon's friends, I found no evidence of its use. Might Thomas Jefferson, a quintessential agricultural experimenter, have considered building an electro-vegetometer? He was familiar with experiments on plant growth and electricity, but he eventually accepted his friend Ingen-Housz's negative conclusions.[80]

Bertholon's second invention allowed him to apply charge to plants from an electrical machine.[81] Standing on a squat, insulated table, the farmer held in his hands a large metal syringe for spraying electrified water (Fig. 38, bottom). A chain connected the syringe to the machine's prime conductor. In another variant of this invention, the farmer stood on a small insulated

Figure 38.   Bertholon's electro-vegetometers.
(Adapted from Bertholon 1783, pl. 2, figs. 1 [bot-
tom] and 2 [top]. Courtesy of the Bakken Library,
Minneapolis.)

cart, which could move along the plant rows (Fig. 38, top). Perched on the
cart, the farmer dispensed charged water from a metal watering can connected by chain to the prime conductor. Presumably, the cart was periodically moved by an unmentioned assistant. Bertholon claimed that it did not
take much longer than usual to water a garden in this manner.

Even if these inventions had been commercialized, few farmers would
have been inspired to acquire them. Although the application of charged
water might have increased crop yields, such gains would have been difficult to demonstrate. Even if farmers believed that the inventions worked,
they would have to invest in an utterly unfamiliar, sometimes temperamental technology, whose use demanded more effort than conventional
watering methods. We may suspect that convincing farmers to switch to
electrical watering technology would have been a tough sell.

According to Bertholon, electricity's use in agriculture was not limited
to accelerating plant growth. Electricity also could make an insecticide for
treating infected fruit trees. Appreciating that the damage was done by in-

sect larvae feeding under the bark, Bertholon sought to end the scourge by "attacking the enemy in his fort."[82] By placing two iron bars in the tree at either end of a suspected infestation, the farmer could administer a charge and kill the insects. Supplied by a single, small Leyden jar, the charge was sufficient to kill larvae but not powerful enough to damage the tree. He further suggested that electricity could be used daily as a prophylaxis against infestation and even discussed how, for this purpose, the trees in an entire orchard could be connected in a single circuit.[83]

The electrical insecticide raised the possibility that an enterprising person could have traveled about the countryside offering, for a price, to help a farmer rid an orchard of insect pests. It is doubtful, however, that anyone seized the opportunity to peddle an electrical bug zapper.

⊣||⊢

Apparently, Bertholon's clerical duties left him much time for experimenting outdoors. In addition to the electro-vegetometer, he created the "earthquake rod," a device for preventing earthquakes and volcanoes that was supposed to work electrically.[84]

Buoyed by the results of experiments with his model village, Bertholon believed that the underlying cause of earthquakes (and volcanoes) was an accumulation of excess charge in the earth. When that surplus electricity discharged abruptly into the atmosphere, thereby restoring the natural equilibrium, an earthquake resulted.[85] These catastrophic discharges could be prevented, argued Bertholon, if charge were vented from the earth gradually. With an appropriate electrical technology, both earthquakes and volcanoes might be prevented.

Earthquake rods were long iron spears that were to be placed in the ground, especially around towns but just about everywhere else too. Both ends of each spear bristled with points to assist in the transfer of charge.[86] If in fact earthquakes and volcanoes were electrical phenomena, as many at the time believed, then earthquake rods made a certain sense. At least one person in the eighteenth century, a king of Spain, took the suggestion seriously and wrote a letter to Bertholon praising his technology, but it was "too flattering to repeat."[87] Despite the king's endorsement, the support of seemingly cogent theoretical arguments, and a lack of competing technologies for discharging temblors, Bertholon's earthquake rod aroused little interest. In contrast to lightning conductors, the earthquake rod remained at the conceptual level, apparently neither copied nor commercialized.[88]

Nonetheless, Bertholon's flurry of inventions underscored his belief, perhaps fostered by an acquaintance with, and respect for, Franklin, that scientific research—even theory—lays a foundation for practical products. Others have repeated this mantra of scientific ideology, but few in the eighteenth century worked as hard as Bertholon to realize it. Even more surprising is Bertholon's hubris, perhaps unmatched until the twentieth century: the conviction that mere mortals could actually control or modify the divine forces of nature. For a cleric to have held this view seems to us almost startling. Apparently, new worldviews were taking shape in many corners of Enlightenment society.

⊣║⊢

When Dalibard successfully carried out in France Franklin's proposed experiment to draw electricity from the heavens, the news of his triumph— and thus Franklin's—took weeks to reach Philadelphia. Even within the colonies, overland travel was exceedingly slow and perilous. In Britain too, the roads connecting major cities were often in poor condition. Nonetheless, over these roads government-subsidized mail coaches began running in the 1780s.[89] In many places, faster and more reliable communication was a long-standing cultural imperative, particularly among merchants, sovereigns, shippers, and militaries.[90] Not surprisingly, experiments with various long-distance communication technologies were a near-constant feature of the late Enlightenment, and many of these systems had centuries-old antecedents rooted in signaling fires, torches, or explosions. The experimental work on technologies for distant communication was so rich that a review of previous efforts, by Johann Bergsträsser in 1784, filled five volumes. (Bergsträsser, by the way, invented a binary code, consisting of 0s and 1s, for transmitting letters of the alphabet.)[91]

Beginning in the 1790s, systems of mechanical telegraphy based on optical signaling were widely adopted.[92] The signaling devices were usually placed on hills between which there was line-of-sight communication, not unlike today's microwave relay towers. Although many countries installed mechanical-optical telegraphs, the French were especially avid adopters, as Napoleon's government sought to consolidate its control over distant territories. Perhaps the best-known French telegraph was the Chappe system, based on a T-shaped apparatus with movable arms. Chappe also receives credit, at least from the *Oxford English Dictionary*, for coining the term *télégraphe* in 1794. By the 1840s, virtually all of France was integrated by mechanical telegraph, making possible rapid communication between Paris and all major cities.[93] On the French network a message could be sent 150 miles in only 15 minutes.[94]

Not satisfied with the "French telegraph," which was useless at night,

Thomas Northmore offered an improved system, his "nocturnal or diurnal telegraph."[95] It employed four large lamps, placed in a row, whose illumination could be seen from front or rear. Letters of the alphabet were formed by raising and lowering different combinations of lamps. For example, if only the lamp on the far left were raised, then the array denoted the letter *A*; if that same lamp were lowered, it was instead an *E*. Northmore's system was probably never built, but his rudimentary digital code (up, middle, down) foreshadowed later contributions to telegraphy. However, the difficulty of nighttime communication would not be solved decisively until the advent of electrical telegraphy.

Many electrophysicists in the first half of the eighteenth century established the rudiments of electrical telegraphy. In an earlier chapter we met Stephen Gray, who carried out early research on the communication of electricity. In an ambitious series of experiments performed in mid-1729, Gray along with his wealthy patron Granville Wheler and the latter's servants showed that electricity could be transmitted over long distances.[96] In the penultimate—and most dramatic—experiment, they rigged up a series of 10-foot poles in the garden of Wheler's country house. The conductor of packthread, suspended from the poles by silk threads, snaked its way back to the house and inside, entering through a window.[97] When Gray, stationed in the garden, placed an excited glass tube or rod near the end of the packthread, Wheler, in the house, was able to discern electrical activity: an ivory ball connected to the packthread attracted a piece of gold leaf. Gray and Wheler then exchanged places, Gray moving to the house and Wheler to the garden, and they repeated the experiment with success. The total distance traversed electrically was 650 feet; a later experiment achieved a distance of 765 feet.[98]

William Watson, an apothecary and gifted experimenter, along with several other members of the Royal Society and assorted servants, carried out in 1747–1748 ingenious studies of electrical conduction.[99] They charged a Leyden jar with a globe machine that had been placed in a nearby house and sent discharges through an iron wire insulated from ground. The experiments, conducted in diverse configurations, took place in and near London, including Shooter's Hill, Highbury Barn, and the Westminster bridge over the Thames; the last location was especially unsatisfactory because a crowd gathered and repeatedly broke the wire.

Using his collaborators as human electrometers—connected in the circuit, they could report when they received a shock—Watson and his crew demonstrated that the "electrical Power" could be transmitted easily over 2 miles.[100] They also showed that either water or the ground could serve as the circuit's second conductor. This latter finding would have significant implications, unappreciated at the time, for constructing telegraph systems economically. In the meantime, additional experiments were in the offing.

An observer passing by Shooter's Hill on August 5, 1748, would have spied a curious spectacle: four grown men in a row, holding hands, periodically jumping or convulsing in unison. Once more, Watson and his colleagues were carrying out experiments on electrical conduction, but this time they were measuring the speed of electricity traveling through a long circuit.

Again Watson used a Leyden jar as the transmitter and his colleagues as receiver-electrometers. The circuit consisted of two iron wires, each about a mile long, with the men in the middle to complete it. Watson recorded exactly when the Leyden jar discharged and when his collaborators convulsed. Unable to detect a time lag in repeated trials, Watson arrived at the remarkable conclusion that the "Velocity of Electricity was instantaneous." [101] As Watson knew, sound passing through air over the same distance would have taken about 11 seconds.[102]

By showing that electricity could travel instantaneously over long distances, Watson and his collaborators had established a basis for electrical telegraphy. Yet these gentlemen were interested only in scientific results, not in building communications technology. In 1748 Watson published an admirably detailed report in the *Philosophical Transactions.* In this most prominent venue, people in many countries could read about Watson's experiments, and some soon drew from them profound implications for communicating electrically over long distances.

The first suggestion that electricity could be used to make a practical telegraph emerged in 1753, in an article in *Scots' Magazine,* signed simply "C. M." [103] This proposal was remarkable, a telegraph in all but name. The idea was to connect two places by a set of parallel wires, one for each letter of the alphabet. People at either end of the wires could send or receive, depending on the configuration of the equipment; a change from receiver to sender and vice versa would be set in motion by a prearranged signal.[104]

At the sending end, a gun-barrel prime conductor, connected to an electrical machine, was to be placed perpendicularly to the ends of the wires, which were fixed upon solid glass insulators. To indicate a letter, the operator would merely bring the end of the wire representing that letter into contact briefly with the prime conductor. At the receiving end, a ball was attached to each wire; almost immediately below the ball was a letter of the alphabet written on paper. When a circuit was closed, the charged ball attracted the paper, indicating which letter had been sent. Appreciating that some people might consider the use of such a receiver "tiresome," C. M. suggested a creative alternative. Instead of balls and paper, letters were represented by bells of varying size. With some practice, C. M. assured the reader, an operator could "come to understand the language of the chimes in whole words, without being put to the trouble of noting down every letter."[105]

Although there is no evidence that C. M., whoever he was, actually built a telegraph system, his blueprint influenced other visionaries and inventors for many decades.[106] Seeing farther ahead than his contemporaries, Louis Odier suggested in 1773 that someday it would be possible to "intercommunicate all that you wish, at a distance of four or five thousand leagues in less than half an hour."[107] Furnishing plans and prototypes of electrical telegraphs almost became a cottage industry.[108]

Beginning in the 1780s, several inventors began to assemble hardware for electrostatic telegraphs. The Frenchman Claude Chappe's system was especially intriguing because it used clocks for selecting and displaying numerals.[109] On the second-hand dials of two clocks, he drew ten wedge-shaped zones, which he numbered from zero to nine. Beginning with perfectly synchronized clocks, the operator at the sending end discharged a Leyden jar when the second hand swept into the zone of the desired numeral. At the receiving end, a vigilant operator awaited the signal of the Leyden jar's discharge, presumably indicated by a bell or electrometer. When the signal was detected, the operator immediately recorded the numeral corresponding to the location of the second hand. Details on the design and operation of Chappe's system are, regrettably, fuzzy. Nonetheless, the use of synchronized clocks, perhaps inspired by methods for measuring longitude, was a clever approach to sending and receiving information.

Although individuals in several countries built prototype electric telegraphs in the 1780s and 1790s, the ambitions of Don Francisco Salvá of Barcelona, Spain, outshone them all. Combining—perhaps reinventing—the ideas of earlier thinkers, Salvá constructed several actual systems.[110] One used two cables comprised of seventeen pairs of insulated wires, each pair indicating an essential letter of the alphabet; powered by Leyden jars, this system employed sparking metal foils on glass to indicate the transmitted letter.[111] There is a reliable report that he used this telegraph in 1796 to communicate "from the Academy of Sciences to the Fort of Atarazanas . . . a distance of about a kilometre."[112] Salvá also demonstrated his system successfully to the king of Spain, Carlos IV, and his court.

Possibly with patronage from the Spanish crown, Salvá constructed a much longer system in 1798. Although sources conflict on the details, it seems to have required but one wire (and, perhaps, a ground connection) along with Leyden jars and an electrical machine whose disk was about a meter in diameter.[113] Extending from Madrid to Aranjuez, the line reached the unprecedented distance of 26 miles.[114] This long line apparently did transmit messages, as the king once reported having received through it interesting news.[115]

Early in the nineteenth century, even after the first experiments with galvanic telegraphs had taken place, inventors continued to work on electrostatic systems. Picking up in 1816 where Chappe left off, a British investigator, Francis Ronalds, employed synchronized clocks to indicate both numerals and twenty letters of the alphabet.[116] In place of a second hand, each clock had a metal disk that rotated once per minute. An aperture on the disk permitted the operator to see, at any given instant, one letter and one numeral. Installed in the spacious garden of Ronalds's house at Hammersmith, the telegraph had several design features that overcame problems in earlier electrostatic systems.

Ronalds buried his single-strand cable, 525 feet long, in a thick glass tube sealed with pitch inside a slender wooden trough.[117] At the receiving end, in addition to the clock, were a pith-ball electrometer and an electric pistol. To alert the operator at the receiving end that a message was imminent, the telegrapher placed a large charge on the line, which caused the pith balls to diverge widely. In case the operator were looking away or daydreaming, the large charge also set off the electric pistol. Thus alerted, the operator awaited the next signal—collapse of the pith balls—and then immediately set his clock to the letter *A*, which synchronized the sending and receiving apparatus. Further collapses of the pith ball, which indicated that the system had been grounded at the transmitting end, were timed to contain information.

To speed up transmissions, Ronalds employed a "telegraphic dictionary," a standardized code developed by mechanical telegraphers.[118] Surprisingly, many words, even sentences, could be sent with only three characters: the first indicated the page of the dictionary, the second the column on that page, the third the row. At the intersection of row and column were common words and sentences.

Although Ronalds's complete system was short, he also built a model using 8 miles of wire, which confirmed the instantaneous transmission of electricity that Watson had reported so many decades earlier. Looking ahead to the design of long-distance installations, Ronalds suggested that the batteries of a large system could be charged with plate machines powered by small steam engines.[119]

In 1816 Ronalds offered his perfected telegraph to the British Admiralty. Unsuccessful in obtaining an interview with the first lord of the Admiralty, he at last received a letter: "Telegraphs of any kind are now wholly unnecessary; and no other than the [mechanical] one now in use will be adopted." [120] This curt dismissal ended Ronalds's efforts, but he was not alone in making an entreaty to the Admiralty on behalf of electric telegraphy.

Ralph Wedgwood, relative of the famous Josiah, started his own pottery, which failed, and also sought jobs teaching chemistry—apparently with as little success. [121] In 1803 at age thirty-nine, Wedgwood moved to London, where he sought to sell several inventions, including a device he had patented for mechanically producing a written document in multiple copies. A few years later he began work on an electrostatic telegraph, which he completed in 1814. His first effort to commercialize the system involved a proposal to the British Admiralty, but it met with cold rejection.

In another effort to attract financial backing, in 1815 Wedgwood issued a pamphlet describing in general terms what his telegraph could offer the government and the public. Specifically, he enumerated its advantages over mechanical systems then in use: "Whilst this invention proposes to remove the usual imperfections and impediments of telegraphs, it gives the rapidity of lightning to correspondence *when* and *wherever* we wish, and renders *null* the *principal disadvantages of distance to correspondents*." [122] Moreover, to furnish a revenue stream, he suggested that telegraph offices could be leased to individuals for transacting private communications. Despite Wedgwood's best efforts, there were no takers: the system was not commercialized, and the "secrets" of its operation died with him.

─┤│├─

The case of telegraphy brings into sharp relief the obstacles faced by inventors working on electrical *systems* at the dawn of the nineteenth century. John Fahie, a nineteenth-century historian of telegraphy, calls attention to these problems when offering an insightful explanation for Wedgwood's failure to commercialize his telegraph. [123] Noting that putting a telegraph system into operation was too "gigantic" a project for an individual to undertake, Fahie assumed that the assistance of "Government or a powerful company" would have been required. Since support from neither source was forthcoming—the government was too conservative and railroads did not yet exist—the project languished.

Why were governments, not just the British, unreceptive to installing electrostatic telegraphs? An answer is not difficult to discern. The period

from 1789 to 1815—encompassing the French Revolution and the preda-
tions of Napoleon's armies—was one of unrest and widespread war. Governments, having an immediate need for reliable and rapid long-distance
communication, invested in far-flung mechanical telegraph systems. Despite its performance deficiencies, mechanical telegraphy was a simple,
proven technology that could be emplaced quickly without extraordinary
initial costs. In contrast, electrostatic systems were obviously still in an experimental stage, their construction was materials- and labor-intensive,
and their far-flung cables would have been susceptible to sabotage. In short,
on the basis of performance characteristics and costs, the choice between
mechanical and electrostatic systems was easy.

Dependent on mechanical systems, which apparently worked well for
limited purposes, governments became "conservative," resisting the heroic
efforts of inventors like Wedgwood who agitated for electrical telegraphy.
Tellingly, inventors of the earliest galvanic telegraphs, which had liabili-
ties like those of electrostatic systems, were also turned away by government agents. In the years after 1820, as the performance shortcomings of mechanical telegraphy were underscored daily in the increasingly
information-hungry industrial nations, governments gradually became
more receptive to arguments in favor of subsidizing the development of
electrical telegraphy.

As Fahie also notes, electrical inventors like Ronalds and Wedgwood
were hampered by a lack of private financial and entrepreneurial institutions to underwrite large-scale undertakings. These institutions would take
shape in the nineteenth century, partly under the impact of the railroads,
making it possible for inventors like Charles Wheatstone and Samuel Morse
and, eventually, Thomas Edison to commercialize large-scale electrical systems. By the time these societal constraints were removed, new electrical
technologies had been invented and commercialized. In many—but not
all—"practical" applications, performance characteristics of the upstart
electrical technologies gave them a decided advantage in head-to-head
competition with electrostatic technologies and they prevailed.[124]

—|ı|⊢

Taken together, the electrical technologies of the eighteenth century,
whether invented by visionaries or members of other communities, comprise a remarkable array of artifacts and potential systems. More than that,
the manipulation of these technologies led to the recognition of myriad
intriguing effects, from the production of light and rotary motion to the
virtually instantaneous transmission of information over long distances.

Perhaps, with further development, these effects could have become the basis of many "practical" products. However, the electrical world of the eighteenth century—Franklin's world—differed greatly from the one in which Edison lived and invented.

By the early nineteenth century the role of independent inventor was emerging in the United States along with the modern patent management system. In that dramatically altered context of burgeoning industrial capitalism, inventors of key components of complex electrical systems, such as the telegraph, telephone, and electric lighting and traction, were able to market their intellectual property, and these technological systems eventually became the cornerstones of a new electrical world.

If Franklin had been born in the middle of the nineteenth century, Thomas Edison might have had a formidable rival. Like Erasmus Darwin, Pierre Bertholon, and others, the Philadelphian had stirring visions of a future brightened by new science and new technologies. In a 1780 letter to Priestley, Franklin reflected:

> The rapid Progress *true* Science now makes, occasions my Regretting
> sometimes that I was born so soon. It is impossible to imagine the
> Height to which may be carried in a 1000 Years the Power of Man over
> Matter. We may perhaps learn to deprive large Masses of their Gravity
> & give them absolute Levity, for the sake of easy Transport. Agricul-
> ture may diminish its Labour & double its Produce. All Diseases may by
> sure means be prevented or cured, not excepting even that of Old Age.[125]

In his eighth decade, Franklin appreciated all too well that he would not be around to help realize such visions.

Nonetheless, as the preceding chapters have shown, the legacy of eighteenth-century electricity was far from insignificant. Franklin and his friends bequeathed to Edison and other nineteenth-century workers fundamental scientific principles and terminology, a plethora of technological effects, many fascinating devices and product ideas, and the fervent and infectious belief that electrical science and technology could create a better world.

12. Technology Transfer: A Behavioral Framework

The preceding chapters have presented a panorama of eighteenth-century electrical artifacts in relation to specific activities and the communities that carried them out. Although this technology had its beginnings as the laboratory apparatus of electrophysicists, from the mid-1740s onward a host of new functional variants was invented—everything from Seiferheld's electric oracle to Bertholon's electro-vegetometer to Volta's electric lamp. I was drawn to this technology in part because its seemingly relentless growth and change conformed to a common pattern, the process of "technological differentiation," which is familiar to most archaeologists.[1] Many technologies, from Neolithic pottery to twentieth-century home electronics, began their developmental trajectories in a limited number of functional variants. Over decades, centuries, even millennia, these technologies became diversified as people in different communities, capitalizing on technological effects, created and acquired new variants, often highly specialized.

Regrettably, scholars studying technological change have available precious few theoretical tools for dealing with large-scale processes such as technological differentiation. An old favorite, diffusion theory—even in its modern quantitative guises—furnishes at best a superficial description of the temporal, geographic, or social "spread" of a particular technology. Because technological differentiation entails not only spread but also *change*, diffusion theory is clearly unhelpful. Suffering defects similar to diffusionism, other theories in history and in the social sciences cannot convincingly handle large-scale processes of technological change in a behaviorally sound manner. By "behaviorally sound," I mean theories grounded in the materiality of human life—in the concrete people-artifact interactions that comprise activities.

In view of the dearth of relevant theory, I crafted an archaeological framework that structured my engagement with eighteenth-century electrical artifacts. Built around technology transfer, this framework can help us explain technological differentiation. The framework's major premise is that, as a technology passes from one community to another, members of recipient communities invent new functional variants, differing in performance characteristics, that are more suitable for participating in their activities. Understanding this kind of technological change, then, requires that we examine the activities carried out in recipient communities, members of which envision—and attempt to realize—a new technology's possibilities. In this chapter I elaborate the technology-transfer framework, drawing upon material in previous chapters as well as on behavioral theory from archaeology.[2]

When studying large-scale processes such as technological differentiation, we must define the technology of interest as an aggregate of a given type of artifact (e.g., stone axes, cave paintings, electrical products) or as a material technology (e.g., ceramics, basketry, thermoplastics) or as a technological system (e.g., cooking technology, weaving technology, ritual technology).[3] In the present study I defined electrical technology expansively to include *all* artifacts that interacted with electricity in human activities; that is as any part, product, or system employed in the collection, generation, storage, distribution, measurement, use, or display of electricity. Thus any object—from a piece of gold leaf to a frog's legs—used to manipulate electricity *is* electrical technology, whether or not it was designed to enable such manipulations. This archaeological move permitted me to transcend treatments of early electrical technology confined to the "instruments" of physics.

As is well known, all technologies incorporate knowledge, usually have ideological correlates, and are embedded in social, political, and economic structures. In the present framework, I temporarily hold these important factors in abeyance in favor of first discerning the proximate, *behavioral* causes of technological differentiation. We need to focus initially on the functions that variants carried out in activities regardless of the latter's nature (i.e., scientific, domestic, religious, recreational, or political). In adopting this methodological strategy, we can establish a secure basis for contextual explanations that invoke relevant "external" factors. The aim is to avoid building contextual explanations on flimsy foundations, unconnected to the actual activities—people-artifact interactions—in which the technologies participated. Any artifact or technology's immediate context is always activities—those that take place throughout its life history.[4] Thus all questions about technology are, first and foremost, questions about human behavior.

⊣||⊢

That technologies are redesigned for different activities when transferred between and among *communities* is hardly a revelation, but that claim has interesting implications in communities defined in very general terms. A community—my basic analytic unit throughout this work—is any group of people whose members take part in one or more activities that incorporate variants of a given technology. In the preceding chapters, I designated the following communities: electrophysicists, disseminators, collectors and hobbyists, electrobiologists, earth scientists, property protectors, new alchemists, and visionary inventors. Although many of these were science-related, this concept of community is applicable to *any* group defined on the basis of activities and artifacts. For example, there are communities of bonsai gardeners, particle physicists, letter carriers, craft potters, Cajun chefs, Internet surfers, rock climbers, philatelists, and Catholic priests; each group's members, regardless of whatever else they do and whether they interact among themselves, conduct certain activities using particular technologies.

Community members may consist of all people in an existing community, may be part of one *or more* existing communities, or may form an entirely new group. Thus, many electrophysicists were members of the larger physics community who had acquired electrical technology for their research. In contrast, most of the early electrotherapists were not physicians but people of diverse occupations who formed a community around this new treatment technology.

This latter example leads me to the conclusion that it is often impossible to predict the composition of a recipient community *before* the technology is actually transferred and modified. We might expect that certain groups would *potentially* adopt and adapt a given technology, but such forecasts can be way off the mark. For example, no one could have predicted that the electrotherapists would include many people having previous experience with neither electrical technology nor the healing arts. Thus, the statement that technologies are transferred between and among communities or from one community to another is an investigator's after-the-fact analytical judgment, since the recipient community as such has no prior existence. The composition of a recipient community can only be identified after people carrying out different activities have begun to obtain and alter the technology. Needless to say, if archaeologists and historians could reliably predict (actually, retrodict) the composition of recipient communities, our services would be avidly sought by marketing researchers.

Given wide latitude in defining analytic units, the investigator can designate communities at various scales, depending on how finely he or she desires to resolve activities and technologies. In chapter 6, for example, I recognized an overarching community of electrobiologists (people who employed electrical technology in biological research activities), which consists of both electrobotanists and electrophysiologists. Yet by narrowing the definition of activities and corresponding technologies, I could have treated both latter groups as communities in their own right (or subdivided the electrophysiologists into two communities, depending on whether the experiments were performed on humans or on other animals). In the present study I sought to furnish proximate explanations of technological differentiation as well as to fashion an engaging narrative; that is why I collapsed many potentially distinguishable communities, fearing that an analysis too fine-grained would obscure the main story lines.

As previous chapters have demonstrated, the memberships of different communities may overlap to any extent, from completely to not at all; conversely, one person may belong to many communities. The lack of community exclusivity may be jarring at first, but such seemingly ephemeral communities faithfully reflect the sorting—and constant resorting—of people among activities in any society. Community members can even be drawn from different nation-states; indeed, most eighteenth-century electrical communities were composed of people from England, France, Germany, the Netherlands, Italy, the American colonies, and other polities. What matters most in this framework is to delineate behaviorally based communities, regardless of how their memberships crosscut existing social, political, or geographic boundaries.

A community defined by activities and associated technologies had a behavioral reality, but in the past its presence may not have been recognized or labeled. I emphasize that community is simply a flexible analytic unit that permits the investigator to construct activity- and technology-based groups.[5] Although not a unit applicable to all research in the history of science and technology, community may be useful to investigators seeking to work above the scale of individuals and "research schools" without being shackled to a particular polity or region.[6] Moreover, community does not imply science-based, or even academically oriented groups, as do "research schools" and "invisible colleges." The technology-transfer framework should be general enough to accommodate any groups that acquire and modify the technology of interest.

-||⊢

Regardless of the nature of the communities (or technologies) involved, technology transfer can be modeled as a six-phase process. In the first phase, *information transfer,* people learn about a technology through one or more information-transfer modes, such as word of mouth, written materials, or examples of the technology itself. The second phase, *experimentation,* involves an assessment of the suitability of the new technology for given activities. The third phase, termed *redesign,* entails modification of the technology so that its performance characteristics—its behavioral capabilities—become better suited for particular activities of the recipient community. In the fourth phase, *replication,* the modified technology is reproduced through one or more replication modes and made available to community members. *Acquisition,* the fifth phase, takes place when some people obtain the new (redesigned) variant through one or more replication modes. Finally, the sixth phase is *use,* incorporation of the new technology into the recipient community's activities. After people acquire and use replicated examples of the modified technology, the investigator is able to specify the composition of the recipient community.

These phases, I hasten to add, are neither rigidly discrete nor sequential in real-world instances of technology transfer. For example, limited acquisition often takes place before redesign and replication. A case in point: many of the earliest electrotherapists acquired off-the-shelf electrical machines and Leyden jars for their clinical practice. Thus, we could easily treat experimentation as an early phase of acquisition. Also, redesign and replication are usually iterative processes that alternate with use. Merely a heuristic device, these phases allow the investigator to break down a complex behavioral process into manageable units of study. I now discuss the phases in more detail.

Information Transfer

This first phase transfers a description of the technology—and, often, its known technological effects—from person to person, directly or indirectly. In this way, people who are potentially the nuclei of new recipient communities learn about the technology. Innumerable modes of information transfer in the eighteenth century alerted people to the existence of electrical technology, such as scientific journals and monographs, physics textbooks and experiment books, newspapers and magazines, meetings of scientific societies and academies, public lectures, instruction from college and university teachers, scientific-instrument shops and catalogs, interpersonal interaction, not subsumed by the above modes, such as corresponding by mail and visits, and nonmarket exchange of the actual technologies through gifts.

Key individuals who are nodes in several communication networks may play pivotal roles in transferring information about a new technology. In the present study, both Benjamin Franklin and Joseph Priestley were important agents of information transfer. Sometimes, such individuals initiate the experimentation phase that can lead to the establishment of a new community. Information about electrical technology also moved with ease across scientific fields as well as international borders by following modes of communication established during the late seventeenth and early eighteenth centuries through societies and academies.

I hasten to add that identifying the precise modes of information transfer is unnecessary. Indeed, because information transfers do not always leave material traces in the archaeological and historical records, efforts to infer *in detail* information flows among communities may fail.[7] In addition, we should always keep in mind that the transfer of *information* about a technology, through any mode, is never a sufficient condition for the transfer of the technology itself. Only a mere fraction of the countless people who learned about electrical technology in the eighteenth century ever acquired it.

Experimentation

The second phase usually begins when a few people try out the new technology in different activities or forecast in "thought experiments" how it might perform. In this way, knowledge is created about the fit, real or imagined, between given activities and the technology's performance characteristics. For example, investigators like Cavendish and Priestley became intrigued with the possibility of using Volta's electric pistol in pneumatic chemistry. However, the pistol was mainly a display device, ill suited for the careful control of gaseous reactants and products. Likewise, earth scientists learned quickly that existing electrometers were not sufficiently sensitive to reliably indicate atmospheric electricity. These kinds of early experiments usually reveal that the technology shows some promise in an activity, but its mix, or weighting, of performance characteristics is unsuitable. In response, people redesign the technology to yield variants having different, and more appropriate, weightings of performance characteristics. Thus, both Cavendish and Priestley designed new electrical reaction chambers, but their variants still enabled insufficient control over reactants and products. These performance deficiencies were at last remedied by Lavoisier's elegant—and costly—version, which allowed one to measure quantities of gases before and after ignition. Similarly Bennet, Cavallo, and others developed ultrasensitive electrometers to detect minute atmospheric charges.

In some ways experimentation is the most intriguing phase of the process because so little of it usually survives in the archaeological and historical records. It is safe to assert that Wedgwood and Watt were not the only industrialists who considered electrical technology as a possible enemy or ally. But perhaps after thought experiments, others went no further. This part of the technology-transfer process is hidden from view but might be as informative as those exceptional cases in which development proceeded. After all, thought experiments with negative outcomes represent rejection decisions, many doubtless made without the slightest verbalization. We can, I suggest, model rejection decisions by comparing the performance characteristics of existing technologies to the likely performance characteristics of the rejected alternative. That is how I handled electric lighting in the previous chapter, arguing that it would have been dismissed without much explicit deliberation because the technology appeared utterly unfeasible without a reliable, constant, and inexpensive source of electricity.

Redesign

When experimentation does goes forward, people recognize a need for redesigning the technology, which can eventuate in the creation of new—and, usually, functionally specialized—variants. An appreciation for this common consequence of technology transfer sensitizes us to seek, in the recipient community's activities, the situational factors that might have influenced how the performance characteristics of new variants were weighted. Situational factors include the composition of the activity's social unit, the specific artifacts employed, the activity's precise interactions among people and artifacts, material flows to and from other activities, and the location and frequency of activity performance.

Situational factors can be regarded as the proximate determinants of a technology's functional requirements. Indeed, potentially relevant "external" variables, such as social inequality, religious ideology, political structure, and exchange networks, affect a technology's design only by acting upon situational factors. For example, as noted in chapter 5, the social and political activities of elite persons often required the conspicuous demonstration of learning to visitors, which was sometimes accomplished by displaying and/or conversing about a collection of scientific instruments. In such activities, impressive visual performance was an important functional requirement that surely influenced the design of electrical machines for wealthy collectors. In this way I was able to link Enlightenment ideology, through the situational factors of activities, to the performance characteristics of electrical artifacts. I emphasize that external variables have causal efficacy only when the investigator succeeds in showing how they affected the situational factors of specific activities in the technology's life history.

The process of design or *re*design—there is no need to distinguish between them in the present framework—has already been the subject of a general behavioral theory.[8] I now present that theory in an abbreviated version.

In the design process, the artisan (a gloss for any person or group that designs a technology) can receive feedback from other people whose activities take place during that technology's life history.[9] From this feedback the artisan can learn about situational factors and functional requirements; that is, which performance characteristics should *ideally* be weighted in particular activities (of manufacture, marketing, use, maintenance, discard, etc.). Often the performance requirements of different activities conflict, and so the artisan usually fashions a design that compromises pertinent performance characteristics. The *actual* weighting of performance characteristics is influenced by external variables as they impinge upon situational factors. For example, van Marum's enormous electrical machine was designed to have stunning visual and electrical performance, as required for its political and scientific functions, but it was rather difficult to operate; ease of use had clearly been sacrificed in favor of the more heavily weighted performance characteristics. I stress that performance characteristics are a technology's activity- and interaction-specific behavioral capabilities, which run the gamut from those enabling mechanical and thermal interactions to others that come into play when someone views another's artifact collection, listens to a lecture, or smells a turkey roasting.[10]

In the absence of experimental data and detailed evidence, the investigator can estimate the actual performance characteristics of a variant on the basis of its materials, form (size, shape, arrangement of parts, etc.), *and* the functional requirements of its activity-specific interactions with people and other technologies. Even when historical or ethnographic evidence is available, however, assessments of actual performance characteristics are inferences. In the preceding chapters, I could not always justify my inferences about performance characteristics on the basis of eighteenth-century sources, but I believe my judgments are generally accurate.

Replication

This fourth phase replicates a variant—its manufacture and distribution to purchasers and users. I enumerate three, somewhat idealized, replication modes prevalent in the eighteenth century.

1. One person makes a new, unique example of the technology, often for his or her own use. Such singular technologies, while seemingly ephemeral, can be of great importance, especially in activities of scientific research. Examples of singular technologies include van Marum's shellac-disk machine that ran in a trough of mercury, Bennet's printing technology, and Dr. Graham's electric throne. Sometimes instrument makers, like Nairne and Cuthbertson, were commissioned to design and build unique electrical technologies.

2. More than one individual constructs copies of the technology for their own use, creating recognizable examples of a particular variant. Many items produced or commissioned by scientists for their own use conform to this replication mode. Countless electrophysicists and members of other communities, for example, assembled their own Leyden jars and prime conductors.

3. One or more people or workshops manufacture a technology and make it available for exchange or purchase. In previous chapters, I have referred to this replication mode as "commercialization." As we have seen, Nairne, Adams, Nollet, Gütle, and other makers of scientific instruments brought to market many kinds of electrical technology.

An appreciation for differences in replication modes helps the investigator understand how people could have acquired examples of particular variants. In the present study, all replication modes seem to have been employed by members of all communities; indeed, there were many ways to acquire electrical technology throughout the eighteenth century.

Acquisition

During acquisition, examples of new variants are obtained by people who thereby become members of the recipient community. Acquisition, which I equate for present purposes with "consumption" or "adoption," is much studied in many disciplines. Commonly, investigators seek to correlate acquisition behavior with cultural and sociodemographic variables. However, from a behavioral perspective, *explanation* of acquisition requires the investigator to estimate and compare the performance characteristics of the adopted technology against alternatives in relevant activities.[11] With knowledge of a technology's replication modes, acquisition patterns, communities, community-specific activities, and alternative technologies, the investigator can rigorously pose and seek answers to the question, "Why did members of a community adopt technology y in preference to technologies u, v, and z?"[12]

However, explaining the acquisition of a *specific* variant goes beyond the goals of the technology-transfer framework because such explanations are concerned with small-scale, not large-scale, technological processes. In dealing with technological differentiation, the investigator need do no more than call attention to the new variants' weightings of performance characteristics, which rendered them more or less suitable for a recipient community's activities. That is why, in the chapters above, I did not systematically compare the performance characteristics of particular electrical technologies with their competitors. However, I did sometimes make very general comparisons, as in emphasizing the performance characteristics of electromedical technologies in relation to many traditional ones.

Use

The sixth and final phase, use is composed of the activity or activities in which the acquired technology interacts with community members; that is, when its use-related performance characteristics come into play. By distinguishing use from acquisition, we acknowledge that a community can include far more people than just those who *directly* acquire the technology. For example, the disseminator community consisted not only of lecturers but also of assistants and spectators. Similarly, in discussing the electromedical community, we must include patients and assistants. In both cases, an appreciation for a community's diverse composition is essential for assessing the ideal weights of use-related performance characteristics. Commonly, the performance requirements of different user groups favor different weightings. In the electromedical community, we might suppose that patients' performance requirements for medical accessories would have emphasized treatments that were brief, effective, safe, and painless. In contrast, the ideal performance requirements favored by electrotherapists doubtless would have centered on convenience for administering treatments, which could have been affected by the availability of assistants.

By considering diversity in a community's membership (and even re-lationships between members and nonmembers), we have an opportunity to explicitly treat issues such as conflict and negotiation, social power, or access to resources and community membership, which can influence the actual weighting of performance characteristics. For example, the perfor-mance characteristics of medical accessories were usually weighted toward the convenience of the electrotherapist, perhaps reflecting the differential in social power between electrotherapist and patient *in that activity*. And, to reassure the patient that shock treatments were safe, the electrothera-pist conspicuously employed an electrometer whose function was entirely symbolic.

An investigator may delineate groups making up a community on the basis of behavioral roles, social roles, age, sex, gender, ethnicity, national-ity, social class, and other sociodemographic variables.

╶╢║╟╴

Although the investigator can employ the technology-transfer framework as an aid for building new theories of technology or as a starting point for studying small-scale processes of technological change (e.g., the adoption of a particular variant), the most important use of the framework, I sug-gest, is to establish the behavioral parameters on which to fashion contex-tualized narratives of technological differentiation.

Once the investigator has identified a technology's new variants and in-ferred which performance characteristics were weighted in light of the ac-tivities and groups of recipient communities, he or she has endless possi-bilities for building narratives. One possibility is to fit the details of a case study into a biographical framework if one or a few individuals were in-volved in transferring the technology and creating its variants. Although I have described various activities of Benjamin Franklin in relation to the many electrical communities in which he participated, I did not use his life history to structure the overarching narrative, for he was a central figure in only a few communities.

Or a purely chronological framework may be useful, emphasizing the sequence in which new variants emerged in the course of transfers from community to community. Yet time can be Procrustean, especially when there are many communities, many transfers, and many new variants, as in eighteenth-century electrical technology. In the past, the use of chronol-ogy as a narrative's structuring principle has often been favored by those who prefer to fashion linear stories of "progress." However, a chronologi-cal framework need not employ such value-laden judgments. In chapters 2 and 3, I began my narrative by setting forth, in a temporal sequence, the electrophysicists' main inventions, many of which were transferred to other communities and redesigned. These chapters also served to introduce, I hope as painlessly as possible, basic principles of electrostatics.

As the backbone of my narrative, after chapter 3, I dispensed with time and employed the communities themselves. These appear to me as more-or-less natural units, given my interest in explaining technological differentiation by situating this process in the activities of recipient communities. Moreover, this structure enabled me easily to fold in biographical information, discussions of *relevant* external variables, and many entertaining digressions. Nonetheless, this structure also fostered a few temporal anomalies, in that chapters sometimes refer to people and variants not yet discussed; I sought to minimize such anomalies by generously cross-referencing—and judiciously ordering—the chapters.

The technology-transfer framework, then, permits the investigator to lay a behavioral foundation for fashioning contextualized narratives about technological differentiation. The precise form and content of any narrative will be tailored not only to the particulars of the case study but also to the investigator's interests and theoretical preferences. As a behavioral archaeologist, I prefer narratives grounded in the materiality of human life—that is, in the people-artifact interactions of activities. Others favor narratives that are dominated by religious, political, or social themes. Regardless of theoretical persuasion, I submit that investigators should ascertain, upfront, the behavioral dimension of technological variants (i.e., relevant groups, activities, artifact functions, and performance characteristics). Thus, the starting point for delineating contexts is the activities in which a technology took part.

On the one hand, I was often able to link major features of the Enlightenment, such as the growing importance of education and the conspicuous display of one's learning, to the situational factors of specific activities and thus the functional requirements of the participating electrical technologies. And I was able to show that the emergence of some electrical variants was influenced by the activities of courtly culture in the old regimes. In addition, I demonstrated that the development of other electrical variants was stimulated by military activities throughout the Western world, not for making advanced weapons but for protecting installations from lightning. George III's investment in Wilson's model of the Purfleet armory was the most flamboyant example, but the varied lightning conductors deployed on powder magazines testify to a wider influence of endemic militarism.

On the other hand, my efforts to discern the effects on electrical technology of an emerging modern capitalism and of the industrial revolution came to little.[13] Neither set of socioeconomic factors seems to have played any *direct* role in the differentiation of eighteenth-century electrical technology. We might argue that a mature industrial capitalism, such as was developing in some nation-states during the nineteenth century, would have led to a more rapid and more far-reaching proliferation of electrical variants. This of course did occur in the nineteenth century, but by that time electrochemical and electromagnetic technologies were also available for development.[14]

Electrical technologies of the eighteenth century were manufactured by traditional processes embedded in equally venerable production organizations. The instrument makers were part of an old craft-based industry with long-established patterns of inventiveness, hardly meriting the appellation of industrialists. In addition, no exotic materials were needed to make electrical artifacts. Although a few technologies, like the van Marum machine at the Teyler Museum (and its clones), did require the highest level of glassworking skill as well as many redesigned components, they employed no radically new industrial processes. Moreover, there was no mass production of electrical technology in factories having the extreme division of labor or standardization of parts that would be seen in later centuries. If the industrial revolution contributed to the proliferation of electrical variants, its influence was indeed subtle or so indirect as to constitute merely a generalized backdrop. Significantly, electrical technology was not acquired for use in new activities created by industrialization.

Perhaps the most telling pattern of all is that electrical technology differentiated in countries whose involvement with modern capitalism and industrialization varied enormously. France, Germany, and the American colonies, for example, were little affected by these two processes during the eighteenth century, in contrast to England, where both had begun to establish an appreciable foothold.[15] Yet in all areas, most electrical communities flourished, their members inventing many new variants of this extraordinary technology. This pattern, it seems to me, favors the conclusion that the common factor promoting the differentiation of electrical technologies was Enlightenment-related activities taking place in polities linked by trade, intermarriage, warfare, and, especially, scientific societies.

A final point: there were few pecuniary incentives to develop new electrical products. With the exceptions of instrument makers, some disseminators, and the people who designed and installed lightning-protection systems, not many people made money by inventing and selling electrical technology. Moreover, there was a conspicuous lack of capitalist backing for the creation of new variants beyond a few sovereigns and investigators, like Franklin and Cavendish, who financed their own projects. Most developmental activities were in fact financed, indirectly through salaries, by colleges and churches. Not until the early nineteenth century would electrical inventors patent their creations and treat the resultant patents as commodities or as the nucleus for building a manufacturing enterprise.

My basic conclusion is that electrical technology was differentiated by literate and clever individuals of diverse socioeconomic, religious, and national backgrounds who took part in varied activities—scientific, political, recreational, or medical. These cross-cutting activities furnished the immediate contexts in which inventive people—Gray, Beccaria, Bertholon, Franklin, Nollet, and countless others—explored opportunities to develop new variants. In these contexts the creative urge flourished, but in a manner highly constrained compared to the opportunities that new electrical technologies in vastly altered socioeconomic contexts would afford inventors of later centuries.[16] Perhaps other investigators, also employing a behavioral approach, will be able to establish convincing causal linkages between external variables and the differentiation of eighteenth-century electrical technology that have eluded me.

Regardless of which, if any, external factors an investigator implicates in the differentiation of a technology, the technology-transfer framework is a useful tool for investigating large-scale patterns of technological change. This framework applies in principle to any case of technological differentiation, such as the proliferation of worked-bone technologies during the Upper Paleolithic, the differentiation of chemical research apparatus in the nineteenth century, or the expansion of plastics-manufacturing technologies in the twentieth century. In any event, the technology-transfer framework helped me construct the story of a fascinating and little-known technology that was at the beginning of our modern electrical world.[17]

Notes

1. THE FRANKLIN PHENOMENON

1. A specific technology may be a unique artifact, such as the Hubble telescope; an artifact type, such as all telescopes; or even diverse artifact aggregates, such as all astronomical instruments, all consumer products, or all electrical devices. On different ways to designate "aggregate" technologies, see Schiffer (2001a). Although many scholars define technology to include social and cultural factors, *for present purposes* I employ the hardware definition, which enables me to bring in *relevant* external factors without blurring boundaries between artifacts, behavior, and knowledge (see chapter 12).

2. The view that technologies take part in all human activities has been most thoroughly developed by behavioral archaeologists (see, for example, LaMotta and Schiffer 2001; Rathje and Murphy 1992; Rathje and Schiffer 1982; Schiffer 1992; Schiffer and Miller 1999; Walker et al. 1995). For recent studies in the anthropology of technology, see Schiffer (2001b).

3. A community (or "technocommunity"), as employed here, is a kind of "community of practice" defined on the basis of activities *and* artifacts (on practice theory, see, e.g., Lave and Wenger 1991; Suchman 1987). In a quite different project, Golinski (1992, 10) also situates scientific activities and technologies within a "community of practitioners." Kuhn (1970, 177) in a discussion of scientific communities mentions that "major subgroups" can be isolated on the basis of "similar techniques." In chapter 12 I elaborate my concept of community.

4. In defense of this practice, which enables the solution of research problems and contributes to the construction of a reader-friendly narrative, I merely quote Jardine (2000, 265): "Use of categories alien to the agents studied is often perfectly legitimate."

5. "Electrophysicist," a term used by Higgins (1961), designates only those physicists who used electrical technology to carry out research.

6. A "performance characteristic" is a behavioral capability that comes into play in a specific activity or activities. Performance characteristics enable

appropriate interactions to take place among people and artifacts. Not only do performance characteristics facilitate electrical, mechanical, and thermal interactions, they can pertain to every human sensory mode—such as visual, acoustic, and tactile. For discussions of performance characteristics by behavioral archaeologists, see Schiffer and Miller (1999, ch. 2), LaMotta and Schiffer (2001), Schiffer (1995, 2000, 2001a), and Schiffer and Skibo (1987, 1997).

7. The terminology of "electrostatic" or "static electricity" is both anachronistic and inaccurate. It is anachronistic because the terms "static electricity" and "electrostatics" did not come into use until the mid nineteenth century (e.g., Mascart 1876); electricity in the eighteenth century was simply "electricity" or, after atmospheric electricity was discovered, "artificial electricity." It is inaccurate because some eighteenth-century electrical technology employed circuits in which current—albeit in tiny quantities—flowed through conductors; not all of its effects were wrought by static charges. However, eighteenth-century investigators did establish the fundamental principles of electrostatics (for a modern introduction to electrostatics and its applications, see Moore [1973, 1997]).

8. Schäffer (1766, 2); my translation.

9. It would be impossible to examine every variant of eighteenth-century electrical technology. Despite its inevitable gaps, the historical record is so rich that I have had to be selective in the present work, focusing on a mix of common and rare types around which to write my narrative.

10. Many recent works consider the place of science in eighteenth-century societies (e.g., Golinski 1992; Hankins 1985).

11. I have obtained information on Franklin's early years from Franklin (1986) and Van Doren (1938).

12. Franklin 1986, 13–14.

13. Franklin 1986, 21.

14. Franklin 1986, 75.

15. Franklin 1986, 91–92.

16. Van Doren 1938, 98.

17. Franklin 1986, 115.

18. Some might regard mathematical skill as prerequisite for scientific research, but much natural philosophy in the eighteenth century was nonmathematical. Electrical science remained essentially without quantification well into the nineteenth century.

2. IN THE BEGINNING

1. The positions advanced here apply mainly to physical science (see, e.g., Galison 1987; Gooding 1990; Gooding et al. 1989; Hankins and Silverman 1995) and, to a lesser extent, biological science. My views on experimentation in history and social science are expressed elsewhere (e.g., Schiffer 1992; Schiffer and Miller 1999). The claim that experiments are necessary for the pursuit of science is, of course, contestable (see, e.g., Barnes et al. 1996, 142).

2. "Activity" is a key concept in behavioral archaeology. For discussions of activities, see Schiffer (1992, ch. 1) and Schiffer and Miller (1999, ch. 2).

3. Warner's (1990) analysis has shown that the term "scientific instrument" came into use in the middle of the nineteenth century carrying a heavy ideological load. Modern usages of the term "scientific instrument," which I believe are too narrow for understanding the behavior of scientists, can be found, for example, in Joerges and Shinn (2001).

4. For an engrossing narrative that highlights the scientist's interaction with laboratory apparatus, see Buchwald (1994).

5. De Solla Price (1984) emphasizes how technologies create novel phenomena (see also Schiffer and Skibo 1987). Many technologies unconnected to scientific activities also create "unnatural" phenomena and materials, from stainless steel to teflon.

6. See Hankins and Silverman (1995, 11) on what physical scientists study.

7. Biographical information about William Gilbert comes from Bordeau (1982, 3–12) and Mottelay's "Biographic memoir" in Gilbert ([1600] 1958, ix–xxvii). On the origins of Gilbert's scientific method, see Zilsel (2000).

8. The history of early electrical theories has been presented in masterly fashion by John Heilbron (1979; see also the abridged version, 1982). This work must be the starting point for anyone interested in eighteenth-century electricity. Other useful works on early electrical theories are furnished by Home (1979, ch. 2; 1981, 1992) and Whittaker ([1910] 1951); for comparative purposes, I also recommend theoretical sections of late-eighteenth- and early-nineteenth-century treatises on electricity (e.g., Adams 1792, 1799; Cavallo 1795; Cuthbertson 1807, 1821; Ferguson 1775, 1778; Ferguson and Partington 1825; all these sources are in English). The unsurpassed primary source for learning where electrical theory stood at the end of the period covered in this chapter is Desaguliers's (1742) elegant dissertation. Mottelay (1922) furnishes a chronology of key events in the early history of electrical science and technology along with rich references. For an introduction to eighteenth-century electricity, see Fara (2002).

9. To demystify descriptions of experimental findings and technologies— and empower the nonspecialist reader—I present essentially modern principles along with accounts of the earliest scientific studies of electricity that inevitably involve anachronistic concepts and terms. For example, I use the concept of "charge" in discussing the works of Gilbert and Guericke, even though this term did not gain currency until the middle of the eighteenth century.

10. See Barnes et al. 1996, ch. 3. Many scientific advances arise from an investigator's crucial insight that there are important differences among the members of a formerly unitary class of phenomena (see Barnes et al. 1996, 62).

11. I follow Nagel (1961) in distinguishing between experimental laws and theories but acknowledge that the formulations of logical positivists like Nagel have fallen out of fashion in many academic quarters. The distinction between theories and experimental laws, though far from unproblematic, is nonetheless useful for the present project.

12. Gilbert 1958, 77.

13. Gilbert 1958, 95.

14. To rub the test materials, Gilbert (1958, 87) used substances that "do not foul the surface, and cause them to shine, e.g., strong silk, and coarse woolen cloth, scrupulously clean, and the dry palm of the hand." He also specified that it is preferable to perform electrical experiments on fair days having low humidity. Later in the seventeenth century, Boyle (1675) extended these experiments, as did countless investigators in the eighteenth century; the culmination of this line of research was the encyclopedic work of Ritter (1805).

15. Gilbert 1958, 77–78.

16. Du Fay's (1737b) extensive experiments were especially influential.

17. The term "revolve" (Gilbert 1958:79) is highly suggestive. Assuming it is not a translation error, the word hints that Gilbert had come close to inventing an electrostatic motor. This seems very unlikely to me, given his apparatus. "Revolve" must be taken to mean that the versorium's needle merely made some movement about its axis. According to Heilbron (1982, 161), Fracastoro had described a versorium in the previous century.

18. The authoritative history of early electrical measuring instruments is Hackmann's (1978b) monograph; Bordeau (1982) is also a useful source but includes many post-1800 developments.

19. Gilbert 1958, 92.

20. Gilbert 1958, 93.

21. Many people in the eighteenth century would propose that electrical matter was in fact Newton's aether. But in the early twentieth century the aether fell on hard times, and Einstein finally killed it.

22. For a hagiographic but still interesting treatment of Guericke's electrical experiments, see Coulson (1943).

23. On the historical and philosophical significance of the air pump, see Shapin and Schaffer (1985).

24. Guericke [1672] 1994, 160.

25. Bazerman 1993, 215.

26. See Bazerman 1993, 216. On natural magic and its relationship to science, see Hankins and Silverman (1995, esp. chs. 1 and 2).

27. On the intellectual context of Guericke's experiments, see Bazerman (1993). Guericke himself never used the term "electrical"; later investigators called attention to the electrical nature of his experiments.

28. Guericke 1994, 227.

29. Guericke 1994, 227–228.

30. Early histories of electricity credit Guericke with inventing the first electrical machine or generator (e.g., *Histoire* 1752, 4–5; Bose 1738). Except Hackmann (1978a), modern historians of technology repeat the error (e.g., Bowers 1982, 4; Dibner 1957, 13–14). Perhaps the first to draw this incorrect inference was du Fay (1737a, 34), who even claimed that Guericke spun his sulfur sphere *avec rapidité* (du Fay makes several small errors of this kind when describing the work of earlier students of electricity). He also refers to Guer-

icke's apparatus as a *machine* (du Fay 1737a, 34). Du Fay was quite familiar with Hauksbee's electrical machines (e.g., du Fay 1737a, 41) and doubtless saw their antecedents in Guericke's sulfur globe.

31. Hackmann (1978a, 20–21) denies credit to Guericke for inventing the electrical machine because Guericke did not understand that the phenomena he observed were electrical in nature. This sort of argument is unconvincing because it requires investigators to have some level of *scientific* understanding of the effects they are producing—knowledge that often does not develop until far in the future. What matters most, I suggest, is the device's design and how its author actually used it to produce effects. On that score, Guericke's sulfur globe was clearly not an electrical machine because it was not cranked to produce a continuous charge as were all later machines.

32. As I noted above, Guericke himself did not generalize the effect in these terms.

33. Boyle 1675, 29–30.

34. Two Francis Hauksbees, uncle and nephew, both played roles in the development of electrical technology. Francis Hauksbee (the elder), discussed in this chapter, carried out original research on electricity and developed new experimental apparatus (for biographic information, see Guerlac 1972a). Francis Hauksbee (the younger), mentioned in chapter 3, manufactured scientific instruments for sale, including electrical apparatus invented by his uncle (for biographic information, see Guerlac 1972b).

35. Like the French academy, many of these institutions were called "academies," but hereafter I use the generic term *society* to encompass both societies and academies. Much information on the scientific societies and academies in this chapter comes from McClellan (1985). For useful histories of the Royal Society, see Lyons (1944) and Weld (1848).

36. Information on the St. Petersburg Academy of Sciences comes from Home (1979, 28–31).

37. Cohen 1941, 14.

38. Van Doren 1938, 138–140; Smyth 1970, 2:277.

39. Home 1979, 144.

40. Some regional societies also offered prizes for noteworthy contributions. The Bordeaux academy, for example, awarded J. T. Desaguliers a gold medal for his dissertation on electricity (Barbot, in Desaguliers 1742, 48–49).

41. McClellan 1985, 250.

42. Although Thomas Whewell is credited for coining the term "scientist" in the early nineteenth century (e.g., McClellan 1985, 233), the term "science" was widely used in the eighteenth century.

43. Heilbron 1982, 168.

44. Heilbron 1979, 229.

45. My information on Hauksbee's electrical apparatus and experiments comes from his 1719 volume, which also contains a section reporting previously unpublished experiments.

46. Hauksbee 1719, 6–12.

47. The "invention" of the vacuum pump is attributed solely to Boyle (Hauksbee 1719, preface), which suggests that Hauksbee was unfamiliar with Guericke's work. If so, then it is doubtful that Guericke's sulfur globe was the inspiration for Hauksbee's electrical machine. The latter is shown by Hauksbee (1719, tb. 2); the electrical parts are described on 21–23. After Hauksbee, Hartmann (1766) carried out the most extensive studies of electrical phenomena in vacuo.

48. Hauksbee 1719, 24–25.

49. In 1600 Gilbert demonstrated that glass was an electric, but that discovery has often been credited to Isaac Newton (e.g., Hackmann 1978a, 25–26). In any event, it was a phenomenon doubtless known to Hauksbee.

50. Hackmann (1978a) is the definitive history of the electrical machine; also useful is Dibner's (1957) earlier effort.

51. Two pulleys were used in some experiments to turn two vessels, one inside the other, in opposite directions. Precisely that configuration is illustrated in Hauksbee's (1719) tb. 3.

52. Hauksbee 1719, 45–46, tb. 3.

53. Hauksbee 1719, 46. This light is produced by the ionization of gases remaining in the partially evacuated vessel.

54. For examples, see Hauksbee (1719, 34, 38, 174–175). He used the term "spark" on 34.

55. Hauksbee 1719, 65. It is this configuration that is illustrated in tb. 3.

56. Hauksbee 1719, 53. Desaguliers (1742, 5–6) provided more detailed instructions on how to excite electricity on the tube.

57. See, for example, Hauksbee (1716).

58. For biographic information on Stephen Gray, see Cohen (1954a), Corrigan (1924–1925), Heilbron (1972, 1979, 242–245), and Hackmann (1978a, 54–61). On an interpretation of the social context and implications of Gray's scientific activities, see Ben-Chaim (1990).

59. Gray 1720, 106. The ox gut was in the form of gold-beater's cloth. In later experiments, he added to the list of electrics many vegetable materials, including grass, wheat, and dried leaves (Gray 1731, 38), and showed that water could be charged (Gray 1732a). These experiments were extended by du Fay (1737b), who was able to electrify many formerly recalcitrant materials, such as hard stones, after heating them.

60. Gray 1731, 19.

61. Gray 1731, 21–22.

62. Gray 1731, 20. Hauksbee (1719, 54) had previously observed alternating attraction and repulsion of small particles by a charged glass tube.

63. To understand induction, recall that charges come in two varieties, positive (a deficiency of electrons) and negative (a surplus of electrons). Opposite charges attract, whereas like charges repel. Now assume you are holding, by an insulated handle, a negatively charged brass wand. As you bring the wand toward an uncharged glass ball (an insulator), the wand's negative field repels the electrons on the ball's surface, forcing them inward and leaving behind on the

ball's surface a net positive charge. In the opposite case, a positively charged wand attracts electrons to the surface of the glass ball, charging it negatively. Because glass is an insulator, it inhibits an internal rearrangement of electrons that would quickly return the ball's charge to its neutral state. As Gray's experiments showed, some materials resist the flow of electrons so greatly that they can hold charges for months, an effect that would be put to good use decades later (see chapter 3, on the electrophorus; and chapter 11, on Lichtenberg figures and Bennet's printing process).

The field around a charged object also affects conductors. If you bring the negatively charged wand near the left end of an iron bar, suspended by insulators, you will discover that the two ends of the bar acquire different charges, the left side is positive (because it has a deficiency of electrons) and the right side, hosting more than its fair share of electrons, is negative. But if the wand is removed, the charges will disappear quickly because the bar, being a good conductor, returns to its neutral state by the internal flow of electrons. (Needless to say, this explanation of induction—employing the concept of field—could not have been offered until the mid eighteenth century. In place of field, investigators used terms such as "electrical atmosphere" and "sphere of influence"; Heilbron [1981] emphasizes that fieldlike concepts were prevalent in the eighteenth century.)

64. Gray 1731, 22–26.

65. Gray 1735a, 19. Gray did not write in terms of the storage of charge; rather, he appreciated that some objects were capable of acquiring a "greater Quantity" of electric effluvia (see previous note).

66. Gray 1735a, 19–20. Sometimes he supported the iron bar on glass or resin insulators.

67. Gray 1735a, 22.

68. Gray 1732a, 228.

69. Gray's (1732b, 291) paper is especially noteworthy for containing a "catalogue"; today we would call this unremarkable page a "table of data" but it was the first to be published in a report of electrical experiments. It is likely that Gray's experiments with these substances were inspired by Hauksbee's (1719, 153–158) previous work on sulfur and resin.

70. Gray 1732b, 287. Desaguliers (1741, 666) adopted the thread-on-a-stick indicator, terming it a "Thread of Trial" (see also Desaguliers 1742, 4, 28).

71. Gray 1732b, 287. Building on this finding, Alessandro Volta decades later would invent the electrophorus (see chapter 3).

72. The importance of this discovery for designing experiments was explicitly recognized later by Franklin and Kinnersley: "By some lucky Accident, Mr. Stephen Gray . . . first discover'd that Wax, Pitch, Silk & several other things, wou'd stop the Electric Virtue from running off, or dissipating in the Common Mass of Matter" (Kinnersley, in Cohen 1941, 411). Gray did not use the term "ground." A few years later Desaguliers (1739, 206), who was intimately familiar with Gray's work, noted that electrified objects in contact with the floor would lose their electrical "Virtue." By 1742, Desaguliers (1742, 2–

3) had clearly formulated the modern concept of grounding or (in British English) earthing, "if a Non-Electrick be touched by another non-Electrick, which touches a third, and so on; all the Electricity received by the first will go to the second, and from the second to the third, and so on, till at last it be lost upon the Ground, or the Earth" (see also Desaguliers 1742, 25).

73. Apparently a prickly personality, Gray maintained a monopoly on electrical research, at least in the Royal Society. Almost immediately after his death, however, Desaguliers resumed Gray's lines of research, using glass tubes, iron bars, and sundry common objects to perform further experiments on the conduction of charge (Desaguliers 1739, 1741, 1742; for an illustration of one apparatus, see 1741, 665). He was the first person to use the term "conductor" (Desaguliers 1739, 193), but in this early work it had a pronounced mechanical as well as electrical connotation. Not long afterward, Desaguliers (1742, 2) reported that conductors need to be "insulated" if they are to retain electricity imparted to them by a charged body. Among the materials that could serve as insulators (his actual term was "supporter" [17], which emphasized its mechanical function of holding up a conductor), Desaguliers (1742, 17) listed "Hair-Ropes, Fiddle-strings, or Cat-guts, ribbons, Strings of Silk, Glass Tubes, long Bodies of Sulphur or of Resin." He also indicated the necessity of keeping insulators dry. Although adding great conceptual clarity to discussions of electricity in what were, essentially, extended commentaries on the works of Hauksbee and Gray, Desaguliers invented no significant electrical technologies.

74. For biographical information on du Fay, see Heilbron (1971a; 1979, 250–252). Maluf (1985, ch. 4) also presents a useful discussion of du Fay's theoretical ideas. His surname is also written Dufay and Du Fay, especially in modern works. Additional publications by du Fay on electricity, not cited below, include du Fay (1738a, 1741, 1745).

75. For an explicit acknowledgment of Nollet's role in his experiments, see du Fay (1737c, 345). Maluf (1985, ch. 4) discusses du Fay's influence on Nollet.

76. Du Fay 1738b, 716–719; regrettably, he did not describe or illustrate this machine. Possibly it was manufactured by Francis Hauksbee the younger or by Nollet, who was an instrument maker (see chapter 3).

77. Du Fay 1737b, 113–115. He had previously employed stands of wood and metal, but the one of glass was the most serviceable; he also made a stand of sealing wax.

78. See, for example, Adams 1799, 141.

79. Du Fay 1737d, 629–630.

80. Du Fay's (1737d, 634) precise terms were *électricité vitrée* and *électricité résineuse* (original emphasis).

81. These experiments suggested to du Fay a technique for ascertaining the kind of charge on any object: simply expose it to a charged piece of glass or amber (du Fay 1737d, 638–640). If it was repelled by the glass or attracted by the amber, the charge was vitreous; in the opposite case, the charge was resinous. For undertaking these assessments of charge, du Fay devised another version of his versoriumlike device in which a counterbalanced glass needle, tipped by

a metal ball, could pivot on a glass stand; sealing wax could be substituted for the glass parts (du Fay 1737d, 640–641).

82. Du Fay's (1737d, 631) exact statement of the law: "Les uns & les autres repoussent les corps qui ont contracté une électricité de même nature que la leur, & ils attirent, au contraire, ceux dont l'électricité est d'une nature differente de la leur." Another version was presented on 634 (see also du Fay 1738b, 722).

83. For a synthesis of his sixteen principles of electricity, see du Fay (1738b, 720–723).

84. See du Fay (1737c, 337) for a description of this experiment. Further experiments (du Fay 1737c, 339) enabled him to describe "la matiere électrique passant librement à travers la verre." Although he believed that all electrics, such as glass or silk, could allow electricity to pass over short distances, such substances were not appropriate for long-distance communication of electricity (du Fay 1738b, 721).

3. A COMING OF AGE

1. Members of other communities propounded theories of electricity that were more or less conformable with the effects produced by their apparatus, but the scope of this book does not allow me to trace out these connections in detail.

2. I also include the inventions of a few instrument makers and others that do not fit well anywhere else.

3. Toward the end of the eighteenth century, a number of specialized scientific journals were established, including *Journal de physique, Annales de chimie, Annalen der Physik,* and *Allgemeines Journal der Chemie.* Research on and with electricity was published in these journals, but not until 1843 was a journal dedicated to electricity founded (*The Electrical Magazine*).

4. In a fascinating discussion, du Fay (1737c, 331–332) illuminated the still-evolving practice of crediting prior workers for their discoveries.

5. For example, see *Mémoires de Mathématique & de Physique de l'Académie Royale des Sciences,* 1747.

6. Heilbron 1971a, 214.

7. A strong pattern of citing previous researchers, sometimes in extensive detail, appears early among electrophysicists, such as du Fay (1737a, 1737c), Desaguliers (1742), and especially Bose (1738, 1744).

8. For example, see Lavoisier 1793.

9. Beginning late in the eighteenth century, it became increasingly difficult for one society to integrate all areas of interest effectively. Thus, a number of more specialized scientific societies emerged (McClellan 1985, 257), and these would become, in the nineteenth and twentieth centuries, the principal venues for reporting new discoveries. Nonetheless, the general societies remained the nexus of scientific action throughout most of the eighteenth century.

10. Daumas (1989) is a good general source on the early scientific instru-

ment makers. For compendia of Scottish and British instrument makers, see, respectively, Bryden (1972) and Clifton (1995). Bedini (1964) treats early American instrument makers.

11. Clifton 1991, fig. 9.

12. On the scale of London instrument shops, see Daumas (1989, 236–241).

13. Webster (1974) gives some biographical information on Edward Nairne. See also D. J. Warner, *Rittenhouse* 12 (1998): 65–93.

14. For examples of eighteenth-century catalogs containing electrical technology (but not also mentioned below), see Adams (1799, end of book, 9–10), Adams and Jones (1799, 5:9–10), Adams et al. (1746, 258–259), Dollond and Dollond (n.d.), Gütle (1790, end of book), Guyot (1770, 173), Nairne (1793, 73–76). A catalog by Benjamin Martin is reproduced by Millburn (1976, 219–222). For examples of papers and experiment books that showcased the electrical wares of instrument makers, see Adams (1784, 1792, 1799), Adams and Jones (1799), Cuthbertson (1807), Guyot (1770), Martin (1746, 1759a, 1759b), Nairne (1773, 1774, 1783, 1793), Neale (1747), Ribright (1779), and Watkins (1747).

15. A brief biography of Francis Hauksbee (the younger) is supplied by Guerlac (1972b). It is possible, even likely, that his uncle also made instruments for sale (cf. Guerlac 1972a, 169), but I have been unable to document to my satisfaction that he actually sold electrical machines. Hackmann (1978a, 39) also questions whether even the younger Hauksbee sold electrical machines, given that they are not included in the list of apparatus advertised in Hauksbee and Whiston (1714). This omission is not significant for several reasons. First, the term "electrical machine" did not come into use in England until the 1740s. Second, Hauksbee and Whiston (1714, 20) illustrate and discuss the electrical machine under the category of "Pneumaticks." The advertisement lists airpumps "with all their appurtenances," which in that context probably included electrical machines employing exhausted glass vessels. Third, the lack of any "electrical" entry in the advertisement may only indicate that the term would have had little meaning to prospective buyers at that time. What seems significant is that the electrical machine was pictured in this book and used in Whiston's lectures. As even Hackmann (1978a, 39) acknowledges, surely Hauksbee—whose income depended on selling instruments—would have made the machine for any spectator who requested it, as the following statement implies: "All the above-mention'd Instruments, according to their Latest and Best Improvements, are made and sold by Francis *Hauksbee*, (the Nephew of the late Mr. *Hauksbee*, deceas'd)" (Hauksbee and Whiston 1714, advertisement; original emphasis).

16. This machine is shown by Hauksbee and Whiston (1714, pl. 6); they also describe the accessories and their mode of operation (20). Whiston delivered lectures at Hauksbee's house; subscribers paid two and a half guineas for an extensive physics course (Hauksbee and Whiston 1714, 3).

17. Hackmann (1978a, 47–48) reports that Johann George Cotta included, in a sales catalog of 1724, a modified Hauksbee-type machine made by Jacob

Leupold of Leipzig. I focus on Musschenbroek and Nollet here to lay a foundation for their other works discussed later in this chapter.

18. On the Musschenbroeks and their workshop, see de Clercq (1997). Some
biographical information on Petrus van Musschenbroek is also provided by
Struik (1974). On the place of Leiden University in eighteenth-century science,
see Ruestow (1973).

19. Musschenbroek (1739b, catalog at end, p. 4; my translation). The same
items were offered again, similarly situated, in 1751 (Musschenbroek 1751,
catalog at end, pp. 3–4).

20. Among the useful biographical sources on Nollet are Heilbron (1974),
Maluf (1985), and Torlais (1954). Benguigui (1984) contains the extensive
Nollet-Jallabert correspondence.

21. On Nollet's collaboration with, and intellectual relationship to, du Fay,
see Maluf (1985, 118–126).

22. Vazquez y Morales, in Nollet (1747, foreword; my translation).

23. The physics text: Nollet (1738, 1745–1765, 1748); electrical books: Nollet (1747, 1749b, 1754, 1764); *L'art des expériences:* Nollet (1770). Volume 1 of
the latter is about the choice and working of materials, such as wood and metal.

24. Paulian 1781, 3:156; my translation.

25. Nollet 1747, 38, 78–79, 82. Early on, Nollet's electrical research came to
the attention of British investigators (e.g., Nollet 1749a, 1751–1752; Watson
1753).

26. Nollet 1747, 82–83.

27. Nollet's theory, developed over many pages in the context of experiments, suffers from ambiguities, inconsistencies, and outright contradictions.
These uncertainties make it possible, in the past as well as today, to support
many interpretations of Nollet's views (compare, for example, renderings by
Heilbron [1979], Home [1981], Maluf [1985], and Torlais [1954, 1956]).

28. Nollet 1747, 40.

29. A charged object induces a field on the near side of the glass; in turn,
that field induces a field of opposite charge on the far side of the glass. Finally,
the latter field effects attractions and/or repulsions of nearby objects, also by
induction.

30. Nollet 1738. The electrical items in Nollet's catalog are numbered 270
to 300 (Nollet 1738, 175–179).

31. This is probably item 270 in the 1738 catalog: "Un gros globe de cristal
ajusté à une machine de rotation" (Nollet 1738, 174–175). Although the catalog was not illustrated, Nollet's first book on electricity does show this machine
(Nollet 1747, fig. 2). Years later Franklin confided in a letter to Joseph Priestley
that Nollet's machine "was a very bad one, requiring three Persons to make the
smallest Experiment . . . And the Effect after all but weak" (Franklin to Priestley, May 4, 1772 [Smyth 1970, 5:395]).

32. Hackmann (1978a, 69) mentions this academic appointment, which
took place in 1737 or 1738.

33. The term "prime conductor" appears in the English literature in 1747

(e.g., Watkins 1747, 5). Among those who used gun-barrel prime conductors were Franklin (e.g., [1769] 1996, 15), Martin (1746, 25), Rackstrow (1748), and Watkins (1747). Swords were also employed as prime conductors in early experiments.

34. Winkler (1748, 11, and pl. 1, fig. 1). Winkler specifically noted the performance problems of the Hauksbee machine that his rubber was supposed to correct: it required dry hands, one person could not both turn the globe and apply constant hand pressure, and it was tiring to use (8).

35. For an amalgam recipe and application instructions, see Cuthbertson (1807, 17–18).

36. Hackmann 1978a, 71–72.

37. Hackmann 1978a, 81.

38. See, for example, Winkler (1748, 7).

39. Winkler (1748, 8–13, 16–17, and pl. 1, figs. 1 and 2, pl. 2, figs. 1 and 2).

40. For example, see Abstract (1744), Bose (1745), and Winkler (1745).

41. Watkins (1747, 75). The machine is described on pp. 76–77, and illustrated in pl. 3.

42. An historical account 1745, 197. This article, which is amply footnoted, twice mentioned a "current of the electric matter." William Watson also employed the phrase "current of the electric aether" (e.g., 1746b, p. 55).

43. An historical account 1745, 194.

44. Basalla (1988, 66–78) discusses the roles of play and fantasy in the invention of technology.

45. Galileo did not invent the telescope (see Turner 1991, ch. 10, 217), but he was one of the first to use it for making systematic astronomical observations.

46. Some readers may find that my arguments flirt with technological determinism. But I simply wish to underscore that the role of laboratory apparatus, and the possibilities it presents for discovering effects and creating new technologies, have been undertheorized by recent historians of science. Whether they consist of kitchen utensils or multimillion-dollar machines, laboratory apparatus present manipulative possibilities. And it is the playful realization of some such possibilities that produces seemingly serendipitous discoveries.

47. Watkins 1747, 32.

48. I have drawn this account of the invention of the Leyden jar from Heilbron (1979, 312–314); Hackmann (1978a, 90–99) and Keithley (1998, ch. 6) are other good sources.

49. On von Kleist's experiment, see Heilbron (1979, 309–312) and Keithley (1998, ch. 6). After failing to replicate his experiment (see Mottelay 1922, 174), even German-speaking investigators ignored von Kleist; not until decades later do references to a *Kleistischen Flasche* appear (e.g., Gälle 1813, 89; Kühn 1783, 2:157).

50. Winkler 1746, 211–212. A more detailed, technical account of the Leyden jar experiments was provided by Needham (1746).

51. *The Gentleman's Magazine* 16 (1746):163.

52. This effect was first reported in 1750 by Benjamin Wilson in England (Gluckman 1996, 15; see also Priestley 1769). However, Home (1979, 124–125) claims that the oscillating discharge of a Leyden jar was "not confirmed experimentally for another hundred years."

53. Cohen 1941, 16.

54. Cohen 1941, 14–15.

55. Here I follow Cohen's (1990, ch. 4) identification of the mysterious "Dr. Spence," the lecturer who appears in Franklin's autobiography (Franklin 1986, 133) as Archibald Spencer.

56. Franklin 1986, 133. Dibner (1976) is a brief and very readable introduction to Franklin's electrical researches.

57. Franklin 1986, 267.

58. For a summary of the arguments and evidence on this point by John Heilbron and I. Bernard Cohen, see Cohen (1990, 64).

59. Franklin 1996, 7–8.

60. On the first use of the terms "positive" and "negative" as well as "plus" and "minus," see Franklin (1996, 8–9).

61. Franklin himself (1996, 14) refers to the "two states of Electricity."

62. This mathematical conception is explicit, with quantitative examples, in Franklin's first discussion of the charges on a Leyden jar (Franklin 1996, 13). Perhaps this mathematical conception is an outgrowth of Franklin's experience in business, where an account balance is formed by the sum of receipts (+) and expenditures (−).

63. Franklin 1996, 13. Regrettably, he did not specify how he measured the magnitudes of the charges.

64. Franklin 1996, 13–14, 26.

65. This novel hypothesis was stated as follows: "Thus, the whole force of the bottle, and power of giving a shock, is in the GLASS ITSELF; the non-electrics [i.e., conductors] in contact with the two surfaces, serving only to *give* and *receive* to and from the several parts of the glass; that is, to give on one side, and take away from the other" (Franklin 1996, 26; original emphasis). Franklin's explanation was incomplete by modern standards, but it sufficed to convey understandings that would be useful in designing electrical apparatus, including new kinds of capacitors (see, esp., Beccaria 1776).

66. Franklin 1996, 27. "Shock" at that time meant a large discharge, whether received by an animate or inanimate object.

67. Franklin 1996, 27–28. Some years later Franklin suggested to friends that one could also make capacitors from layers of wax-impregnated paper and metal foils (Franklin to Kinnersley, July 13, 1771, in Labaree et al. 1974, 18: 182)—a design that was used in the twentieth century.

68. For series configurations of capacitors, see Franklin (1996, 25, 28). His first use of the term "electrical battery" occurs on 28.

69. Franklin 1996, 94.

70. For battery prices, see Adams (1799, catalog at end, p. 9) and Nairne (1793, catalog at end, p. 74).

71. Cohen (1941, 139–148) presents details on the publication of Franklin's electrical writings.

72. Watson 1751–1752a, 211. Heilbron (1976a) furnishes a brief biography of Watson; among his early works were Watson (1745, 1746a, 1746b, 1747).

73. Franklin 1986, 173. On the controversy between Nollet and Franklin, see Torlais (1956).

74. Franklin 1986, 173–174. Although Franklin generally preferred to avoid scientific controversy, he departed from this pattern by avidly participating in debates on the nature of clouds and rainfall (see the many discussions in Franklin [1996]) and in vigorously defending his lightning conductor (see chapter 9).

75. In Franklin's own words, Nollet "has laid himself extremely open, by attempting to impose false accounts of experiments on the world, to support his doctrine" (Franklin to Cadwallader Colden, January 1, 1753 [Smyth 1970, 3:105]; see also Franklin to James Bowdoin, December 13, 1753 [Smyth 1970, 3:193]).

76. Franklin 1986, 174.

77. Cohen 1995, 140.

78. Franklin 1986, 175.

79. Cohen has made this point since his earliest work (e.g., 1941), but he has more recently presented an expanded case (Cohen 1995, ch. 3).

80. Franklin 1996, 11.

81. The best modern source on eighteenth-century electrical machines, which comes close to being exhaustive, is Hackmann's (1978a) thoughtful and well-illustrated monograph. Dibner's (1957) thin volume is a good read and a fine starting point, but it treats only major developments.

82. Hackmann (1978a, 143–147) discusses the plate machine's plausible inventors. For later developments of the plate machine, focused on the contributions of the Cuthbertsons, see Hackmann (1973).

83. For biographical information on Ingen-Housz, see van der Pas (1973) and Wiesner (1905). Ingen-Housz is written several ways both in eighteenth-century and modern literatures (e.g., "Ingenhousz"—Cohen 1941, 137).

84. Ingen-Housz 1779a, 61–63; see also Hackmann (1978a, 143–145).

85. Ramsden (1766, figs. 1 and 2) illustrated his machine and some accessories.

86. Hackmann 1978a, 147.

87. Finn 1971.

88. According to the 1791 inventory of his laboratory, Joseph Priestley owned five electrical machines, including cylinder and plate varieties (Bolton 1892, 226–227), but he preferred globe and cylinder machines (Priestley to Franklin, June 13, 1772, in Labaree et al. [1975, 1:175]). In another example, Jacquet (1775) mentions both globe and plate machines, but the experiments, as illustrated, all employed a Nollet-type globe machine. New varieties of cylinder machines continued to be developed, even at the end of the eighteenth century (e.g., Hoffmann 1798).

89. Dal Negro 1799; the completed machine and its components are well il-

lustrated in 10 figures. Nearly two decades earlier, Maggiotto (1781) had originated a nearly identical design.

90. On the insulated grips, see Gütle (1790, 117–121, and tb. 6, figs. 3 and 4); on the copper-disk machine, see Gütle (1790, 122–124, and tb. 6, fig. 2); on the woolen-disk machine, see Gütle (1790, 105, tb. 3, fig. 1, and catalog at end).

91. Ingen-Housz 1779a, 668–669.

92. Franklin to Ingen-Housz, April 26, 1777 (Smyth 1970, 50).

93. Ingen-Housz 1779a, 673. Another machine Ingen-Housz invented, with Franklin's encouragement, employed silk ribbons held rigid in a frame with a sliding rubber of copper or cat's fur (Ingen-Housz 1785, 99–135, pl. 1).

94. For examples of drum machines, see Bohnenberger (1784–1791), Gütle (1790, tb. 1, fig. 5), and Rouland (1785). These sources are particularly rich in design details.

95. Gütle 1790, catalog at end of volume.

96. For instructions on how to convert an old treadmill into a belt machine, contact the author.

97. Biographical information on Volta comes from Heilbron (1976b), whom I have shamelessly paraphrased in many places. It is possible that Franklin and Volta met in 1782, for both attended a meeting in France hosted by La Blancherie (Labaree et al. 2001, 36:lxiii, 358).

98. Beccaria to Franklin, February 20, 1767 (Labaree et al. 1970, 14:49–57). This version is an English translation from the original Latin.

99. Volta [1775] 1926.

100. Ingen-Housz to Franklin, November 15, 1776 (Labaree et al. 1983, 23: 7–12).

101. Franklin to Ingen-Housz, April 26, 1777 (Smyth 1970, 7:50).

102. Ingen-Housz 1778b; see also Henly (1778). A lucid modern description of the electrophorus's operation is found in Baird and Nordmann (1994). For early experiments with the electrophorus, see *Schreiben* (1776) and Socin (1777).

103. Because the electrophorus was not "discharged"—it communicated charge to the metal cover by induction—Volta's experiments were irrelevant to explaining the persistence of charge in insulators after an actual discharge. Beccaria's arguments, however inelegantly phrased, actually drew attention to the enigmatic processes underlying the behavior of dielectrics.

104. Gütle 1790, 127–128, and tb. 4, fig. 4. Lacking metal accoutrements, this device was billed as an "air electrophorus"; it merely induced a charge on nearby objects. Gütle (1790, 139–141) also furnished recipes for an electrophorus's resin disk.

105. Webers 1781. Schäffer (1780) also designed a series of demonstrations around the electrophorus.

106. Baird and Nordmann 1994, 46.

107. A very brief account of the empress's electrophorus is furnished by "T. C." in *The Gentleman's Magazine* 51 (1781):355. Gütle (1790, catalog at end of volume) commercialized some huge examples of the electrophorus.

108. The 1797 edition of the *Encyclopaedia Britannica* commented that "the science of electricity seems to be at a standstill" (quoted in Heilbron 1979, 490).

109. Biographical information on Coulomb is furnished by Gillmor (1971).

110. According to Hackmann (1978a, 89), as early as the 1740s Waitz and Winkler had claimed that the force of electricity decreases away from an electrified "body with the square of the distance." Heilbron (1979, ch. 19) furnishes a history of attempts at quantifying electrical force; see esp. 462–464 on Stanhope.

111. The pith-ball electrometer, which often used cork balls suspended by dry silk threads, was invented by John Canton in 1753 (reprinted in Franklin 1996, 144).

112. The experiments and arguments are presented by Mahon (1779, 56–65); this work is sometimes cited as "Stanhope," but the name on the title page is Mahon.

113. I have adapted my simplified, even oversimplified, description of the torsion balance from Heilbron's dense discussion (1979, 471–473).

114. Among the few other eighteenth-century investigators who also attempted to quantify electrical phenomena were Aepinus (Home 1979) and Cavendish (Maxwell [1879] 1967).

115. For descriptions and illustrations of electrometers, see Hackmann (1978b) and Keithley (1998). An older but still useful source is Walker (1936). Several kinds of electrometers were commonly advertised in instrument catalogs (e.g., Adams 1799, catalog at end, p. 10; Nairne 1793, catalog at end, p. 75).

116. On this point, see Ruestow (1973, 153).

117. Roberts (1999) has furnished an engaging survey of electricity and the electrical machine in the lives of the Dutch during the late eighteenth century.

118. I have obtained biographical details on Martinus van Marum from Muntendam (1969). Van Marum's writings, published and unpublished, have been catalogued by Bruijn (1969).

119. For a catalog of van Marum's scientific instruments in the Teyler Museum, see Turner (1973).

120. Martinus van Marum, quoted by Heilbron (1979, 441).

121. John had a brother, Jonathan, who was also an instrument maker. For information on the Cuthbertsons, see Hackmann (1973).

122. For information on the Teyler Museum's large electrical machine, I have relied on Hackmann (1973, 1978a, 159–161). Another useful source is Dibner (1957, 43–49).

123. Heilbron 1979, 441–443.

124. For an intriguing essay on late-eighteenth-century voyages of discovery as "big science," see Raj (2000).

125. On the Landriani visit, see Hackmann (1978a, 167).

126. As Hackmann (1978a, 167–168) points out, the rift between Cuthbertson and van Marum was also rooted in their differing approaches to measuring the performance characteristics of electrical machines.

127. Hackmann (1971) provides an overview of the wide-ranging experiments conducted with the Teyler machine.

128. Muntendam 1969, 22–23.

129. Sutton (1995) does furnish examples of women scientists of the Enlightenment. I do not wish to imply that the scientific elite consisted only of those who had made major contributions, that everyone who did stellar science received commensurate recognition, or that scientific hierarchies then (or today) depended solely on merit. Rather, I emphasize that the highest ranks of science's social hierarchy were sufficiently flexible to accommodate people whose only credentials to join the elite were their scientific achievements.

130. Although Latin was ostensibly the universal language of literate folk, hence of scientists, few important works of science were published in Latin after the middle of the eighteenth century. Scientists wrote in their native tongue, and their books, if popular, might be translated into several languages. Commonly, the proceedings of elite societies published the most significant findings or at least a secondhand synopsis of them.

131. Cohen (1995) has explored the role of science in the thought of America's founding fathers, but his near-exclusive focus on ideas rather than on material, behavioral, and social factors leaves room for many future analyses.

4. GOING PUBLIC

1. Hankins and Silverman (1995, 12, 43) distinguish among several kinds of demonstration, which I doubtless conflate in my use of the term.

2. On the prevalence of such beliefs, see Fara (1996, 50).

3. Illustrated by Dibner 1957, 23.

4. Franklin 1996, 10. A lucid description of the electric kiss is furnished by Cuthbertson (1807, 33); for an early example, see Watkins (1747, 30).

5. Needham 1746, 252. Early versions of this demonstration were carried out by Stephen Gray using two people (Gray 1732c) or three (Gray 1735b, 168–169).

6. Heilbron 1979, 318.

7. Benz (1989) explores at length possible connections between religious theology and electrical theory and discourse.

8. Crediting Bose, Jallabert (1748a, 50–51) furnished an early description of the electric beatification.

9. Franklin 1996, 30.

10. See Rattansi 1972, 22.

11. On the adoption of the Newtonian paradigm of experimental physics in colleges and universities in the seventeenth and eighteenth centuries, see Ruestow (1973) and Heilbron (1979, 134–147). The importance of technology-intensive demonstrations for illustrating principles and phenomena in early Newtonian physics is especially evident in the textbooks of Desaguliers (1719) and 's Gravesande (1731).

12. See, for example, Adams (1771, catalog at end), Adams et al. (1746, catalog at end).

13. Information in this paragraph comes from Heilbron (1979, 147–152).

14. See 's Gravesande 1731, 2:2–13.

15. Musschenbroek 1739b, 363–379 (for the Dutch version, see Musschenbroek 1739a). Electricity was no better represented in the English edition of 1744: 11 pages or 1.7 percent (Musschenbroek 1744). In Musschenbroek (1751), 17 pages were devoted to electricity, a discussion that took up less than 2 percent of the book.

16. Musschenbroek 1769. Musschenbroek's English counterpart, gifted lecturer and Newtonian J. T. Desaguliers, also published a physics textbook in several editions. The second edition (Desaguliers 1719) contained nothing on electricity. The third (corrected) edition (Desaguliers 1763) contained very little on electricity, nothing beyond that present in his dissertation (Desaguliers 1742). In addition, Pierre Polinière in the 1741 edition of his textbook allotted electricity a mere five pages in 2 vols., and Helsham (1755) makes no mention of electricity.

17. Nicholson 1782.

18. Quoted in Brinitzer (1960, 120).

19. Muntendam 1969, 14.

20. Bazerman 1991. Although Priestley's book was the first lengthy history of electricity published in English, a much earlier, two-volume treatise had appeared in French (*Histoire* 1752).

21. Biographical information on Joseph Priestley comes from Schofield (1975) and Schofield's introduction to Priestley ([1775] 1966). For a creative attempt to show the relation of Priestley's science to his theology and politics, see Eshet (2001). Other useful sources on Priestley's scientific activities are Golinski (1992, chs. 3 and 4) and McEvoy (1979).

22. On the Birmingham Lunar Society, see Schofield (1963).

23. See Cohen 1941, 20.

24. Van Doren 1938, 418.

25. Van Doren 1938, 521.

26. In seeking recompense for his losses, Priestley prepared an inventory of the apparatus in his laboratory; the long list, which includes many electrical items, is reprinted in Bolton (1892).

27. Early electrical experiment books include Boullanger (1750), Desaguliers (1742), Doppelmayr (1744), Gordon (1745), Martin (1746), Morin (1748), Rackstrow (1748), Sguario (1746), Turner (1746), Watkins (1747), and Watson (1746a).

28. Typical of later eighteenth-century experiment books are Adams (1792, 1799), Aragão (1800), Bauer (1770), Cavallo (1795), Cuthbertson (1807), Donndorff (1784), Faulwetter (1791), Langenbucher (1780, 1788), Marat (1782), Rabiqueau (1785), Sigaud de La Fond (1776), and Webers (1781). Although not typical of its time, Milner's (1783) small experiment book is of interest because he emphasized the ease with which his devices could be constructed at home.

One can still buy experiment books featuring eighteenth-century demonstrations (e.g., Bakken 1996).

29. Adams et al. 1746, 258–259.

30. The most useful original source on eighteenth-century demonstration devices in English is Adams (1799). Not only is it lavishly illustrated with five plates consisting of a total of 105 figures, but each figure is discussed individually in a list (139–154).

31. Webers (1781) designed his experiments around the electrophorus.

32. For examples, see Follini (1791), Jacquet (1775), and Mazzolari (1772).

33. Priestley (1966, 1:106–118) compared various machines (excepting plate machines) on the basis of several utilitarian performance characteristics.

34. On the interpretation, by modern investigators, of symbolic artifact functions, see Schiffer and Miller (1999).

35. Electric fountains were based on early experiments by Gray (1732a, 229) as well as Desaguliers's (1742, 16) "artificial Fountain." Electric fountains were described and illustrated, for example, by Langenbucher (1780, 89–90, tb. 4, fig. 4); on the urinating fountain, see Ribright (1779, pl. 2, fig. 8). For examples of the electric horse race, see Faulwetter (1791, tb. 2, fig. 11) and Langenbucher (1788, tb. 4, fig. 182). Beneath the circular track, the horses were attached to the rotor of an electrostatic motor. A seesaw—essentially a rocking beam motor—is illustrated by Veau Delaunay (1809, pl. 3, 27). The tightrope walker is shown by Faulwetter (1791, tb. 1, fig. 20) and Langenbucher (1780, tb. 4, fig. 159). The "small head with hair" was available from Adams (1799, catalog at end, p. 9) for 7s. 6d. (see also Ribright 1779, pl. 2, fig. 11).

36. Langenbucher (1780, tb. 7, figs. 2 and 3). Among early investigators reporting the electrical ignition of gunpowder was Jallabert (1748a, 48).

37. Adams (1799, catalog at end, p. 9) sold electric pistols (7s. 6d.) and cannons (16s.), both using hydrogen.

38. For an illustration of the exploding wine glass, see, for example, Langenbucher (1780, tb. 3, fig. 10).

39. This effect will not be seen in a perfect vacuum because it requires that some gas be present. Examples of zig-zag tubes can be seen in Langenbucher (1780, tb. 6, figs. 15 and 16).

40. For discharge tubes and vessels, see Adams (1799, pl. 3, fig. 49-CD; pl. 4, figs. 59–62). Adams (1799, catalog at end, p. 9) also sold a set of accessories for creating these lighting effects: "An useful and illustrative apparatus, compounded of the luminous conductor, exhausted flask, two jars, exhausting syringe, insulated stand, and wires with balls, &c"; it sold for £3. Luminous conductors themselves fetched 12s. to £1 5s. For an example of a vacuum pump used to evacuate glass tubes and vessels, see Langenbucher (1780, tb. 6, fig. 1).

41. Adams (1799, catalog at end, p. 9) brought the "electrical shooter and mark" to market for 5s.

42. For an example of the "luminous palace," see Veau Delaunay (1809, 57–58, pl. 4, fig. 44).

43. Foil-on-glass devices were sold, for example, by Adams (1799, catalog

at end, p. 9): "a spotted jar" sold for 6s. to 10s. 6d.; "spiral tubes to illuminate by the spark," which cost from 6s. to 10s. 6d.; and "luminous names, or words," ranging in price from 10s. 6d. to £1 11s. 6d. Luminous words were also illustrated by Langenbucher (1780, tb. 8, fig. 5). On the use of electric lighting to communicate words in the late nineteenth century, see Marvin (1988).

44. Cuthbertson 1807, 75–76.

45. Rabiqueau 1785.

46. Portable electrical apparatus was sold, for example, by Adams et al. (1746, 259) and Benjamin Martin (Millburn 1976, 222). A portable demonstration outfit, complete with case, can be seen today at the Burndy Library, Dibner Institute, Massachusetts Institute of Technology.

47. Recent sources describing and interpreting the activities of demonstrator-lecturers include Golinski (1992), Millburn (1976), Schaffer (1983), and Sutton (1995). Rueger (1997) draws a connection between aesthetics and public demonstrations, especially of electricity.

48. On eighteenth-century amusements in England, most of which had counterparts on the continent and in the colonies, see, for example, Altick (1978) and Hibbert (1987, 408–427).

49. On natural magic and its technologies, see Hankins and Silverman (1995); on early automata, see Chapuis and Droz (1958).

50. Gütle (1791) was one electrical demonstrator and instrument maker who explicitly situated his experiments in the natural-magic tradition. Demonstrators working in the tradition of natural magic also assimilated electrical displays into their repertoires (e.g., Hooper 1774, vol. 3; see also Ozanam 1790, vol. 4, revised posthumously).

51. Sutton (1995) explores lucidly and at great length the emphasis, in eighteenth-century scientific lectures, on demonstrating phenomena or effects. Morus (1998) carries the story into the early nineteenth century, focusing on London.

52. An historical account 1745, 194.

53. Nollet 1745–1765, vol. 6, pl. 2 and pl. 4, fig. 14.

54. Sguario 1746, frontispiece.

55. An historical account 1745, 194.

56. An historical account 1745, 194.

57. Émilie du Chatelet, quoted in Sutton (1995, 225).

58. Nollet 1754, frontispiece.

59. Martin 1743, preface; original emphasis. For a recent biography of Martin, see Millburn (1976).

60. Martin 1759b. Dialogues 6 and 7 are on electricity, 299–327.

61. Martin 1759b, pl. 25.

62. On the participation of women in science education, see Schiebinger (1989).

63. Franklin to Collinson, March 28, 1747, in Smyth (1970, vol. 2).

64. Versions of these motors were offered, for example, by Adams (1799, catalog at end, p. 9; see also pl. 3, fig. 34).

65. For examples of electric carillons, see Langenbucher (1788, tb. 4, fig. 182) and Adams (1799, pl.2, fig. 19).

66. Franklin 1996, 31–32.

67. Among those constructing and/or illustrating Franklin-type motors are Jacquet (1775, 185–187, pl. 6, fig. 54) and Langenbucher (1780, tb. 3, fig. 9).

68. The "self-moving wheel" is described by Franklin (1996, 32–34).

69. Franklin 1996, 29.

70. Franklin 1996, 40.

71. Adams 1799, catalog at end, p. 9.

72. Biographical information on Ebenezer Kinnersley comes from Cohen (1941, 401–408) and Lemay (1964).

73. Kinnersley, quoted in Lemay (1964, 20–21).

74. Lemay 1964, 92.

75. Lemay 1964, 5. Cohen (1990, 57) reports that the first public lectures on science in the colonies were delivered, beginning in 1727, by Isaac Greenwood, a native American, who a year later became the first Hollis Professor of Mathematics and Philosophy at Harvard.

76. This advertisement, of August 24, is reproduced by Lemay (1964, 60).

77. Lemay 1964, 61.

78. For information on Dömjén, see Lemay 1964, 61–62.

79. It had been believed that Kinnersley's first lectures were those advertised in the *Pennsylvania Gazette* on April 11, 1751; however, Lemay's (1964, 63) evidence for the earlier Annapolis lectures is definitive.

80. Morse 1934, 374. Morse (1934) reports that the following men performed electrical demonstrations in the Boston area: William Claggett, a clockmaker; Captain John Williams; Daniel King, in Salem; Ebenezer Kinnersley; Joseph Hiller, jeweler; David Mason; William Jones; and William Johnson.

81. The first Kinnersley lecture, excepting two missing pages, is reproduced by Cohen (1941, 409–421).

82. For a description of the artificial spider, see Franklin (1996, 10–11) and Kinnersley in Cohen (1941, 416–417).

83. The artificial spider was sold, for example, by Adams (1799, catalog at end, p. 9) for 1s. 6d.

84. Kinnersley in Cohen (1941, 421).

85. I have used the advertisement in the *New York Gazette* of June 1, 1752; it is reproduced, at a scale almost too small to read, by Lemay (1964, frontispiece).

86. Kinnersley in Lemay (1964, frontispiece).

87. Langenbucher (1780, tb. 5, fig. 3) had a model village, with the vulnerable church in the middle. He also displayed a ship (1780, tb. 5, fig. 4) and other structures (1780, tb. 5, fig. 1; tb. 8, figs. 6 and 14). Model structures for illustrating the use of lightning conductors were commercialized, for example, by Adams (1799, catalog at end, p. 9), who offered a house (6s.), a powder house (16s.), and "obelisk or pyramid" (10s. 6d.).

88. Lemay (1964, 109–110) reproduced Kinnersley's last advertisement in the *Pennsylvania Gazette*, published on December 29, 1773.

89. This incident is described at length by Van Doren (1938, 467–478).

90. Lemay 1964, 110–111.

91. Lemay 1964, 82–83.

92. Lemay 1964, 87. Two centuries later Cohen (1956) would compare Franklin and Newton.

93. This price comes from an advertisement in the *Maryland Gazette*, transcribed by Lemay (1964, 68–70).

94. Kinnersley in Lemay (1964, frontispiece).

95. Lemay 1964, 67.

96. Musschenbroek 1744, 1:x.

97. *Pennsylvania Gazette*, April 11, 1751, quoted by Lemay (1964, 77). William Johnson, another American electrical lecturer, appropriated Kinnersley's rationale almost verbatim: "As the Knowledge of Nature tends to enlarge the human Heart, and give us more noble and exalted Ideas of the God of Nature, it is hoped that this Course will prove, to many, an agreeable entertainment" (from a Johnson advertisement, quoted in Morse [1934, 372]; see also Cohen [1952, 424]). Benjamin Martin, also an itinerant lecturer, offered a similar argument: "*Does not every one know, that* RELIGION *is founded wholly in* PHILOSOPHY, *or a true Knowledge of Nature? Is it not from thence that we have the first and truest Notions of the Divine and Adorable* CREATOR *of all Things, and the most direct and cogent Arguments for his Existence and Perfections?*" (Martin 1743, preface; original emphasis).

98. *Pennsylvania Gazette*, April 11, 1751, quoted by Lemay (1964, 77; original emphasis).

99. *New York Gazette* of June 1, 1752. The testimonial is reproduced in a readable form by Lemay (1964, 77).

100. Lemay 1964, 75.

101. Paulian 1781, vii–viii; my translation.

102. On the archaeological discoveries, see Grayson (1983).

103. On the importance of electrical demonstrations for underscoring tenets of elite Enlightenment ideology, see Sutton (1995).

5. POWER TO THE PEOPLE

1. This engraving is reproduced by Multhauf and Davies (1961, fig. 1). Their caption reads: "An engraving by Cornelius Tiebout, thought to be taken from a lost portrait painted by Rembrandt Peale in 1800. The static electrical machine on the table and the globe beneath it are probably items 58-39 and 58-40 from the collection catalogued here [i.e., by the American Philosophical Society]."

2. On the influence of science on Thomas Jefferson's construction of the Declaration of Independence, see Cohen (1995, ch. 2). Cohen (1995, 321, note 5) also furnishes additional bibliographic sources on Thomas Jefferson as scientist.

3. Boyd 1950–1972, 7:303.

4. Boyd 1950–1972, 7:622. The actual entry is "little electrical machine," but Lucia C. Stanton, Senior Research Historian at Monticello, suggests that it was a "pocket shocker" (personal communication, 2000).

5. There is precious little evidence that women in the eighteenth century purchased electrical equipment. However, at the beginning of the electrical craze of the 1740s, the following curious account appeared: "The Hanoverian ladies of quality did yet more, they procured machines, and try'd the experiments themselves, and electricity took the place of quadrille" (An historical account 1745, 194).

6. Other elite individuals sometimes made their equipment available for use by scientists. For example, Priestley, in many of his experiments on gases, used "a very powerful" Nairne electrical machine that had been built for his benefactor, Lord Shelburne (Priestley 1779, 285). Another interesting case is van Marum, who had at his disposal electrical apparatus, which he helped design, that had been paid for by Pieter Teyler's estate (see chapter 3). Turner (1987) discusses collections of scientific instruments made by Enlightenment elite.

7. In a survey of collecting behavior in Tucson, Arizona, we found that 62.5 percent of the sampled households contained one or more collectors (Schiffer et al. 1981).

8. On the prices of mechanical orreries, see Adams (1771, catalog at end, pp. 12–13); on air pumps and accessories, see Adams (1771, catalog at end, pp. 10–11).

9. Fortescue [1927–1928] 1967, 1:333. Barometers would have been used to determine elevation.

10. On the other hand, collections sometimes take control of the collector's life. As folklorist Henry Glassie (1999, 84) has put it, "Collecting can be seen as a neurotic disorder or as a heroic attempt to create some order, some place of personal control and satisfaction, in a world gone haywire."

11. On the contribution of collections to the establishment of science museums, see Bedini (1965).

12. On the history of George III's instrument collection, see Morton and Wess (1993, 5–37).

13. Biographical information on Stephen Demainbray can be found in Morton and Wess (1993, 89–119); on his lecturing activities, see Morton (1990).

14. The 24-page text of Demainbray's lectures is reproduced by Morton and Wess (1993, 128–133); they also describe and illustrate the portion of George III's instrument collection that came from Demainbray (135–241).

15. This phrase appears on the title page of Adams's science books and instrument catalogs.

16. For the listing of eighteenth-century electrical apparatus, see Chaldecott (1951, 68–76) and Morton and Wess (1993, 505–520); the latter publication includes beautiful images of the electrical machines.

17. Similar machines could be bought from Nairne in the late eighteenth century.

18. Fortescue 1967, vols. 1 and 2. However, Pain (1975, 33) claims that "George was more at home with the sciences and with handicrafts," than with art or literature, but that is hardly a ringing endorsement of his scientific enthusiasm. Perhaps the king's scientific correspondence, if such existed, was not saved or is still to be discovered.

19. Pain 1975, 31.

20. It is possible that the earl of Bute (himself an avid collector of scientific instruments, see below) had, in his role of tutoring the prince and preparing him for the role of king (see McKelvey 1973, 84), counseled him to collect.

21. Millburn 2000, 102.

22. The inventory of Bute's instrument collection, along with information on his diverse collecting interests, comes from Turner (1991, ch. 11). For additional biographic information on Bute, see McKelvey (1973) and the essays in Schweizer (1988).

23. On Bute's scientific interests and his relationship with Peter Collinson, see Miller (1988).

24. Pain 1975, 12.

25. The two eighteenth-century Dollond and Dollond catalogs I have consulted are regrettably undated, but they mention the availability of telescopes up to 8 feet long with mahogany tubes. They were sold by the foot, with prices ranging from £2 per foot for a 2-foot model to £10 per foot for one 4 feet long. At this rate of increase, Bute's 10-foot model would have had a base price of several hundred pounds; upgrades in optics and so forth could have raised the price still further.

26. Turner 1991, 238.

27. Daumas 1989, 142.

28. On this machine, see Hackmann (1978a, 151 and pl. 24); however, its maker is unknown.

29. Cf. Daumas 1989, 137.

30. Adams 1799, catalog at end, p. 9.

31. Gütle 1790, catalog at end.

32. Nairne 1793, 74.

33. Unless otherwise noted, these economic data come from Hibbert (1987, 308, 320, 352, 409, 411, 451). I am relying on Hibbert's conversion factor of 60.

34. On the wages of bishops and other high clerics, see Fortescue (1967, 1:33–44); on salaries at the bottom rung of the clergy, see Hibbert (1987, 310–311).

35. Hobsbawm ([1962] 1996, 28) places the beginning of the industrial *revolution* in Britain at about 1780, but his criterion is a "take-off" in economic growth. The process of industrialization had, of course, begun rather earlier.

36. Hibbert 1987, ch. 42.

37. Fortescue (1967, 1:42, 44) lists six Crown-supported professors at Oxford and Cambridge who earned in 1762 between £100 and £350 per year. Not

surprisingly, administrators earned vastly more. The master of Trinity College at Cambridge, for example, took in £800 (44).

38. Early on, 's Gravesande (1735) had aimed a book on Newtonian physics at "the youth of Leyden." By the middle of the eighteenth century, natural philosophy was regarded as an appropriate subject in which to instruct children (Secord 1985). By the end of the eighteenth century, there was already a genre of books targeted at young, amateur scientists (e.g., Telescope 1798).

39. Priestley 1786, 6–7, 71–72.

40. Priestley 1966, 2:119–133. By the end of the century, the existence of "young electricians" was taken for granted, as in Imison's (1796, 81) advice to them on diagnosing the causes of failed experiments.

41. A few biographical details on Seiferheld can be found in Poggendorff (1863, 2:898).

42. Seiferheld 1791a, 6.

43. He explicitly acknowledged his debt to the works of Langenbucher, Cavallo, G. Adams and others (Seiferheld 1802b, preface).

44. Seiferheld 1802a, 51–86, and tbs. 6–9.

45. On assembling a Leyden jar, see Seiferheld (1791a, 7–11, and tbs. 1–2).

46. Seiferheld 1796, 27–32, and tb. 3, figs. 18–22, and tb. 4.

47. Seiferheld 1796, 15–21, and tb. 1.

48. Electric cannon: Seiferheld (1791c, 41–42, 87, and tb. 2, figs. 12, 38); pistols (1791c, 80–83, 85–86, and tb. 4, figs. 32–34, 36–37); flame-thrower/mortar (1791b, 76–77, and fig. 39; 1791c, 88, and tb. 4, figs. 39–40); flintlock (1791c, 89–94, and tb. 4, figs. 41–43).

49. The hydrogen generator is presented in Seiferheld (1791c, 78–80, and tb. 4, fig. 31); on obtaining a good mix of hydrogen and air, see Seiferheld (1791c, 82–83).

50. For discussions of this process, see Schiffer (1991, 35–37, 153–160, 162). It would be instructive to examine the youthful activities of the people who developed commercial electrical technologies in the early and middle of the nineteenth century.

51. Biographical information on Percy Shelley is taken from Cameron (1950) and King-Hele (1984).

52. Quoted by King-Hele (1984, 3).

53. Walker's lectures were published as a physics text (Walker 1799); typical for its time, the book devoted 12.4 percent of its contents to electricity; it also contained 60 figures of electrical apparatus in several plates, including many display devices, from carillon to thunderhouse.

54. King-Hele 1984, 4.

55. Leigh Hunt, quoted in King-Hele (1984, 12).

56. Biographical information on Erasmus Darwin comes from King-Hele (1977).

57. E. Darwin to M. Boulton, 11 March 1766, in King-Hele (1981, 39).

58. King-Hele 1984, 160–161.

59. Thomas Jefferson Hogg, quoted in King-Hele (1984, 161).

60. King-Hele 1984, 159.

61. As today, the percentage of Enlightenment youth who tinkered with electrical things was small, but such people sometimes have a disproportionate influence on human affairs. They can bring into adulthood visions of desirable products and sometimes invest resources into bringing them to fruition. See, for example, my discussion of Norman Krim and the first shirt-pocket radio (Schiffer 1991, 161–165, 1993).

62. Biographical information on Mary Shelley comes from Grylls (1938) and Nitchie (1953).

63. Shelley [1831] 1995, 226.

64. Shelley 1995, 227.

65. Shelley 1995, 228.

66. Shelley 1995, 228.

67. See, for example, Small (1973).

6. LIFE AND DEATH

1. Poggendorff (1863, 1:1188) supplies a few biographical details on Jallabert.

2. Benguigui (1984) has published the correspondence between Nollet and Jallabert.

3. Jallabert's experiments on plants are described in Jallabert (1748a, 80–85).

4. Browning (1747, 374) had also reported the former effect, which Bertholon (1783, 136–137) referred to as "beatification."

5. On the sensitive plant, see Freke (1752, 117–119).

6. Nollet 1749b, 358–361. Pl. 2 indicates that the pots were placed on a metal shelf, suspended by a silk cord, and charged by a chain dangling just above the globe on a large Nollet machine.

7. Dulieu (1970, 82–83) presents a few biographic details on Bertholon.

8. A decade later Read (1793) also discussed the influence of atmospheric electricity on plants (and animals).

9. Bertholon 1783a, 11; my translation.

10. Bertholon 1783a, 17–18.

11. For his arguments on atmospheric electricity and plants, see Bertholon (1783a, 31, 41, 49–58, 81–91, 96–97, 216).

12. Bertholon 1783a, 98–99.

13. Bertholon 1783a, 99; my translation.

14. Bertholon 1783a, 100–102.

15. Bertholon 1783a, 111–113; my translation.

16. Bertholon 1783a, 28–29.

17. Bertholon 1783a, 29; my translation.

18. These experiments are described by Bertholon (1783a, 33–34).

19. Bertholon 1783a, 145–146. Later experiments by other investigators supported these findings; see, for example, Beck (1787, 131–147).

20. Bertholon 1783a, 151–165.

21. Bertholon 1783a, 258–259.

22. Priestley 1966, 1:178.

23. Information on van Marum's shellac-and-mercury-trough machine comes from Hackmann (1978a, 165–167). Beccaria (1776, 9) describes a plate machine that also rotated in a trough of mercury. The difficulty with this design was the tendency for the mercury to splash out of the trough when the plate rotated rapidly (Beccaria 1776, 66–67).

24. On the spring-driven electrical machine, see Toaldo's (1783) brief note. For an early biography of Toaldo, see Life of Joseph Toaldo (1801).

25. I have relied on Thomson's (1840, 575) account of Ingen-Housz's experiment.

26. See, for example, Sidaway (1975).

27. Nollet 1747, 77; my translation.

28. These experiments on animals are described by Nollet (1749b, 366–372).

29. Nollet made no mention of weight loss through respiration, but it is possible that his term *transpiration* subsumed both perspiration *and* respiration.

30. Priestley (1966, 1:173–176), for example, discussed them at length.

31. Barneveld 1787, 48–52. For the original Dutch version, see Barneveld (1785). For a very early use of a data table, showing the extension of the electric virtue from various rubbed substances, see Giovanni (1759, 16).

32. I have inferred the apparatus from Barneveld's illustrations (1787, tbs. 1 and 2) and his mention of the "electric bath," a benign mode of electrifying a person (see chapter 7). Some later investigators, conducting similar experiments, obtained no change in pulse (e.g., Rabiqueau 1785, 123; van Marum [1795] 1974e, 160–165).

33. Martin 1746, 31. Jallabert (1748a, 75–78) also carried out this experiment on human subjects.

34. A description of Priestley's experiments is furnished in Priestley (1966, 2:253–259).

35. Priestley 1966, 2:255. The dog experiment is described on 255–257.

36. I have obtained this account from Read (1793, 71–72). He gave the name of the experimenter as "Richard Hen. Alex. Bennet," a name that does appear in the membership roll of the Royal Society in 1781 (*Philosophical Transactions of the Royal Society* 71, list). I presume that this person is one and the same as the Reverend Abraham Bennet, who was a very active electrical experimenter in the 1780s (see chapter 9).

37. Franklin, quoted in Van Doren (1938, 161).

38. Franklin, quoted in Van Doren (1938, 161); another account of this incident can be found in Franklin (1996, 153–154).

39. Franklin 1996, 153.

40. Priestley 1966, 2:132.

41. *Gymnotus* is a ray, not a fish, but I employ the less precise term as it was used by Cavendish and others. On early accounts of the torpedo, see Falconer (1790, 280–283).

42. Walsh 1773.

43. For biographical information on Cavendish, see Jungnickel and Mc-Cormmach (1996) and McCormmach (1971).

44. Beccaria (1776, 97–101) also used Leyden jars in various series/parallel combinations as voltage and current dividers.

45. Cavendish 1776, 203.

46. Prior to the invention of the galvanometer, early in the nineteenth century, there was no instrument, per se, available for measuring current. That is why so many investigators, including Cavendish and the great Volta, used their own bodies and senses to indicate relative amounts of current. On the growing importance of current in early nineteenth-century electrical studies, see Brown (1969).

47. These experiments are described, and arguments based on them presented, by Cavendish (1776, 197–200).

48. Franklin's views are encapsulated in the following statement: "When this [electric] fluid has an opportunity of passing through two conductors, one good, and sufficient, as of metal, the other not so good, it passes in the best, and will follow it in any direction" (Franklin 1996, 480).

49. Cavendish actually made two models of the torpedo. The first, which had a wooden backbone, did not serve well, and so I do not describe it here.

50. This statement is in Cavendish's words (1776, 219), but it presumably paraphrases John Walsh's comment.

51. Cavendish 1776, 223–224.

52. For more extensive discussion of the limitations of verbal communication, see Schiffer and Miller (1999).

53. For a very brief account of this event, from Cavendish's notes, see Maxwell (1967, xxxviii). This kind of scientific show-and-tell remains important in paleontology and archaeology, where the integrity of highly controversial deposits and associations of finds must be demonstrated in the field to skeptics (on the history of such demonstrations in archaeology, see Grayson [1983]).

54. Cohen (1954b, 25) mentions such experiments conducted by Beccaria and Caldani.

55. This problem was framed, for example, by the physician Robert Whytt (Whytt 1768, ch. 1).

56. In 1755 Haller outlined six criteria that had to be met by candidates for the nervous fluid (quoted in Cohen 1954b, 24). On various eighteenth-century theories of the nervous system, see Pera (1992, 55–60).

57. See Cohen (1954b, 23–24).

58. Beccaria, quoted in Pera (1992, 58).

59. See Cohen (1954b, 25–26).

60. The quote is from Fothergill (1796, 79). Berdoe (1771, 79–80) was another early exponent of the view that nerves convey electricity. An even earlier suggestion of this possibility came from Hales (1733, 59), who was familiar with Stephen Gray's experiments on the communication of electricity.

61. Biographical information on Galvani can be found in Bresadola (1999),

Brown (1972), Cohen (1954b), Dibner (1971), Pera (1992, ch. 2), and Simili (1999).

62. On Galvani's home laboratory, see Bresadola (1999).

63. Bresadola (1999, 73–74) calls attention to the connection between Galvani and Veratti, a Bolognese physician who also carried out diverse electrical experiments (Veratti 1748, 1750). In discussing Galvani's research on electricity and muscular motion, I have relied mainly on Galvani ([1791] 1953) and Pera (1992). Galvani's research has been the subject of a huge modern literature in the history of science (e.g., Hoff 1936; Home 1970; Kipnis 1987; and see other notes below).

64. Galvani also carried out experiments using apparatus that drew charge from atmospheric electricity; I discuss similar apparatus in chapter 8.

65. Galvani (1953, 74), in the conclusion to his monograph, argued that "perhaps the hypothesis is not absurd and wholly speculative which compares a muscle fibre to something like a small Leyden jar or to some other similar electrical body charged with a twofold and opposite electricity, and by comparing a nerve in some measure to the conductor of the jar; in this way one likens the whole muscle, as it were, to a large group of Leyden jars."

66. Galvani 1953, 79. Seligardi (1999, 94–95) points out that Priestley also believed that the brain played a role in forming electric fluid.

67. See, for example, Fowler (1793). Jacyna (1999) discusses the early British investigations of animal electricity. See Trumpler (1999) for discussions of the galvanic research undertaken in Germany by Johann Ritter and Alexander von Humboldt.

68. On the French reactions to Galvani's research and the assorted commissions, see Blondel (1999).

69. Valli 1793, 11–28.

70. Valli 1793, 44, 48.

71. Valli 1793, 43.

72. Valli 1793, xv.

73. Valli 1793, 86–88, 109.

74. Valli 1793, xv–xvi.

75. In England, an animal protection act was enacted by Parliament in 1822, and the Royal Society for the Prevention of Cruelty to Animals was founded in 1824. Needless to say, much scientific research remained beyond the reach of such protections.

76. Valli 1793, 104, 83, 86–87.

77. Valli 1793, 85–87, 95–104.

78. Valli 1793, 94.

79. I have relied on Pera's (1992, 102–103) account of this experiment, which quotes extensively from Volta.

80. In exploring galvanic phenomena, British investigators were especially creative in placing pieces of metal on their person. Richard Fowler, for example, accustomed an eye to tolerate contact with a silver pencil. When this pencil touched a piece of tin-foil on his tongue, Fowler reported perceiving "a pale

flash of light diffuse itself over the whole of my eye" (Fowler, quoted in Jacyna 1999, 172). Fowler also produced optical effects by a less direct route. With a piece of zinc on his tongue, he put a silver rod into a nostril and pushed it up as far as possible. When he brought the two metals into contact he saw a flash of light (Jacyna 1999, 173). Perhaps the most daring self-experimenter was John Robison, who did not fail to insert metal pieces into every bodily orifice (Jacyna 1999, 176). Jacyna (1999, 176) suggests that Robison crossed the "fine line between self experiment and sado-masochism."

81. Pera 1992, 107.

82. Volta, quoted in Pera (1992, 110).

83. Pera 1992, 113.

84. Pera 1992, 119.

85. Pera 1992, 120–122.

86. Pera 1992, 122–123.

87. Pera 1992, 142–143.

88. I have relied on Volta (1800) for discussions of these new technologies. More detailed descriptions of how he actually carried out the experiments are presented in Volta (1802). An English translation of the epochal paper of 1800 is furnished by Dibner (1964, 111–131) alongside his more popular account of the experiments. Pancaldi (1990) provides a useful account of activities leading up to Volta's invention of the electrochemical battery, stressing that Nicholson's suggestions may have made a crucial contribution.

89. Volta 1800, 406; see tb. 17, fig. 3.

90. Volta 1800, 418; see tb. 17, fig. 2.

91. For an early example, see Moratelli (1805, 222–223).

92. Volta 1800, 416–417.

93. Volta 1800, 411–412; see tb. 17, fig. 1.

94. Volta 1800, 426.

95. Volta 1800, 427.

96. Volta 1800, 428–429.

97. He also sometimes called his inventions "electromotors" (Volta 1802, 135), but that name did not catch on either.

98. Volta 1800, 421.

99. Priestley 1802; van Marum 1802.

100. The earliest instrument catalog I have found containing a voltaic pile is appended to Adams (1799, end of volume, p. 10). It is listed as follows: "Volta's newly-invented metallic pile of zinc, silver, &c. plates, that produces spontaneous and repeated electricity, decomposes water, &c." It sold for £6 16s. Although experimenters in England were privy to Volta's paper in 1799, before it was published in the *Philosophical Transactions*, and thus had the jump on others in building piles and carrying out experiments, we can be skeptical that the pile was commercialized quite this early. Most likely, a later catalog—but not much later—was bound with pages from the 1799 book.

101. Volta 1802, 141–142.

102. A very brief biography of Aldini is furnished by Dibner (1970). Ful-

ton and Cushing (1936) supply an annotated bibliography of Aldini's writings on animal electricity and galvanism.

103. Aldini 1819, 16.

104. Aldini 1819, 18.

105. Aldini 1819, 20.

106. For descriptions and sources, see Schechter (1983, 42 – 45).

107. Aldini 1819, 27.

108. Aldini 1819, 26.

109. Pt. 1, *The Gentleman's Magazine* 73 (1803):238–239.

110. Hogg's eyewitness description (quoted by Cameron 1950, 80) mentions a "galvanic trough." Most likely this apparatus was a commercial version of Cruikshank's invention, which employed metal plates, immersed in electrolyte, separated by insulators in a wooden trough.

111. Shelley 1995, 33. The remaining quotes from *Frankenstein* in this section occur, in order, on 34 (two), 45, 46 (two), 47, and 51 (two).

112. Shelley 1995, 228–229. The preface to the first edition was written entirely by Percy Shelley (Shelley 1995, 230).

113. There are many discussions of the science in *Frankenstein;* see, for example, Seymour (2000) and Turney (1998, 19–23).

7. FIRST, DO NO HARM

1. Washington, quoted by Irving (1800, 2:447).

2. For accounts of Washington's fatal illness and medical treatment, I relied on Carroll and Ashworth (1957), Flexner (1972), Irving (1800), and Wells (1927). The statement on Washington's death issued by Drs. Craik and Dick, published in the *Alexandria Times* (December 19, 1799), is reprinted in full by Carroll and Ashworth (1957, 640–641).

3. "George Washington, who developed quinsy (tonsillar abscess) . . . was probably done in by excessive bleeding, 2.5–2.8 quarts within twelve hours, in addition to further dehydration from both fever and calomel cathartics, tartar emetic, and other ineffective remedies" (Shapiro and Shapiro 1997, 18).

4. The term for tracheotomy in the eighteenth century was "bronchotomy" (Fothergill 1796, 170), which had been known since before midcentury. Indeed, one instrument maker (Musschenbroek 1739b, catalog on 2:8) sold a cannula for bronchotomies.

5. For examples of home-remedy handbooks, see Chambers (1800) and Smith (1746); additional references to such books are found in Riddle (1997).

6. Washington's old nemesis, King George III, had an unfortunate but not terminal encounter with accepted medical treatments in 1788. Suffering from the (undiagnosed) psychological and physical effects of porphyria, the king in 1788 was eventually admitted to the madhouse of Rev. Francis Willis. The good reverend "tied his royal patient to his bed, enclosed him in a straitjacket, stuffed handkerchiefs in his mouth to keep him quiet when he was being reprimanded, blistered him when it was considered necessary 'to divert the mor-

bid humours' from his head, and doctored him with a formidable variety of medicines . . . [including] calomel and camphor, digitalis, quinine, and, as an emetic, tartarized antimony which made him so sick that he knelt on his chair fervently praying that he might either be restored to his senses or allowed to die" (Hibbert 1987, 430).

7. Buchan 1809, ix. He goes on to say that changes in medical practice "have been usually opposed by the faculty, till every one else was convinced of their importance. An implicit faith in the opinions of teachers, an attachment to systems and established forms, and the dread of reflections, will always operate upon those who follow medicine as a trade. Few improvements are to be expected from a man who might ruin his character and family, by even the smallest deviation from an established rule." On alternatives to orthodox medicine, see Bynum and Porter (1987).

8. For reference-rich modern discussions of Enlightenment electromedicine, see Licht (1967), Rowbottom and Susskind (1984, 15–30), and Schechter (1983, 26–36). The latter author, however, is unduly dismissive of eighteenth-century electrotherapies. Another useful source is Sutton (1981).

9. The term "nervous juice" comes from Wesley (1771, penultimate page of preface).

10. One of the first to suggest that electricity could be used therapeutically was Krüger (1744, 23). For theoretical justifications of electromedical practice, see, e.g., Becket (1773, 49–55), Bertholon (1786, 1:1–206), Graham (1779), Sigaud de La Fond (1802), and Wesley (1771, 1–41).

11. Information on this case comes from Jallabert (1748a, 127–136).

12. For early accounts of electrical treatments, see Allaman (1758), Baker (1748), Hart (1754), Himsel (1759), Jallabert (1748b), Louis (1747), Mangin (1752), Nollet (1747, 407–424), Pivati (1749), Kratzenstein's 1744 writings in Snorrason (1974), Spengler (1754), Veratti (1748, 1750), and Watt (1751).

13. A brief account of Franklin's electromedical experiences is furnished in a letter of December 21, 1757, to John Pringle, M.D. (Franklin 1996, 359–361).

14. Examples of such lists, often supported by case histories, appear in, e.g., Becket (1773, 65–104; his cases are mostly derived from other practitioners), Bertholon (1786, vols. 1 and 2), Gale (1802), Lovett (1756, 1760), Lowndes (1787), and Wesley (1771, 42–43; 1792, 119). For a somewhat negative assessment of electrotherapies, see Marat (1784). For additional sources on electrotherapy that also contain case histories, original or derivative, see Adams (1799), Cavallo (1780a), Gardane (1768), Gardini (1780), Hall (1806), Hart (1754), Hartmann (1770), Masars de Cazeles (1780–1785), Mauduyt (1784), Molenier (1768), Sans (1772), Schäffer (1766), Sigaud de La Fond (1802), Symes (1771), Veratti (1748, 1750), and Wilkinson (1799, 60–81).

15. Wesley 1771, 42–43.

16. These conditions were highly gendered: women usually suffered from hysteria, whereas melancholy mostly beset men. A valuable source on these treatments is La Beaume (1820), who also provided case histories.

17. Nairne 1793, 75.

18. Bertholon 1786, vol. 2, pl. 2, fig. 9.

19. Schmidt 1784, 72–73, tb. 4, fig. 14.

20. For examples of wood-tipped conductors, see Mauduyt (1784, 70–71, pl. 2, fig. 11), as applied to the eye, see Nairne (1793, 29).

21. Mauduyt 1784, 94–100.

22. For examples, see Adams (1792, pl. 5, fig. 84), Mauduyt (1784, pl. 2, figs. 13, 14).

23. "In most cases it may be proper to begin with drawing of sparks from the part affected, and then proceeding to moderate discharges" of the Leyden jar (Becket 1773, 61).

24. For example, see Cullen (1790, 2:248) and Gale (1802).

25. Becket 1773, 63.

26. The use of handheld directors is illustrated by Adams (1792, pl. 5, fig. 85), Bertholon (1786, vol. 2, pl. 2, fig. 14), Mauduyt (1784, pl. 1, fig. 10), and Nairne (1793, pl. 3, fig. 6, pl. 5, figs. 1, 2).

27. Nairne 1793, pl. 3, figs. 1–6. Pointed conductors are also shown by Bertholon (1786, 2:230–237, pl. 5, figs. 35, 36).

28. Bertholon 1786, 2:185–186; pl. 6, figs. 45–46.

29. Bertholon (1786, 2:186) credits this invention to both Steiglehner and Hiotberg, which suggests that the electric sandal may have been widely copied.

30. The electric bandage (my term) is described and illustrated by Bertholon (1786, 2:182–185, pl. 6, figs. 38–43); he attributes its invention to Steiglehner. A less elegant version is illustrated by Nairne (1793, pl. 5, fig. 1).

31. Bertholon (1786, 2:184; my translation).

32. Lane 1767. For illustrations of the Lane electrometer's use in electrotherapy, see, e.g., Bertholon (1786, vol. 2, pl. 4, fig. 30), Blunt (1797, pl. 2), Mauduyt (1784, pl. 1, fig. 10), and Nairne (1793, pl. 5, figs. 1–2).

33. At best, the Lane electrometer was an indicator of intensity (voltage). However, as Cavendish had demonstrated (chapter 6), a shock's severity was strongly related to quantity (current).

34. Bertholon 1786, 2:188–191.

35. Bertholon (1786, 2:191) suggested calculating the areas covered by metal foils to give a measure of a Leyden jar's *capacité*. This was a common though inaccurate method, for it was known that the thickness of the glass also inversely influenced its capacity (e.g., Cavendish 1776, 206).

36. This machine is illustrated by Lane (1767). Hackmann (1978a, 127–128) discusses four surviving examples of the Read machine. Watkins's compact, portable electrical machine of 1747 might have been purchased for medical applications; he even mentioned this as one possible use (Watkins 1747, end of book, n.p.).

37. Bertholon 1786, 2:230–237, pl. 5, figs. 34–35. Indeed, Nairne (1784) furnished in French a description of his machine along with directions for treating specific diseases. For a detailed description in English, see Nairne (1783; 1793, 1–39) and Blunt (1797, 1–24).

38. The cheapest package cost little more than the ordinary machine for

"common electrical experiments" (£4 4s.), and was vastly cheaper than the large machines, which ranged from £16 6s. to £168 (Nairne 1793, 74). Other instrument makers also sold machines in packages for medical applications. Dollond and Dollond (n.d.), for example, offered two machine packages with cylinders 6.5 and 8 inches in diameter. Both came with "the necessary Apparatus for Medical Uses" and cost £7 7s. and £8 8s.

39. Nairne 1793, 74.

40. The electrotherapist John Wesley (1771, preface) appreciated this problem: "In some Cases, where there was no Hope of Help, it [electrical treatment] will succeed beyond all expectation. In others, where we had the greatest Hope, it will have no Effect at all. Again, in some Experiments, it helps at the very first, and promises a speedy Cure: But presently the good Effect ceases, and the Patient is as he was before. On the contrary, in others it has no Effect at first: It does no good; perhaps seems to do hurt. Yet all this Time it is striking at the Root of the Disease, which in a while it totally removes."

41. As early as 1759, John Wesley (1771, penultimate page of preface) argued that electricity was "the general and rarely failing Remedy, in nervous Cases of Every Kind (Palsies excepted)."

42. Graham 1779, 185.

43. John Wesley's medical activities are but briefly mentioned in most biographies, as if they were an embarrassment (e.g., Bowen 1937, 318; Collier 1928, 32–34; Simon 1927, 222–223; Tyerman 1870, 2:161–162; Vulliamy 1931, 184–185; Winchester 1906, 186–187).

44. Willcox 1966, 128.

45. Quotations in this paragraph are from Wesley (1792, 28, 80, 119; original emphasis).

46. As Wesley (1771, last page of preface) himself admitted, the case histories were mostly derived from the practices of others.

47. Bowen 1937, 318; Tyerman 1870, 2:162; Vulliamy 1931, 185. Vulliamy erroneously claims that Wesley was using a "galvanic apparatus" in 1756. A photograph of one of Wesley's electrical machines, with crucial parts missing, is furnished by Collier (1928, 35).

48. The London Electrical Dispensary, founded in 1793, was a charity operation that served the "Afflicted Poor" (La Beaume 1821, 82). Chambers (1800, 284–285) in his *Pocket Herbal* lists dozens of diseases amenable to electrotherapy; his list is taken, almost word for word, from Wesley (1792, 119).

49. Graham 1779, 37.

50. Graham 1779, 159–160.

51. Graham 1779, 160–161.

52. This room is described in Graham 1779, 161–166.

53. Graham 1779, 161.

54. Graham 1779, 162.

55. See Franklin 1996, 82–85; Watson 1749, 1751–1752b.

56. Graham 1779, 164. The phrase "delicate, nervous and irritable consti-

tutions" was among the jargon used then to describe what are today called psychosomatic or psychogenic conditions.

57. Graham 1779, 164.

58. This room is described briefly by Graham (1779, 166).

59. For a description of this room, see Graham (1779, 166–168).

60. Quotations in this paragraph are from Graham (1779, 166, 167).

61. A description of the electrical apparatus in the Great Apollo Apartment is given by Graham (1779, 168–177).

62. Graham 1779, 168.

63. See *Random House Dictionary*, 1971 ed., s.v. "Apollo," 70. Obviously he was a very busy god.

64. Graham 1779, 170.

65. Quotations in this paragraph are from Graham (1779, 170, 171; original emphasis).

66. Graham 1779, 173–175. The throne is described on 175–177.

67. Graham 1779, 173.

68. The dragon is described by Graham (1779, 175).

69. Graham 1779, 177.

70. I have relied on Schecter's (1983, 34–35) account and illustration of the Celestial Bed; Hibbert (1987, 430) also provides a brief description.

71. James Graham, quoted in Schecter (1983, 34).

72. Schecter (1983, 35) has reproduced one broadside.

73. Graham (1779, 204).

74. Hibbert (1987, 429), writing about the eighteenth century, remarks that "of all quacks none was more celebrated than James Graham" (see also Jameson 1961). Thompson (1929, 333) refers to Graham as the "Emperor of Quacks . . . one of the most extraordinary charlatans of his day."

75. See Huisman's (1999) thoughtful paper on the social construction of quackery in Dutch historiography.

76. Quotations in this paragraph are from Graham (1779, 50).

77. Graham 1779, 51–52.

78. Shapiro and Shapiro 1997, 21–22, 30–31.

79. In the following discussion of factors that promote the placebo effect, I draw upon Ader (1997), Fields and Price (1997), Shapiro and Shapiro (1997), and Spiro (1997), but I emphasize the role of artifacts more than these authors (see Schiffer and Miller 1999, chs. 5 and 7). One of the earliest experiments to demonstrate the power of the placebo effect took place in England in the 1790s. It was the inadvertent by-product of a study designed by Richard Smith to expose metallic tractors—Galvanian metal conductors—as a medical fraud. Patients were deceived into believing that they were being treated by the tractors when in fact the investigator had substituted wooden sticks. Nonetheless, many patients reported improvement. Smith was surprised that "mere imagination" could have a "powerful Influence upon diseases" (Smith, quoted in Jacyna 1999, 182). Smith's letter reporting this experiment is included in Haygarth (1800).

80. Buchan 1809, 552; see also Mauduyt 1784, 174.

81. Hibbert (1987, 405) reports that "Illegitimate children were accepted quite casually both in the upper and in the professional classes." In the lower middle class and indeed throughout the ranks of society, attitudes toward their mothers were probably far less forgiving.

82. William Hunter, in Hawes (1794, 423 – 429); see also Jackson (1996).

83. On herbal technologies for inducing abortions in Western societies, see Riddle (1997). Wood and Suitters (1970, 86) mention abortifacients employed by desperate women in England.

84. Anonymous, quoted in Stopes (1928, 293).

85. Riddle 1997. In seeking additional support for my claim that women in the eighteenth century obtained electrical abortions, I suggest that historians with some knowledge of Enlightenment electricity scrutinize women's diaries; perhaps previously obscure passages would now make sense.

86. Hawes (1794, 51) is my source for the case of Sophia Greenhill. Another account of this case was furnished by Kite (1788, 165–166); although his account differs in details from that of Hawes, the essentials are the same: a young girl falls, apparently to her death, but is revived by electrical stimulation.

87. "Apparent death" was the term applied optimistically in the late eighteenth century (and beyond) to people showing diminished life signs, such as an absence of pulse and respiration. In Fothergill's words (1796, 56; original emphasis), "apparent death . . . may be defined as A TEMPORARY suspension of the vital motions."

88. For a general history of emergency medicine that acknowledges the early role of electrical technology, see Eisenberg (1997).

89. For a brief, early history, see Royal Humane Society (1809, 1–4).

90. Hawes 1794, 6; see Royal Humane Society (1809, 1–4).

91. On the history and activities of the Massachusetts Humane Society, see Howe (1918).

92. The Royal Humane Society also offered, beginning in the mid-1780s, "prize medals for the best dissertations on subjects which had a direct reference to suspended animation" (Royal Humane Society 1809, 6).

93. Fothergill (1796, 169–173) furnished in detail an ideal list of apparatus, which makes fascinating reading today. For a discussion of the medical assistants of the London Humane Society, see Hawes (1794, 13–17).

94. Use of this evocative term can be seen, for example, in Hawes (1794, 51) and Royal Humane Society (1809, 6).

95. An example of a broadside from the Humane Society of Philadelphia, "Directions for recovering persons who are supposed to be dead," is on deposit at the Bakken Library, Minneapolis, Minnesota.

96. The term "artificial respiration" was used by Fothergill (1796, 112).

97. Fothergill 1796, 170.

98. Fothergill 1796, 123.

99. Curry 1792, 56.

100. D. Stephenson made the suggestion in *The Gentleman's Magazine* 17

(1747):183: "As the signs of death are uncertain, so long as none of the vital organs are destroy'd, nor any indications of a beginning of general putrefaction, and as there are instances of persons reputed irrecoverably dead, who have been restored to life; among other proper methods for that purpose, will not the operation of bronchotomy, and injecting the ethereal vapour, together with air into the lungs . . . be of real use for restoring to life persons newly dead of syncopes, apoplexies, cold, hunger, damps, hard-drinking, over-doses of opium &c?" By "ethereal vapour" he meant electricity, with which he had considerable familiarity (see chapter 11).

101. The quote is from Hawes (1794, 51); for Henly's letter, see Hawes (1794, 63–65).

102. Eisenberg 1997, 66–67.

103. Fothergill 1796, 128. Curry (1792, 57) recommended that "In order to more certainly pass the shock through the heart, place the knob of one discharging rod above the collar-bone of the right side, and the knob of the other above the short ribs of the left."

104. The body could be insulated by placing it on a door supported by quart bottles (Curry 1792, 57).

105. Hawes 1794, 51.

106. Accounts of successful electrical resuscitations: a man struck by lightning in his home and revived electrically at Guy's Hospital (reported in a letter from Anthony Fothergill, reproduced in Hawes 1794, 282–285); a young man who fell from a window and was revived electrically by a neighbor (as retold by Curry 1792, 83).

107. For a complaint about unskilled operators, see Hawes (1794, 488). Kite (1788, 164) lamented that electricity had not been employed more often "in the recovery of those who have suffered from drowning and other causes, as there is certainly the fairest prospect of its being attended with the most happy and successful event."

108. Fothergill 1796, 171; original emphasis. Curry (1792, 58) describes a more elaborate pair of directors or "discharging rods" as follows: "Two thick brass wires, each about eighteen inches long, passed through the two glass tubes, or wooden cases well varnished, and having at one end a knob, and at the other a ring to fasten the brass chain to." Such conductors were also described and illustrated by Kite (1788, 192, pl. 1).

109. Mr. Fell's machine was described and illustrated in *The Gentleman's Magazine* 62 (1792):299–300, pl. 2, fig. 4.

110. *The Gentleman's Magazine* 62 (1792):299. See Kite's (1788) book, *An Essay on the Recovery of the Apparently Dead.*

111. Hawes 1794, vii. A record of rescues this stellar does raise a red flag. Perhaps the pecuniary incentives given assistants caused them sometimes to attempt a rescue when none was needed. Nonetheless, the statistic was not disputed at that time and helped establish the good reputation of the Royal Humane Society. On favorable reports of rescue work, see, for example, *The Gentleman's Magazine* 58 (1788):706, and 60 (1790):1109.

112. I have not found a primary source in which an objection is raised to reviving the apparently dead. However, it is clear from humane society literature that its members were reacting to criticism, perhaps unpublished. For example, one report noted that there were times "when the tide of public prejudice ran high against our well-intended exertions for the preservation of life" (Hawes 1794, 11). Another report mentioned that Dr. Hawes, in his duties as an officer of the society, had to "refute the falsehoods which were industriously circulated against him and the Society, to expose the calumnies with which they were continually assailed" (Royal Humane Society 1809, 5).

113. For examples, see Clarke (1793) and Robbins (1796).

114. In his own words, "organic life further requires [in addition to irritability] an harmonious arrangement of parts, and the influence of stimulating fluids specifically adapted to the respective organs, in order to produce the functions of an animated being. When these circumstances combine, the action that results appears to me to constitute the IMMEDIATE CAUSE of that condition, which we call VITALITY or LIFE, in its first or simple state of existence" (Fothergill 1796, 13; original emphasis). For a compendium of Fothergill's important medical works, see Fothergill (1783).

115. For the clock and pendulum discussion, see Fothergill (1796, 56–57).

116. Fothergill 1796, 15. A telling quote: "To restore a person from a temporary suspension of vital action, is within the province of the physician: But to restore life, after it has entirely vanished, is an act of OMNIPOTENCE, and belongs only to HIM, who gave it" (Fothergill 1796, 55–56; original emphasis).

117. Questions had already been raised by midcentury about the accepted signs of death (e.g., Sketch of treatise 1745).

118. Mentioned in Royal Humane Society (1809, 7). For a recent history of premature burial, which includes eighteenth-century case material, see Bondeson (2001).

119. Eisenberg 1997, 72.

8. AN ELECTRICAL WORLD

1. Hauksbee 1719, 34; original emphasis; see also 11. In 1735, Stephen Gray also suggested that electricity was "of the same Nature . . . as Thunder and Lightning" (Gray 1735a, 24). As this phrase was the last in his paper, I suspect that Gray attached some significance to this insight, but later workers seem to have ignored it.

2. Martin 1746, 33; original emphasis. He also furnished a list of similarities between lightning and electricity (17–18). A few researchers had maintained that lightning was the result of chemical processes (see Casati 1998, 494), but their arguments carried no weight after midcentury.

3. There were many arrangements for creating artificial lightning. For example, Watkins (1747, 56–57) used a "piece of gilt leather," hung from the prime conductor, to "produce a representation of thunder accompanied with

lightning." When a charged Leyden jar was applied to the leather, it "will flash quite up to the conductor with a very vivid sparkling, and the noise will be like the discharge of a pocket-pistol." The use of laboratory models to mimic large-scale natural processes did have precedents in the seventeenth century: Gilbert ([1600] 1958) experimented with a miniature globe, his "terrella," to argue that the earth was a magnet; and Guericke ([1672] 1994) employed his sulfur sphere to demonstrate earth's "attractive virtue" (see chapter 2).

4. Van Marum, quoted by Muntendam (1969, 14). Among the early advocates for the involvement of electricity in almost all environmental processes were Poncelet (1766) and La Perrière du Roiffé (1761).

5. "Earth scientist" is a decidedly modern construct that has no Enlightenment equivalent. In the present work it is merely a label for designating investigators who used electrical technologies to study geological and atmospheric processes. The closest eighteenth-century equivalent is "meteorologist" and "meteorology," which are present in the French literature (e.g., Bertholon 1787; Hervieu 1835, but written in 1794). Indeed, Bertholon (1787) and others included as "meteors" even geological processes such as earthquakes and volcanoes. However, because its usage in the twentieth century is far narrower than that in the eighteenth, I eschew meteorology in favor of the anachronistic "earth science." In addition, many eighteenth-century meteorologists—in the modern sense of the term—did not propound electrical theories and used no electrical technology in their studies.

6. The quote is from Franklin (1996, 64); the abbreviated early list is from 51–52. The complete list, from notes that were not published until a few years later, reads as follows: "1. Giving light. 2. Colour of the light. 3. Crooked direction. 4. Swift motion. 5. Being conducted by metals. 6. Crack or noise in exploding. 7. Subsisting in water or ice. 8. Rending bodies it passes through. 9. Destroying animals. 10. Melting metals. 11. Firing inflammable substances. 12. Sulphureous smell" (Franklin 1996, 323).

7. Franklin 1996, 52. I have telescoped Franklin's labored arguments, perhaps rendering them an injustice. To read his ideas on the involvement of electricity in clouds and rainfall, see Franklin (1996, 39–45). It should also be noted that Franklin himself developed doubts about his theory and, after further experiments, abandoned it (see Franklin 1996, 111–112, 175–176).

8. The conjecture is presented by Franklin (1996, 45–46); the experiment is on 46.

9. The plan for the experiment is described by Franklin (1996, 66).

10. Franklin actually says "insulating stand," but from his drawing I infer that the object referred to is a stool (Franklin 1996, pl. 1, fig. 9).

11. For information on Dalibard's translation of Franklin and the ensuing events and controversies, I have relied on Heilbron (1979, 348–349), Maluf (1985, 139–140), and Torlais (1956).

12. For a description of the Marly experiment, I drew upon Heilbron (1979, 349) and G. Mazeas's 1752 letter in Franklin (1996, 106–107).

13. On the verification of the Marly experiment in England, see Franklin (1996, 108–110). Bertholon (1787, 1:14–29) supplies references to investigators throughout Europe who repeated the experiment.

14. Heilbron 1979, 349.

15. Cohen 1952, 432. Beccaria also made this observation in 1756 (Beccaria 1776, 449).

16. Biographical information on Richmann, excepting the manner of his death and its aftermath, comes from Heilbron (1975, 1979, 84–85). Details on Richmann's apparatus are given by Watson (1754).

17. My sources for the events of that day are An account (1755), Labaree et al. (1962, 5:219–221), and Watson (1754).

18. This was the view of the South Carolinian John Lining (1754, 757) and most everyone who has since studied the question. Properly speaking, such a connection is called a "shunt"; it would come into play in case of a sudden surge of electricity.

19. Labaree et al. 1962, 5:221. I assume, as the editors do, that Franklin penned the last paragraph of the article.

20. Priestley 1966, 1:107.

21. Home (1979) furnishes biographical information on Aepinus, an English translation of his monograph, and an assessment of its significance.

22. This rendering of Kinnersley's mention of the Richmann tragedy is a close paraphrase adapted from Lemay (1964, 91).

23. Cohen (1990, 67–68) reproduced the original letter, as found in the files of the Royal Society; it differs in style but not in substance from the version included in Franklin's book. Unless otherwise noted, all quotes and information on Franklin's kite experiment presented below come from Franklin (1996, 111–112). Some years later Priestley, in his *History and Present State of Electricity,* also published an account (Priestley 1966, 1:215–217). Because Franklin furnished detailed advice to Priestley while the latter wrote his book, we can reliably take Priestley's account as the gospel according to Franklin. Indeed, in adding details not present in either of Franklin's own accounts, Priestley boasts that he obtained these "few particulars . . . from the best authority." That could only have been Franklin.

24. There is a reasonably credible claim that Jacques de Romas, a Frenchman, preceded Franklin in proposing the use of a kite for drawing down atmospheric electricity (Bertholon 1787, 1:33; Heilbron 1979, 351). Bertholon (1787, 1:32) mentioned the obvious height advantage of kites.

25. The outlines of this controversy are set forth by Van Doren (1938, 164–170) and Cohen (1990, ch. 6), but it has a great antiquity (e.g., Bertholon 1787, 1:33).

26. After assessing the evidence in detail, Cohen (1990, 93) concludes that "there is no reason to doubt that Franklin had conceived and executed the kite experiment before hearing the news of the French performance of the sentry-box experiment."

27. This view of Franklin as little concerned with details of scientific prior-

ity is supported by his positive reaction to the fate of his first letter on electricity sent to his friend Peter Collinson in London. The letter was neither read to, nor published by, the Royal Society. Rather, Collinson circulated it among electrophysicists, including William Watson. Watson liked what he saw and included, in his own paper read to the Royal Society, a large excerpt from Franklin's letter presenting the basic theory of one electricity. Watson's paper was published by the Royal Society. Franklin, rather than being miffed that Watson had stolen his thunder (so to speak), expressed pleasure that his ideas were being taken seriously (this episode is detailed by Cohen 1941, 78–79).

28. Van Doren 1938, 17.

29. The apparent foolhardiness of this experiment has led some critics to suggest that Franklin never carried it out (Cohen 1990, 93). However, Franklin could have believed that it was safe. Given Franklin's nascent concept of electrical resistance, he might have supposed that a thin kite string would have lacked the ability to convey a fatal charge to the experimenter. He also contended that lightning "does not usually enter houses by the doors, windows, or chimneys, as open passages, in the manner that air enters them" (Franklin 1996, 423).

30. For example, Beccaria (1776, 461–462). Bertholon (1787, 1:33–62) discusses other investigators who used kites to explore atmospheric electricity.

31. Information on de Romas's kite and experiment comes from Bertholon (1787, 1:35–37); see also Cohen (1990, 100–106).

32. Torlais 1956, 346.

33. On the kite and reel system, see Bertholon (1787, 2:322–325).

34. Cavallo 1795, 2:8–9; pl. 4, fig. 9.

35. For information on the balloon flights and Franklin's witnessing of these experiments, I have relied on Cavallo (1785, 84–92) and Van Doren (1938, 700–703). These sources differ in some inconsequential details, and I have steered a middle course between them. I have also drawn information from Gillispie (1983), the definitive modern account of the Montgolfier brothers' activities.

36. The term "aeronaut" was in use by 1785 (e.g., Cavallo 1785, iii).

37. Franklin's quotations are from Van Doren (1938, 700, 702).

38. Cavallo 1785, 135. On Bertholon's priority, see Bertholon (1787, 2:333) and Cavallo (1785, 110).

39. Information on the construction of this balloon comes from Veau Delaunay (1809, 186–187; see also pl. 11, fig. 124).

40. Veau Delaunay 1809, 224–226; and pl. 12, fig. 146. Lichtenberg also used hydrogen-filled balloons to study atmospheric electricity (Brinitzer 1960, 122–126).

41. Franklin 1996, 112. This description of Franklin's home lightning conductor is taken from Van Doren (1938, 168–169), who quoted Franklin at length.

42. Franklin 1996, 114–116, 128–129.

43. For biographical information on Beccaria, see Heilbron (1970).

44. Beccaria 1776, iv.

45. Franklin 1996, 427–428.

46. Franklin 1996, 433. On the Armonica, see Franklin (1996, 428–433). Inexplicably, Franklin omitted the drawing of the Armonica from his book when he published the letter to Beccaria. Both items have been published elsewhere (e.g., Smyth 1970, 4:163–169). Editorial notes accompanying Franklin's letter to Beccaria (in Labaree et al. 1966, 10:116–132) supply information about the adoption of the Armonica in Europe.

47. This collecting system is described by Beccaria (1776, 422–423). Beccaria (1776, 2) referred to his electrometer as an "electroscope," a term little used by others in the eighteenth century but now in fashion.

48. A description of this apparatus is given by Bertholon (1787, 2:318–321). Employing a similar apparatus, Galvani (1953) powered some of his electrobiological experiments with atmospheric electricity (see his tb. 2).

49. For biographic information on Tiberius Cavallo, see Bertucci (1999) and Heilbron (1971b). Among his published contributions to electrical research are Cavallo (1776, 1777a, b, c, 1780a, b, 1788, 1795).

50. Cavallo 1795, 2:47–48, pl. 3, figs. 4, 5, and 6.

51. Cavallo 1780b, 21–26, tb. 1, figs. 1–3.

52. Nairne (1793, 75) listed "Cavallo's Electrometer" for 14s.; Adams and Jones (1799, catalog at end, 5:10) also offered it, but for 12s.

53. Bennet 1789, 104. Bertholon (1787, 2:17) also believed that shooting stars are electrical discharges.

54. Bennet 1789, 108–112.

55. Information on Bennet's gold-leaf electrometer comes from Bennet (1789, 18–20, and pl. 1).

56. For the water experiments, see Bennet (1789, 57–74); for the powder experiments, 22–29.

57. The gold-leaf electrometer was brought to market by, for example, Adams (1789, catalog at end, p. 10) at 18s.; Nairne (1793, 75), where it was listed at 14s.; and Adams and Jones (1799, catalog at end, 5:10) who offered it for 18s. Its lack of portability was pointed out by Cavallo (1788, 6).

58. Volta 1782, viii.

59. Bennet 1789, 75–80; he furnished a more detailed description of his "electrical doubler" in Bennet (1787).

60. For a description of how Nicholson's doubler was used in Bennet's new experiments, see Bennet (1787, 82–90); for the original description of this apparatus, see Nicholson (1788); for further experiments, see Read (1794); it was commercialized by Adams (1789, catalog at end). In effect, once given a small charge—even that present in ambient air—Nicholson's doubler functioned as an electrical machine that was based not on friction but on induction. In the nineteenth century, many new induction machines would be brought to market.

61. Beccaria describes his investigations of electricity in fog (1776, 441–445) and dew (1776, 459–475).

62. Desaguliers 1741, 666–667.

63. Desaguliers 1742, 42–43.

64. For a description of this apparatus, see Cavallo (1795, 2:45–46, pl. 3, fig. 2).

65. Landriani's recorder is mentioned briefly by Bennet (1789, 49–50). Regrettably, the actual mechanics of the recording process are unclear from Bennet's description. We could easily jump to the conclusion that the resin plate rotated, but it is also possible that the mechanism for imprinting the charge was the mobile element. A similar—if not identical—device for recording variation in atmospheric electricity is attributed to Lichtenberg. His apparatus is cryptically described as follows: "A clockwork motor was to rotate a resin ball slowly, letting sparks transmitted through the metal wire move a needle on a drum so that the strength, sign and change of sign could be registered automatically and read off any time" (Brinitzer 1960, 119).

66. Information about the ceraunograph comes from Beccaria (1780, 1–11; see also pl. 1).

67. For problems with the first model of the ceraunograph and a description of the second, see Beccaria (1780, 12–18).

68. On early meteorological instruments, see Frisinger (1977, 47–98) and Daumas (1989).

69. Records of atmospheric electricity were published by Achard (1792) and Royal Society (1776).

70. For examples, see Bennet (1789, 114–141), Cavallo (1776, 1777a, 1795, 2:15–36), Lampadius (1793), Mazeas (1753), Read (1793), and Ronayne (1772).

71. Some investigators, such as Bertholon, also worked deductively at times, using models to raise the plausibility of extant theories.

72. Musschenbroek 1744, 2:287–289.

73. Cavallo 1795, 1:239.

74. Bertholon 1787, 2:31–82.

75. Bertholon 1787, 2:77; my translation. This device is discussed on 77 and illustrated in pl. 4, fig. 9.

76. Bertholon 1787, 2:80; my translation. This device is discussed on 80 and illustrated in pl. 4, fig. 13.

77. Bertholon 1787, 2:81; my translation. This device is discussed on 81–82 and illustrated in pl. 4, fig. 14.

78. Adams (1799, catalog at end, p. 9) lists "Exhausted flasks, called Aurora Borealis" at 6s. 6d.

79. Bertholon 1787, 1:390–391. Silva (1756) early on took seriously the theory that earthquakes involve electricity. Many of Bertholon's theoretical arguments were derivative; the source most likely was de la Cepède's two-volume work (1781), a wide-ranging synthesis that included neither new experiments nor new technology; other possible sources were Poncelet (1766), Scudery (1775), and Tressan (1786; supposedly written in 1747). In a volume penned in 1794, Hervieu (1835) built a comprehensive theory to explain, electrically, all meteorological phenomena and invoked the sun as the prime cause.

80. Bertholon 1787, 1:369, 395, 414–415. Decades earlier, the Harvard

scholar John Winthrop had written a pamphlet dismissing electrical theories of earthquakes. Intimately familiar with Franklin's theories, and one of his countless correspondents, Winthrop constructed arguments that seem sound *today*. If the earth were one large and continuous conductor, which he assumed, then localized charge imbalances—of the sort later postulated by Bertholon—could not arise (see Cohen 1952, 430–431). In the eighteenth century, however, Winthrop's assumption-laden theoretical arguments were eminently disputable, even by other Franklinians.

81. Bertholon 1787, 1:409; my translation.

82. Bertholon 1787, 2:7; my translation.

83. Bertholon 1787, 2:83–267.

84. Veau Delaunay (1809, 37–38, pl. 3, fig. 25).

85. Gray and Mortimer 1736, 403; original emphasis. See also Gray 1736.

86. Wheler 1739, 125; original emphasis.

87. Near the end of Kinnersley's first lecture, he conjectured that electricity is responsible for the attractions of gravity that hold the planets in orbit around the sun (Cohen 1941, 420–421).

88. Consult, for example, Troostwyk and Krayenhoff's (1788) nicely balanced treatment of environmental processes.

89. For the development of modern theories on "hydrometeors," see Middleton (1965).

9. PROPERTY PROTECTORS

1. Bertholon 1787, 1:268.

2. Cook 1951, 38.

3. See, e.g., Cohen 1990, 6. Cohen's claim is eminently disputable. It could be argued, for example, that electrical equipment used for medical purposes was also a practical technology derived from science. Other examples could be provided, depending on how we define "science" and "practical."

4. Franklin may well have been *the* first electrical engineer, but I do not wish to develop that argument here. In this context, I define "engineer" behaviorally as someone hired to engage in technology-design activities. Those who define engineer in terms of membership in a professional community recognize no electrical engineers until the mid nineteenth century or later (e.g., McMahon 1984).

5. On these cases, see Cohen (1952, 396, 399).

6. For the following account of bell ringing in churches to ward off thunderstorms, I draw heavily on Cohen (1952).

7. On ritual technologies, see, e.g., LaMotta and Schiffer (2001) and Walker (1995, 2001).

8. Cohen 1952, 396.

9. Bertholon 1787, 1:270; my translation.

10. Franklin 1996, 50.

11. Franklin 1996, 50.

12. Franklin 1996, 61; original emphasis. Watkins (1747, 12) had earlier reported that points were good for receiving the "electric virtue."

13. Franklin 1996, 62–63.

14. Franklin 1996, 63.

15. Franklin 1996, 63.

16. Franklin 1996, 64. Much earlier, Stephen Gray, in experimenting with pointed and blunt iron rods (as prime conductors), showed that a shock from a blunt end was more painful (Gray 1735a, 22).

17. Franklin 1996, 65. Although I have left out some parts of the experiment and simplified the interpretation (below), I do not believe that the omissions are significant.

18. Franklin 1996, 65.

19. Franklin 1996, 66.

20. Franklin 1996, 484.

21. Cohen 1990, 125. Van Doren (1938, 166–167) quotes in its entirety Franklin's one-paragraph announcement of the lightning rod in the 1753 *Poor Richard's Almanack*. A facsimile reproduction of Franklin's momentous invention was also furnished by Schonland (1952, 386), from which I quote extensively: "It has pleased God in His goodness to Mankind, at length to discover to them the means of securing their Habitations and other Buildings from Mischief by Thunder and Lightning. The Method is this: Provide a small Iron Rod (it may be made of the Rod-iron used by the Nailers) but of such a Length, that one End being three or four Feet in the moist Ground, the other may be six or eight Feet above the highest Part of the Building. To the upper End of the Rod fasten about a Foot of Brass Wire, the Size of a common Knitting-needle, sharpened to a fine Point; the Rod may be secured to the House by a few small Staples. If the House or Barn be long, there may be a Rod and Point at each End, and a middling Wire along the Ridge from one to the other. A House thus furnished will not be damaged by Lightning, it being attracted by the Points, and passing thro the Metal into the Ground without hurting any Thing." It is noteworthy that this description of the lightning conductor reads more like a recipe or engineering design than the scientific description that Franklin furnished in other media.

22. A curious remark 1750, 208.

23. I draw heavily on Lemay's (1964, 77–78) discussion of the religious objections to lightning rods as indicated by the topics in Kinnersley's lectures (which Franklin had prepared). Cohen (1952, 423) points out that Kinnersley was lecturing about lightning rods as early as the fall of 1751—before the confirmation of Franklin's prediction at Marly—so confident was Franklin in his new technology.

24. Lemay 1964, 78.

25. Adams, quoted in Cohen (1952, 435).

26. Adams, quoted in Cohen (1952, 434).

27. The Junto Minute Book, in the American Philosophical Library, quoted by Cohen (1952, 422).

28. On a Kinnersley advertisement of April 1753, quoted in Cohen (1952, 424).

29. Stevenson 1959, 176.

30. Cohen 1952, 413.

31. Cohen 1952, 420.

32. Cohen 1952, 436–437.

33. Figuier 1867, 1:569.

34. Veau Delaunay 1809, 189–191, pl. 11, fig. 123.

35. Cohen (1990, 134–136) supplies appropriate quotes from Nollet (see also Cohen 1952, 396–397, 414–415).

36. Franklin 1996, 162. The claim of a preventive function for lightning conductors is not generally accepted today.

37. Franklin 1996, 486.

38. Franklin 1996, 486.

39. Franklin, quoted by Van Doren (1938, 169).

40. Kinnersley, in Franklin (1996, 394).

41. Kinnersley, in Franklin (1996, 396–397).

42. Franklin 1996, 488–489. In a letter to de Saussure in 1772, Franklin reported that he was aware of five authentic cases where struck houses had been saved by lightning conductors (Cohen 1952, 413). And in a letter to Landriani many years later, he noted other structures in Philadelphia that had been struck by lightning but not damaged, including his own house (Franklin to Landriani, October 14, 1787 [Smyth 1970, 9:617–618]).

43. For Franklin's "engineering" discussions of lightning conductors, see Franklin (1996, 124–127, 162–164, 414–417, 420–425, 441–442, 479–485). Of special note is his calculation of the diameter needed for the rods (125).

44. Dalibard, quoted in Kryzhanovsky (1990, 817).

45. Kryzhanovsky 1990, 816–817.

46. Biographical information on Benjamin Wilson was obtained from Turner (1976).

47. Van Doren 1938, 430. A mezzotint based on this painting is the frontispiece of Labaree et al. (1966, vol. 9).

48. Hoadly and Wilson 1759 (see also Wilson 1752).

49. Watson et al. 1769, 160. The letter from the dean of St. Paul's Cathedral prefaces the committee's report. This report serves as the basis for my discussion.

50. On the analysis of the Purfleet buildings, see the committee report (Cavendish et al. 1773).

51. Wilson 1773, 55; original emphasis.

52. On the Eddystone lighthouse's conductor, see Mainstone (1981, 92) and Stevenson (1959, 18).

53. Wilson 1773, 58. He erroneously claimed that a pointed rod had been judged "highly improper" (58); in all likelihood, Smeaton had the copper ball serve a dual purpose as an ornament and as the top of the lightning conductor. Curiously, Cohen (1952, 412) names William Watson's house as the first struc-

ture in England to be outfitted with a lightning conductor; the date given is 1762, long after the Eddystone installation.

54. Henly 1774, 152.

55. Cohen 1952, 413.

56. Henly et al. 1778.

57. Wilson 1778a.

58. Wilson 1778b, 247.

59. A description of the prime conductor's construction is provided by Wilson (1778b, 251–253); its dimensions are given on 311.

60. On the spiral wire, see Wilson (1778b, 252–253, 311).

61. Detailed dimensions of the boarding house model are furnished by Wilson (1778b, 313). Wilson (1778b, 251) claimed that the model's scale was "one-third of an inch to a foot," but it was actually closer to .5-inch per foot (based on the drawing of the actual boarding house [Nickson 1778, tb. 2]).

62. Wilson 1778b, 254.

63. Wilson does not mention how long the experiments took, but he does admit that enough dust had eventually settled on the silk suspenders to make them somewhat conductive (Wilson 1778b, 304). The large number of experiments and their variations, along with shakedown time and the inevitable repairs, suggest that this project lasted many weeks, if not months.

64. Wilson 1778b, 256.

65. Heilbron 1979, 382.

66. Franklin, quoted in Heilbron (1979, 382).

67. Wilson (1778c) also printed a slightly different version of the report, perhaps for more general circulation; it contains some details of the apparatus not present in Wilson (1778b).

68. The remaining committee members were S. Pringle (president of the Royal Society), W. Henly, S. Horsley, T. Lane, and E. Nairne (Pringle et al. 1778).

69. Pringle et al. 1778, 317.

70. In a later paper and employing new apparatus, Nairne (1778) did address the substantive issues raised by Wilson's (1778b, c) experiments. Not surprisingly, Nairne's experiments resulted again in a decisive victory for pointed conductors.

71. There are some minor exceptions here, as in the case of very tall buildings, but they do not detract from the main point. Remarkably, a points-vs.-knobs controversy has been recently resurrected (see Moore et al. 2000).

72. If we assume for the sake of argument that a house and cloud are no fewer than 4 miles apart, and that this is represented in the model by a distance of 5 feet (based on Wilson's drawing), the scale would be 1:4224. For the house to conform to this scale, it should have been about .10-inch across, with rods correspondingly smaller! As Heilbron (1979, 382) has perceptively noted, all rods—even knobby ones—appear as points "On nature's scale."

73. As two obvious examples, consider Thomas Edison's small-scale electric lighting system at Menlo Park and his construction of a miniature electric railway (see Friedel et al. 1986).

74. Pain 1975, 39.

75. Franklin, quoted in Cohen 1952, 413.

76. Toaldo 1779. This edition also contains a lengthy general section on lightning conductors by Barbier de Tinan, the French translator (183–241). Monographs on specific lightning-conductor installations include Bartaloni (1777?), Calandrelli (1789), and Toderini (1771); for other works relating to the design and installation of lightning conductors, see Bertholon (1783b), de Romas (1776), Landriani (1784), Langenbucher (1783), Mahon (1779), Meredith (1789), and Reimarus (1794).

77. Winn 1770, 188.

78. Captain Cook, quoted in Barrow (1993, 101).

79. Harris 1831, 426–428.

80. Landriani 1784, 78.

81. Landriani 1784, 121–136.

82. Landriani 1784, 98.

83. Landriani 1784, 181–284.

84. Franklin to Landriani, October 14, 1787 (Smyth 1970, 9:617).

85. Brief mention of the latter installations occurs in a 1773 letter by Franklin to John Winthrop (quoted in Cohen 1952, 413). Henly (1777, 85) notes that the ship *Generous Friends* was protected by a lightning conductor. Apparently, the pace of installing lightning conductors in England had accelerated between 1773 and 1784.

86. Kryzhanovsky 1990, 816.

87. Raw numbers for structure types, based on Landriani's (1784, 285–304) appendix, are 174 houses, 46 religious structures, 26 castles and palaces, 21 military buildings, 20 public buildings, 6 schools, and 4 factories. There are also 26 unknown or unidentified entries. In my typology, "religious" structures include church, cathedral, convent, basilica, and monastery; among the "public" buildings are archive, art gallery, theater, bank, library, lighthouse, and tower. In the "school" category I place structures at colleges, universities, and academies but exclude religious structures (tallied above).

88. Palaces and castles have many nonresidential functions, and so I include them in this generalization.

89. Adams 1799, catalog at end, p. 10; Nairne 1793, 75.

90. Blake-Coleman 1992, 124.

91. Reimarus (1794, tb. 2) illustrates two straw-thatched buildings that may have been farm houses or barns.

92. This sort of contagious adoption process may account for the clusters of lightning conductors apparent in certain towns, even small ones, in Landriani's compendium; simply having a pushy instrument maker in town might not have sufficed.

93. I do not mean to imply that the lightning conductor had become a social *necessity* for the elite; its adoption was far too infrequent to support that claim. Rather, it seems to have been merely socially appropriate, one of many consumption options. Obviously, the lightning rod—especially if gilt—could

also indicate the homeowner's wealth, but that message would have been re-
dundant, assuming that the house itself was recognizably pricey.

94. Remarkable effects 1787.

95. Cohen 1952, 401.

96. See Henly (1772) for a very detailed case study.

97. Franklin (1996, 441–442) offered advice on the design of lightning conductors for powder magazines.

10. A NEW ALCHEMY

1. Hull 1956, 306. The huge contributions of craft traditions can be readily
seen in the chemical dictionaries and encyclopedias that appeared around the
turn of the nineteenth century (e.g., Chaptal 1807; Fourcroy 1800–1802; Imison 1796, 1803; Nicholson 1795).

2. For a lengthy and detailed history of chemistry, see Partington (1961);
on the eighteenth century in particular, see Holmes (1989). Jaffe (1976) is
a readable history oriented around biographies of major figures. Hull's (1956,
ch. 11) dense monograph contains many important insights; also useful are the
early chapters of Bensaude-Vincent and Stengers (1996). Golinski (1992) furnishes a fascinating treatment of the social context of chemistry in Britain.

3. Lavoisier [1789] 1952, 2. On the importance of laboratory apparatus in
eighteenth-century chemistry, see Eklund (1975).

4. Materials science, per se, emerged as a distinct discipline only in the
twentieth century. Previously, much materials-science research was carried out
in chemistry and engineering. In the latter part of the twentieth century, materials science and chemistry significantly coalesced, often in schools of engineering (W. David Kingery, personal communications, 1988–1999).

5. On the cold-fusion controversy, see Huizenga (1992) and Park (2000).

6. Franklin 1996, 51, 52.

7. Franklin (1996, 68) did use the term "electrical circuit" (1750), which
suggests that he was familiar with William Watson's work. The latter had employed "electrical circuit" in print as early as 1748 (Watson 1748, 49–51). Finn
(1969) discusses Watson's probable influence on Franklin's ideas.

8. Franklin 1996, 69.

9. Maluf 1985, 143–144.

10. Knight 1759.

11. Kinnersley's description of the "electrical air thermometer" was published in a long letter, dated March 12, 1761, that Franklin included in later editions of his book (Franklin 1996, 389–393, and pl. 6). Heilbron (1979, 370) as-
cribes the invention of the air thermometer to Beccaria, in 1753; the latter's
much simpler device differed greatly in design from Kinnersley's instrument
and, moreover, was not used for testing materials, much less for refuting Franklin's cold fusion. Nonetheless, it did show that sparks in the air create heat.

12. Kinnersley, in Franklin (1996, 391).

13. Kinnersley, in Franklin (1996, 392).

14. Most such products were first brought to market toward the end of the nineteenth century, after the advent of commercial electrical systems in the early 1880s (see Schiffer et al. 1994). I have not found the precise term "resistance heating" in the eighteenth-century literature, but it doubtless would have made sense to Kinnersley and others.

15. Franklin 1996, 411.

16. By 1750 other inventors, such as René Réaumur, had developed thermometers that did measure the temperature of liquids and gases (see Frisinger 1977, ch. 4).

17. Eventually it was brought to market by Adams (1799, catalog at end, p. 9). The price of "Kinnersley's electrical air thermometer" was £1 1s.

18. Franklin 1996, 69.

19. Beccaria 1776, 298–302. Sometime later Singer (1814, 175) reported, "The colours produced by the explosion of metals have been applied to impress letters on ornaments of silk and paper." Possibly this was a commercial process, but as yet I have found no additional evidence. A good description of the process is given by Veau Delaunay (1809, 133–134, pl. 9, fig. 101).

20. Lavoisier 1952, 4–5.

21. Cuthbertson (1807, fig. 82) built—and doubtless also sold—an elaborate apparatus for measuring the volume of air consumed in the formation of a metal oxide. A metal wire was placed inside a well-sealed glass jar, its ends terminating in external conductors. After the discharge, which caused a partial vacuum, the investigator could admit measured quantities of air as needed to restore atmospheric pressure. On Cuthbertson's experiments for producing "metallic oxides" through electrical discharge, see Cuthbertson (1807, 197–224).

22. These experiments are described by Beccaria (1776, 310–311).

23. This hypothesis is offered by Beccaria (1776, 305–306). Beccaria (1776, 292–295) was aware that the measured electrical resistance of a metal depended also on the length and cross-section of the wire specimen tested.

24. Priestley, cited by Beccaria (1776, 305). On other early experiments to rank the electrical resistance of different metals, see Blake-Coleman (1992, 125–133).

25. Beccaria 1776, 312.

26. Beccaria 1776, 258, 293–294.

27. Kinnersley, in a letter of February 3, 1952, to Franklin (Franklin 1996, 99): "I placed the needle of a compass on the point of a long pin, and holding it in the atmosphere of the prime conductor, at the distance of about three inches, found it to whirl round . . . with great rapidity." It is likely that this effect was actually produced by electrostatic induction, not electromagnetism.

28. Beccaria 1776, 305–307.

29. Beccaria 1776, 310.

30. For examples, see Aepinus in Home (1979), Haüy (1787), and Swinden (1784).

31. Purrington 1997, 41.

32. On the history of the energy principle, see Purrington (1997, ch. 4).

33. Kinnersley, in Franklin (1996, 385).

34. He also affirmed this result with a much more complex apparatus (Beccaria 1776, 126–129, tb. 6, figs. 5 and 6).

35. Such an apparatus is described and illustrated by Cuthbertson (1807, 113, fig. 59). He also discussed an experiment for showing that hot air becomes a conductor (Cuthbertson 1807, 114).

36. Franklin 1996, 92, 94.

37. Hull 1956, 327. I have drawn extensively on his discussion of the phlogiston theory.

38. Holmes (1998) is an excellent source on the development of Lavoisier's "balance-sheet method."

39. For biographical information on Lavoisier, see Poirer (1998).

40. Lavoisier, quoted in Hull (1956, 332).

41. Holmes 1998; Hull 1956, 332.

42. For a constructivist interpretation of Priestley's chemistry, see Golinski (1992, chs. 3 and 4).

43. Information on Priestley's work with carbonated water comes from Jaffe (1976, 41–42).

44. See Jaffe 1976, 42–43.

45. Jaffe 1976, 47.

46. Hull 1956, 333. Somewhat later Lavoisier (1783, 54) accorded Priestley credit for his experimental research in a monograph that thoroughly refuted the Englishman's phlogiston-based interpretations.

47. Hull 1956, 334.

48. Van Doren 1938, 657.

49. Musschenbroek 1739b, 1:417. Nollet (1745–1765, 4:2) regarded water as an element created by the wise God.

50. Jaffe 1976, 57.

51. Partington 1961, 3:325.

52. Volta mentioned (and sometimes described and illustrated) to several correspondents, including Francesco Castelli, Lord Cowper, and Marsilio Landriani, devices variously called pistols and eudiometers with which it was possible to ignite inflammable gases (see *Le opere di Alessandro Volta*, vol. 6, 1928). On the eudiometer, see Roberts (1998).

53. Schäffer 1778, 9–18.

54. For example, Priestley (1781, 1:61). Veau Delaunay (1809, 213–221) discusses several devices for studying electrical discharges in gases.

55. Nicholson 1790, 51. Priestley (1781) himself published accounts of copious experiments using electrified reaction chambers.

56. My concept of "discovery machine" is similar to Shapin's (1996, 96) "fact-making machine" and Golinski's (1992, 9) "engine of discovery."

57. Among the earliest investigators who ignited hydrogen (in air or sometimes in oxygen) were Volta; John Warltire, a lecturer who had assisted Priestley; Priestley himself; Pierre Joseph Macquer, a French scientist in the Jardin

des Plantes; James Watt, of steam-engine fame (Jaffe 1976, 59, 63; Partington 1961, 3:325–328); and the French chemist Gaspard Monge (Perrin 1973).

58. Partington 1961, 3:329–338.

59. Cavendish (1784, 119) posed the general research questions as follows: "experiments were made principally with a view to find out the cause of the diminution which common air is well known to suffer by all the various ways in which it is phlogisticated, and to discover what becomes of the air thus lost or condensed."

60. Cavendish 1784, 129. Hereafter I usually employ the modern terms "hydrogen" and "oxygen."

61. The description of this experiment is furnished by Cavendish (1784, 129–132). A reaction vessel, claimed to be Cavendish's original, is preserved in the chemistry department of the University of Manchester (Partington 1961, 3:333; it is illustrated in fig. 32). However, Farrar (1963) has shown that the device dates to the early nineteenth century.

62. Cavendish 1784, 133.

63. On the use of Lavoisier's reaction vessel for forming water, see Lavoisier (1952, 32–33, pl. 4, fig. 5); Description (1798a) also furnishes a nice account and illustration (pl. 10, fig. 1). For another demonstration on forming water, see Description 1798b.

64. Cavendish 1784, 152.

65. Description 1798a, 303.

66. Henry (1800, 212) stated most clearly the performance characteristics of electrified reaction-vessels that made them so well suited for initiating reactions in gases: "this series of experiments . . . employed the electric fluid, as an agent much preferable to artificial heat. This mode of operating enables us to confine accurately the gases submitted to experiment; the phaenomena that occur during the process may be distinctly observed; and the comparison of the products, with the original gases, may be instituted with great exactness. [Also,] The action of the electric fluid itself, as a decomponent, is extremely powerful."

67. Accum 1803, 68. About this time Heidmann (1799, vol. 2) published a textbook on electricity that had a large, well-illustrated section on electricity as an investigative tool of chemists.

68. Chemical experiments carried out with the large Teyler machine are reported by van Marum ([1785] 1974a, [1787] 1974b). Swinden collaborated on the magnetization experiments (van Marum 1974a, 46).

69. The new machine designed for chemical experiments is described by van Marum ([1791] 1974c). Other apparatus and experiments using electricity in chemical research are described by van Marum ([1798] 1974d).

70. Cavendish (1785) soon demonstrated, for example, that the nitrous acid seen in the early version of the crucial experiment was actually a product of the sparking of phlogisticated air (nitrogen) in the presence of oxygen. As Partington (1961, 3:338–342) points out, it was some years before other investigators were able to confirm these findings.

71. Ingen-Housz, who acknowledged Volta's idea (1779b, 385), discoursed on this subject at length.

72. Lavoisier 1952, 30–32.

73. I have drawn the description of this apparatus from Pearson (1797); see also Description (1798c).

74. In Nicholson's chemical dictionary, for example, the Dutch experiment is lauded for having clinched the argument on the composition of water (Nicholson 1795, 1023–1024).

75. Pearson (1797) reported his own experiments at great length and illustrated several new apparatus.

76. Wollaston 1801. Cuthbertson (1807, 129–130, pl. 5, fig. 67) showed how this experiment could be easily performed with a wine glass, two glass capillary tubes, and conductors from an electrical machine.

77. Pearson 1797.

78. Davy 1807. In the first half of the nineteenth century, the ability to accumulate a metal at one pole of an electrolytic cell was the foundation of an entirely new industry: electroplating. However, that metals could be plated electrically had been discovered prior to 1800 by Ritter using electrostatic technology (Ostwald [1896] 1980).

79. Singer (1814, 352), in a very early textbook treating electrochemistry, asserted that "chemical and electrical attraction are identical."

80. Schiffer 2001a.

11. VISIONARY INVENTORS

1. Biographical information on Edison is provided by Conot (1979), Josephson (1959), and Israel (1998).

2. Franklin was strongly opposed to the granting of patents, even refusing a patent on his stove offered by the governor of Pennsylvania. His elegant argument concluded, *"That as we enjoy great Advantages from the Inventions of others, we should be glad of an Opportunity to serve others by any Invention of ours, and this we should do freely and generously"* (Franklin 1986, 130; original emphasis). And when others patented his inventions, on occasion earning large sums, he did not contest them (Franklin 1986, 130–131). At the end of a long letter on electrical matters to Dr. Lining, of South Carolina, in 1755 Franklin did however express some bitterness at the treatment of inventors: "One would not, therefore, of all faculties, or qualities of the mind, wish, for a friend, or a child, that he should have that of invention. For his attempts to benefit mankind in that way, however well imagined, if they do not succeed, expose him, though very unjustly, to general ridicule and contempt; and, if they do succeed, to envy, robbery, and abuse" (Franklin 1996, 328).

3. Cooper 1991.

4. On the changing social construction of invention during the nineteenth century in the United States, see Cooper (1991) and Post (1976). Lienhard (1979) found that rates of change in a number of technological systems accelerated

dramatically after the mid nineteenth century, underscoring the claim that various societal changes converged to stimulate invention across a broad front.

5. I hypothesize that, in common usage during the eighteenth century, the term "inventor" meant the author of an invention or someone possessing a disposition to be inventive, not a social role or occupational specialization.

6. This group, whose activities and technologies were highly diverse, cannot be considered a community in the same sense as, say, electrophysicists or electrotherapists. Although I could have treated many of these inventions in earlier chapters, my purpose here is to showcase a range of "forward-looking" creations; many of them did not come to fruition as electrical technologies until more than a century later.

7. Franklin 1996, 37.

8. Franklin 1996, 38; original emphasis. He also once suggested that electrocution "would certainly . . . be the easiest of all deaths" (Franklin 1996, 325).

9. Stephenson's letter appears in *The Gentleman's Magazine* 17 (1747):183.

10. See, for example, Bakewell (1853, 191–192).

11. In another letter (*The Gentleman's Magazine* 17 [1747]:262) Stephenson reiterated that a variety of prime movers could be harnessed to drive electrical machines.

12. Although neon and other noble gases were not discovered until the nineteenth century, gases already known in the eighteenth century—even air at low pressure—give off light in the presence of an electric charge.

13. Hauksbee 1719, 46.

14. As archaeologists have demonstrated, the fostering of new technologies by elite personages is an ancient and widespread process (see Hayden 1998).

15. Friedel et al. 1986, 140–141.

16. Volta [1787] 1929.

17. Ingen-Housz 1784, 1:224. The lamp is illustrated in vol. 1, tb. 2, fig. 11.

18. On the disadvantages of the phosphorus match, see Nicholson (1795, 656).

19. Adams and Jones (1799, catalog at end, 5:10); see also Adams (1799, catalog at end, p. 10). Nairne (1793, catalog at end, p. 75) also sold an "Electrical Tinder-Box" for inflammable air (hydrogen), one version of which came with an electrophorus. For examples in texts and experiment books, see Gütle (1791, tb. 5); Langenbucher (1780, tb. 8, figs. 16 and 17); Veau Delaunay (1809, pl. 8, fig. 88).

20. Brinitzer (1960, 118) mentions that Klindworth in Germany brought these lamps to market; most likely Gütle did as well.

21. Ingen-Housz 1778a.

22. Stephen Gray (1736; Gray and Mortimer 1736) was the first to demonstrate that electricity could produce rotary motion. Although Desaguliers (1742, 38) disputed this effect, Wheler and Mortimer (1739) independently confirmed it.

23. Nairne 1793, 75. Also offered, on the same page, was an "Electrical Planetarium to be set in motion by the electric fluid." Adams (1799, pl. 4, fig. 79)

illustrated an electrical orrery but omitted it from the catalog; presumably customers could order it.

24. Langenbucher 1780, 55–56, tb. 3, fig. 4.

25. Volta did invent a device, based on a balance, for measuring the strength of a charge (described by Chaldecott 1951, 76). One scale of a balance was attracted to a charged metal plate. The weight required to counterbalance the charge, placed on the other plate of the balance, was a measure of the charge's intensity. However, with an independent measure of the charge that could be varied continuously, the device might have been used to weigh very tiny objects.

26. Such electric motors are discussed and illustrated, for example, by Seiferheld (1791c, 39–42, and tb. 1, figs. 14–15).

27. Ferguson 1778, 24–33; Ferguson and Partington 1825, 19–22. For biographical information on Ferguson, see Millburn and King (1988).

28. Nairne and Blunt (Nairne 1793, 75) offered "A set of Paper Models to be set in motion by the electric fluid," which, in all likelihood was Ferguson's machine models. Likewise, Adams (1799, catalog at end, p. 9) sold an assemblage of "working models, to be set in motion by the electrical fluid, consisting of a corn mill and a three-barrelled water pump, worked by one crank only; an orrery, shewing the diurnal motion of the earth, age and phases of the moon . . . an astronomical clock, shewing the aspects of the sun and moon, age, phases, &c. all delicately made of card paper, cork, and wire only, packed in a deal case," all for just £2 12s. 6d.

29. Franklin 1996, 32; original emphasis.

30. Lemay 1964, 73.

31. Seiferheld (1802a, 1–27, and tbs. 1–3) is my source on his electric clock.

32. See Leone and Shackel's (1987) study of probate records for eighteenth-century Annapolis and Anne Arundel County, Maryland. These records underestimate the adoption of clocks because the probate data do not represent younger families, which would have had higher adoption rates.

33. G. C. Bohnenberger was also a prolific electrical inventor. In one case, Bohnenberger (1784–1791, vol. 3, tb. 5, fig. 2) illustrated an electrostatically powered escapement that would have permitted the construction of a somewhat more accurate timepiece. He also wrote extensively on electricity (see Bohnenberger 1788, 1793–1795, 1798).

34. Seiferheld 1802a, 27–33, and tb. 4. It is likely that "color tubes" and perhaps other Seiferheld inventions were inspired by some of Gütle's (1790, catalog at end) electrical amusements.

35. Seiferheld 1799, 13–19, and tb. 4.

36. Seiferheld 1795, 19–25, and tb. 3, figs. 7–10.

37. Evidence for such linkages should be sought in the biographies and autobiographies of nineteenth-century scientists, inventors, and engineers, which might reveal the inspiration that literature and toys provided to young technology enthusiasts. I would be astounded if this process, so common in the twentieth century, were not already present in the Enlightenment.

38. Schiffer 1991.

39. Cavallo 1781, 310.

40. Cavallo 1781, 314; Ingen-Housz 1779b, 381.

41. Ingen-Housz 1779b; it is illustrated opposite 418. His electric pistol, with piston, was modeled after one he had seen demonstrated in Amsterdam in 1777 by Aeneae and Cuthbertson (Ingen-Housz 1779b, 380–381). Both versions could use ether in place of hydrogen.

42. Cavallo 1781, 312, and figs. 19–20.

43. Another spark plug—literally a plug-in, not screw-in variety—was illustrated by Seiferheld (1791c, 84–85, and tb. 4, fig. 35).

44. Partington 1961, 3:327.

45. Ingen-Housz 1779b, 390.

46. Ingen-Housz 1779b, 416.

47. Pertinent histories of the earliest internal combustion engines include Clerk (1886, 1–22), Donkin (1900, 19–29), Evans (1932, 1–11), Hardenberg (1992), MacGregor (1885, 1–9), and Schottler (1902, 10). Hardenberg (1992) is particularly well referenced.

48. Jones 1828, 18.

49. Donkin 1900, 20; Schottler 1902, 4–5.

50. Evans 1932, 7; Hardenberg 1992, 58, 71.

51. Clerk 1886, 13–16.

52. MacGregor 1885, 7.

53. I am aware that this argument contains holes, but the hypothesis of a linkage between eighteenth-century electrical technology and the internal combustion engine is intriguing and merits additional research.

54. For an explicit statement of this principle, see Ingen-Housz (1778b, 1034–35). It is also clear that Desaguliers (1742, 6–7), decades earlier, appreciated that a large insulator—a glass tube in particular—could be charged and discharged in a localized manner.

55. However, there may also be inductive effects in that a charge will induce an opposite charge in the adjacent area, which could in turn induce another charge next to it, and so on. These possible effects, which are of minor importance here, are ignored to keep this presentation manageable.

56. Brinitzer (1960) presents detailed biographical information on Lichtenberg.

57. Takahashi 1979, 3. I have relied on Takahashi for the translation of crucial passages from Lichtenberg 1778. And I consulted both Lichtenberg 1778 and 1779; ostensibly identical, these two versions of his account contain different examples of Lichtenberg figures. For additional discussions of Lichtenberg's discovery, see Baird and Nordmann (1994) and Brinitzer (1960, 117–121).

58. The setup for this experiment is illustrated by Lichtenberg (1778, tb. 4).

59. Lichtenberg figures have remained a curiosity, enchanting generations of electrical researchers and furnishing a means for studying the actual shapes of discharges (Takahashi 1979).

60. Takahashi 1979, 4; Baird and Nordmann 1994, 48.

61. Baird and Nordmann 1994, 48.

62. Employing small rods of clay that shrank as the temperature rose, Wedgwood thermometry was widely adopted in the early nineteenth century, for there were no viable alternatives to taking a kiln's temperature beyond visually inspecting the color of its contents. On Wedgwood's pyrometer, see Chaldecott (1975); on his contributions to science, see McKendrick (1973).

63. Enamel glazes offer a palette of vibrant colors, but at high temperatures they burn out or fade. Enamels must be painted onto pots that have already been fired once at high temperature. The pots are then fired again at a lower temperature, thus retaining the colorful, often intricate decoration.

64. Cavallo (1780b, 17–18) had previously shown that merely pouring a powder could electrify it.

65. Bennet's (1789, 34) acknowledgment: "The following experiments are intended as improvements on M. Lichtenberg's beautiful configurations."

66. This experiment is described by Bennet (1789, 42). He also supplied the recipe for making resin plates: "five pounds of rosin, half a pound of bees wax, and two ounces of lampblack, melted together and poured upon a board sixteen inches square, with ribs upon the edges at least half an inch high, to confine the composition whilst fluid, thus the resinous plate was half an inch thick, which is better than a thinner plate, the figures being more distinct. After the composition is cold, it will be found covered with small blisters, which may be taken out by holding the plate before the fire, till the surface be melted, then let it cool again, and upon holding it a second time to the fire, more blisters will appear; but by thus repeatedly heating and cooling the surface, it will at last become perfectly smooth" (Bennet 1789, 47).

67. Bennet 1789, 49; the detailed ink recipe is on 46.

68. On the electrical pen see Bennet (1789, 54); on Edison's electric pen see Conot (1979, 74); Edison's pen was actually electromechanical and became, among other products, the first electric tattooing pen.

69. Bennet 1789, 54.

70. Millburn 2000, 261. It is curious that Adams did not apparently bring the electric pen or printing process to market.

71. Bennet 1789, 50.

72. Bennet 1789, 50. Unfortunately, descriptions of the ceramic experiments are exceedingly sketchy.

73. Carlson himself acknowledged Lichtenberg as the inventor of "the first electrostatic recording process" (quoted in Baird and Nordmann 1994, 51).

74. Loesser 1954.

75. Delaborde 1761, 1.

76. Delaborde 1761, 14–15.

77. Delaborde 1761, 5; my translation.

78. Electro-vegetometer is a literal translation of *électro-végéto-metre* (Bertholon 1783a, 393).

79. Bertholon 1783a, 393–394; my translation.

80. Jefferson letter to Rev. James Madison, 1788, in Sowerby (1952, 377).

81. This invention is described and illustrated by Bertholon (1783a, 405–411, pl. 2).

82. Bertholon 1783a, 420; my translation.

83. Bertholon 1783a, 423–424, pl. 3. The word he used for circuit was *chaîne* (423).

84. "Earthquake rod" is my liberal translation of *para-tremblemens de terre* (Bertholon 1787, 1:399).

85. Bertholon 1787, 1:369, 395.

86. Bertholon 1787, 1:405–406.

87. Bertholon 1787, 1:401; my translation.

88. Scudery (1775), who accepted the electrical nature of earthquakes, described earthquake rods prior to Bertholon. In Scudery's (1775, 85–90) version, tall iron rods towering over buildings and buried deeply in the ground did dual service, protecting buildings from lightning and towns from earthquakes. In one charming illustration, he showed a small walled town boasting a forest of rods (Scudery 1775, fig. 5). The best place to look more deeply for evidence that someone actually installed earthquake rods is Italy, which had both electrical expertise and a high frequency of earthquakes and volcanic eruptions.

89. On the history of the British post office, see Tegg (1878).

90. The concept of "cultural imperative" is elaborated elsewhere (Schiffer 1993).

91. Bergsträsser's *Sinthematografie* is cited by Wilson (1976, 5); the binary code is reproduced by Aschoff (1984, 158).

92. A worldwide review of these early mechanical signaling systems is furnished by Wilson (1976). For a general history of telegraphy, including electrostatic systems, see Aschoff (1984).

93. Aschoff 1984, 161.

94. Dibner 1967, 452.

95. Northmore 1799. The code is on 12.

96. These experiments are described by Gray (1731).

97. Curiously, Jarvis (1956, 130) erroneously claimed that these experiments used brass wire and an electrical machine. I suspect that iron wire was chosen for many experiments because it was both cheaper and stronger than other metals. Indeed, many nineteenth-century electrical telegraphs used iron wire despite its lower electrical conductivity than copper.

98. Gray 1731, 29–31. In 1733 du Fay, inspired by Gray's publications, conducted electricity through 1,256 feet (French measure) of moistened silk (du Fay 1737c, 347–349); Desaguliers (1742, 16–17) repeated this feat.

99. On these experiments, see Watson 1748.

100. Watson (1748, 49–51) used the term "circuit" in its modern sense, as the path traveled by the "electrical Power" through a continuous "Line of substances non-electrical." The term "non-electrical" was used interchangeably with "good conductor" (51–52) and particularly described metals such as the iron wire. Watson's circuit diagram on 92 is of special interest.

101. Watson 1748, 91. Jallabert (1748a, 86) also reported that he could detect no delay in the transmission of electricity through a long chain.

102. Watson (1748, 81–83) summarized previous experimental studies on the speed of sound, adopting a value of 4.77 seconds per mile—close to that accepted today.

103. *Scots' Magazine*, February 17, 1753; cited by Fahie (1884, 68–71), who also reproduced the letter.

104. C. M. was unfamiliar with the works of Gray and Watson, since he held that electricity had not previously been conducted more than 30 or 40 yards. In fact, he believed that over greater distances, losses of "electrical fire" to the air would require that the conducting wires be coated with insulation—specifically, jeweler's cement.

105. C. M., quoted in Fahie 1884, 70. The Morse telegraph system, widely adopted more than a century later, also used a code that could be "read" acoustically.

106. Debates on the identity of C. M. began immediately and continue to this day (e.g., Fahie 1884, 72–77; Garratt 1958, 649–650).

107. Quoted in Fahie (1884, 79).

108. For histories of telegraphy that include discussions of electrostatic systems, see Aschoff (1984), Fahie (1884), Karras (1909), Marland (1964), Sabine (1869), and Shaffner (1859). Of these, Fahie's book, which draws on interviews and archival sources, is the most authoritative and useful.

109. Chappe's system is described by Fahie (1884, 93–95).

110. Salvá's efforts are described by Fahie (1884, 101–108) and Karrass (1909).

111. The use of sparking metal foils to indicate the transmitted letter had been invented by an anonymous writer to the *Journal de Paris* in 1782 (Fahie 1884, 85).

112. Saavedra, quoted by Fahie (1884, 106 n.).

113. Karrass 1909, 47.

114. Fahie 1884, 108.

115. The account of the successful use of Salvá's long telegraph comes from the *Magazin de Voight,* quoted by Noad (1857, 748).

116. See Ronalds (1823) for a detailed description of his telegraph.

117. The use of buried conductors in pitch-filled wooden troughs had been proposed in 1782 (Fahie 1884, 85).

118. Use of the telegraphic dictionary is described by Ronalds (1823, 8–9).

119. Ronalds 1823, 21.

120. Quoted in Marland (1964, 23–24).

121. On Ralph Wedgwood's telegraph, see Fahie (1884, 123–127).

122. Ralph Wedgwood, quoted in Fahie (1884, 126; original emphasis).

123. Fahie 1884, 127.

124. On the competitions among these electrical technologies, see Schiffer (2001).

125. Franklin to Priestley, February 8, 1780 (Labaree et al. 1995, 31:455–456); original emphasis.

12. TECHNOLOGY TRANSFER

1. In a previous work, I had noted and named the process of technological differentiation (Schiffer 1992, ch. 5).

2. For introductions to behavioral theory, see LaMotta and Schiffer (2001) and Schiffer (1992); for a recent review of behavioral studies of technology, see Schiffer et al. (2001).

3. A more extended discussion of "aggregate" technologies is presented by Schiffer (2001a). On ritual technologies, see Walker (2001).

4. For more on the life-history construct, which is widely used in archaeological and anthropological studies of technology, see LaMotta and Schiffer (2001), Schiffer and Skibo (1997), and Schiffer et al. (2001).

5. Johnson (2000, 2001) has also employed a version of "community" based on technology-related groups.

6. On research schools, see Geison (1981).

7. On the formation processes of historical evidence, see Schiffer (1996a, ch. 3, 1996b, 1996c).

8. The behavioral theory of design is presented by Schiffer and Skibo (1997); for a more reader-friendly version, consult Skibo and Schiffer (2001).

9. Feedback can of course be flawed or incomplete or lack timeliness, depending on the information links and social relationships among people in different activities.

10. On van Marum's machine and its operation, see chapters 3 and 10. For additional theory and case studies that elucidate performance characteristics, especially those based on human senses, see LaMotta and Schiffer (2001), Schiffer (1995, 2001a), Schiffer and Miller (1999).

11. Such comparisons are usually carried out with "performance matrices" (Schiffer 1995, 2000).

12. For an example based on a comparison of early electric versus gasoline automobiles, see Schiffer (1995).

13. On the relations between science and the industrial revolution, see Jacob (1997).

14. On the competitions among these electrical technologies, see Schiffer (2001a).

15. See, for example, Hobsbawm (1996) and Jacob (1997).

16. My views echo those of Friedel et al. (1986, 232), "We will not and cannot understand the true origins of novelty—in technology or in any other part of our lives—unless we acknowledge the personal, human, creative impulse at its root."

17. How are we to apply this framework to other cases of technological differentiation? I recommend that research begin with a provisional definition of the technology being studied. I emphasize *provisional*, because the research it-

self refines the conception of the technology as it progresses. On the basis of the provisional definition, the investigator delineates the functional varieties and maps out their time-space distribution. After these preliminaries have delimited the project's scope, the following questions arise: (1) In what community did the technology originate and in which activities did it take part? (2) To which communities was the technology transferred? (3) What groups made up each recipient community? (4) In which activities of recipient communities did the technology participate? (5) What new variants of the technology arose in recipient communities, and which performance characteristics were weighted in the new designs? (6) How did social, political, economic, and ideological factors, for example, influence—through their effects on the situational factors of activities—actual weights of performance characteristics? (7) Which replication modes were used by members of the recipient community to acquire examples of the new variants? At last, the investigator constructs upon this behavioral foundation a richly textured and theoretically informed historical narrative.

References Cited

Abstract of what is contained in a book concerning electricity, just published at Leipzic, 1744, by John Henry Wintler [sic]. 1744. *Philosophical Transactions of the Royal Society* 43:166–169.

An account of the death of Mr. George William Richman [sic]. 1755. *Philosophical Transactions of the Royal Society* 45:61–69.

Accum, Friedrich C. 1803. *A System of Theoretical and Practical Chemistry.* 2 vols. London.

Achard, Franz K. 1792. Observations sur l'électricité terrestre. *Mémoires de l'Académie des sciences et belles-lettres, classe de philosophie expérimentale* (1786), 13–16. Berlin.

Adams, George. 1771. *Micrographia illustrata; or, The Microscope Explained, in Several New Inventions, Particularly of a New Variable Microscope for Examining All Sorts of Minute Objects.* 4th ed. London.

———. 1784. *An Essay on Electricity.* London.

———. 1789. *An Essay on Vision.* London.

———. 1792. *An Essay on Electricity.* 4th ed. London.

———. 1799. *An Essay on Electricity.* 5th ed. London.

Adams, George, Louis Joblot, and Abraham Trembley. 1746. *Micrographia illustrata.* London.

Adams, George, and William Jones. 1799. *Lectures on Natural and Experimental Philosophy.* 5 vols. 2d ed. London.

Ader, Robert. 1997. The role of conditioning in pharmacotherapy. In *The Placebo Effect: An Interdisciplinary Exploration,* edited by Anne Harrington, 138–165. Cambridge, Massachusetts.

Aldini, John. 1819. *General Views on the Application of Galvanism to Medical Purposes; Principally in Cases of Suspended Animation.* London.

Allaman, M. 1758. Account of the cure of an extraordinary kind of palsey. *The Gentleman's Magazine* 28:467–468.

Altick, Richard D. 1978. *The Shows of London.* Cambridge, Massachusetts.

Aragão, Francisco de. 1800. Breve compendio ou tratado sobre a electricidade. Lisbon.

Aschoff, V. 1984. *Geschichte der Nachrichtentechnik.* Berlin.

Baird, Davis, and Alfred Nordmann. 1994. Facts-well-put. *British Journal for the Philosophy of Science* 45:37–77.

Baker, Henry. 1748. A letter, concerning several medical experiments of electricity. *Philosophical Transactions of the Royal Society* 45:270–275.

Bakewell, F. C. 1853. *Electric Science; Its History, Phenomena, and Applications.* London.

Bakken Library and Museum. 1996. *Sparks and Shocks: Experiments from the Golden Age of Static Electricity.* Dubuque, Iowa.

Barnes, Barry, David Bloor, and John Henry. 1996. *Scientific Knowledge: A Sociological Analysis.* Chicago.

Barneveld, Willem van. 1785. *Geneeskundige electriciteit.* Amsterdam.

———. 1787. *Medizinische Elektrizität.* Leipzig.

Barrow, John, compiler. 1993. *Voyages of Discovery: Captain Cook.* Chicago.

Bartaloni, Domenico. 1777 [?]. *Sul conduttore elettrico della torre della piazza di Siena.* Siena [?].

Basalla, George. 1988. *The Evolution of Technology.* Cambridge.

Bauer, Fulgenz. 1770. *Experimental* [sic] *Abhandlung von der Theorie und dem Nutzen der Elektricität.* Thur und Lindau.

Bazerman, Charles. 1991. How natural philosophers can cooperate: the literary technology of coordinated investigation in Joseph Priestley's *History and Present State of Electricity* (1767). In *Textual Dynamics of the Professions: Historical and Contemporary Studies of Writing in Professional Communities,* edited by Charles Bazerman and James Paradis, 13–44. Madison.

———. 1993. Forums of validation and forms of knowledge: the magical rhetoric of Otto von Guericke's sulfur globe. *Configurations* 2:201–228.

Beccaria, Giambatista. 1776. *A Treatise upon Artificial Electricity.* London.

———. 1780. *Di un ceraunografo e della cagione de'tremuoti.* Turin.

Beck, Dominikus. 1787. *Kurzer Entwurf der Lehrer von der Elektricität.* Salzburg.

Becket, J. B. 1773. *An Essay on Electricity.* Bristol.

Bedini, Silvio A. 1964. *Early American Scientific Instruments and Their Makers.* Washington, D.C.

———. 1965. The evolution of science museums. *Technology and Culture* 6:1–29.

Ben-Chaim, Michael. 1990. Social mobility and scientific change: Stephen Gray's contribution to electrical research. *British Journal for the History of Science* 22:3–24.

Benguigui, Isaac. 1984. *Théories électriques du XVIIIe siècle: correspondance entre l'abbé Nollet (1700–1770) et le physicien genevois Jean Jallabert (1712–1768).* Geneva.

Bennet, Abraham. 1787. An account of a doubler of electricity. *Philosophical Transactions of the Royal Society* 77:288–296.

———. 1789. *New Experiments on Electricity.* Derby.

Bensaude-Vincent, Bernadette, and Isabelle Stengers. 1996. *A History of Chemistry.* Cambridge, Massachusetts.

Benz, Ernst. 1989. *The Theology of Electricity: On the Encounter and Explanation of Theology and Science in the 17th and 18th Centuries.* Allison Park, Pennsylvania.

Berdoe, Marmaduke. 1771. *An Inquiry into the Influence of the Electric-fluid, in the Structure and Formation of Animated Beings.* Bath.

Bertholon, Pierre. 1783a. *De l'électricité des végétaux.* Lyons.

———. 1783b. *Nouvelle preuves de l'efficacité des para-tonnerres.* Montpellier.

———. 1786. *De l'électricité du corps humain, dans l'état de santé et de maladie.* 2 vols. Paris.

———. 1787. *De l'électricité des météores.* 2 vols. Lyons.

Bertucci, Paola. 1999. Medical and animal electricity in the work of Tiberius Cavallo, 1780–1795. In Luigi Galvani International Workshop, Proceedings, edited by Marco Bresadola and Guiliano Pancaldi, 147–166. *Bologna Studies in History of Science,* no. 7.

Blake-Coleman, B. C. 1992. *Copper Wire and Electrical Conductors — the Shaping of a Technology.* Chur, Switzerland.

Blondel, Christine. 1999. Animal electricity in Paris: from initial support, to its discredit and eventual rehabilitation. In Luigi Galvani International Workshop, Proceedings, edited by Marco Bresadola and Guiliano Pancaldi, 187–209. *Bologna Studies in History of Science,* no. 7.

Blunt, Thomas. 1797. *Description and Use of Blunt's Medical Electric Machine, and a New Method of Applying Metallic Conductors to Buildings, &c. for Their Preservation from Lightning.* London.

Bohnenberger, Gottlieb C. 1784–1791. *Beschreibung einiger Elektrisirmaschinen und elektrischer Bersuche.* 7 vols. Stuttgart.

———. 1788. *Beschreibung einiger Elektrisiermaschinen und elektrisher Bersuche.* 3d ed. Stuttgart.

———. 1793–1795. *Beyträge zur theoretischen und praktischen Elektrizitätslehre.* Stuttgart.

———. 1798. *Beschreibung unterschiedlicher Elektrizitätsverdoppler von einer neuen Einrichtung, nebst einer Anzahl von Bersuchen über verschiedene Gegenstande der Elektrizitätslehre.* Tübingen.

Bolton, Henry C. 1892. *Scientific Correspondence of Joseph Priestley: Ninety-seven Letters addressed to Josiah Wedgwood, Sir Joseph Banks, Capt. James Keir, James Watt, Dr. William Withering, Dr. Benjamin Rush, and Others.* New York.

Bondeson, Jan. 2001. *Buried Alive: The Terrifying History of Our Most Primal Fear.* New York.

Bordeau, Sanford. 1982. *Volts to Hertz . . . the Rise of Electricity from the Compass to the Radio through the Works of Sixteen Great Men of Science Whose Names Are Used in Measuring Electricity and Magnetism.* Minneapolis, Minnesota.

Bose, George Mathias. 1738. *De attractione et electricitate.* Wittenberg.

———. 1744. *Tentamina electrica.* Wittenberg.

———. 1745. Abstract of a letter. *Philosophical Transactions of the Royal Society* 43:419–421.

Boullanger, Nicolas. 1750. *Traité de la cause et des phénomènes de l'électricité.* Paris.

Bowen, Marjorie. 1937. *Wrestling Jacob: A Study of the Life of John Wesley and Some Members of the Family.* London.

Bowers, Brian. 1982. *A History of Electric Light & Power.* London.

Boyd, Julian P. 1950–1972. *The Papers of Thomas Jefferson.* 18 vols. Princeton, New Jersey.

Boyle, Robert. 1675. *Experiments and Notes about the Mechanical Origins or Production of Electricity.* London.

Bresadola, Marco. 1999. Exploring Galvani's room for experiments. In Luigi Galvani International Workshop, Proceedings, edited by Marco Bresadola and Guiliano Pancaldi, 65–82. *Bologna Studies in History of Science,* no. 7.

Brinitzer, Carl. 1960. *A Reasonable Rebel: Georg Christoph Lichtenberg.* London.

Brown, Theodore M. 1969. The electric current in early nineteenth-century French physics. *Historical Studies in the Physical Sciences* 1:61–102.

———. 1972. Luigi Galvani. In *Dictionary of Scientific Biography,* edited by Charles C. Gillispie, 5:267–269. New York.

Browning, John. 1747. Part of a letter, concerning the effect of electricity on vegetables. *Philosophical Transactions of the Royal Society* 44:373–375.

Bruijn, J. G. de. 1969. Van Marum bibliography. In *Martinus van Marum: Life and Work,* edited by R. J. Forbes, 1:287–360. Haarlem.

Bryden, D. J. 1972. Scottish scientific instrument-makers, 1600–1900. *Royal Scottish Museum, Information Series, Technology,* no. 1.

Buchan, William. 1809. *Domestic Medicine: A Treatise on the Prevention and Cure of Diseases, Regimen and Simple Medicines.* New ed. London.

Buchwald, Jed Z. 1994. *The Creation of Scientific Effects: Heinrich Hertz and Electric Waves.* Chicago.

Bynum, W. F., and Roy Porter, editors. 1987. *Medical Fringe and Medical Orthodoxy, 1750–1850.* London.

Calandrelli, Guiseppe. 1789. *Ragionamento sopra il conduttore elettrico Quirinale.* Rome.

Cameron, K. N. 1950. *The Young Shelley: Genesis of a Radical.* New York.

Carroll, J. A., and M. W. Ashworth. 1957. *George Washington.* Vol. 7. New York.

Casati, Stefano. 1998. Storie di folgori: il dibattito Italiano sui conduttori elettrici nel settecento. *Istituto e museo di storia della scienza di Firenze — Nuncius: Annali di storia della scienza* 13:493–512.

Cavallo, Tiberius. 1776. Extraordinary electricity of the atmosphere observed at Islington on the Month of October, 1775. *Philosophical Transactions of the Royal Society* 66:407–411.

————. 1777a. *A Complete Treatise of Electricity in Theory and Practice; with Original Experiments.* London.

————. 1777b. An account of some new electrical experiments. *Philosophical Transactions of the Royal Society* 67:48–55.

————. 1777c. New electrical experiments and observations; with an improvement of Mr. Canton's electrometer. *Philosophical Transactions of the Royal Society* 67:388–400.

————. 1780a. *An Essay on the Theory and Practice of Medical Electricity.* London.

————. 1780b. An account of some new experiments in electricity, with the description and use of two new electrical instruments. *Philosophical Transactions of the Royal Society* 70:15–28.

————. 1781. *A Treatise on the Nature and Properties of Air, and Other Permanently Elastic Fluids; to which Is Prefixed, an Introduction to Chymistry.* London.

————. 1785. *The History and Practice of Aerostation.* London.

————. 1788. Of the methods of manifesting the presence, and ascertaining the quantity, of small quantities of natural or artificial electricity. *Philosophical Transactions of the Royal Society* 78:1–22.

————. 1795. *A Complete Treatise on Electricity, in Theory and Practice; with Original Experiments.* 3 vols. 4th ed. London.

Cavendish, Henry. 1776. An account of some attempts to imitate the effects of the torpedo by electricity. *Philosophical Transactions of the Royal Society* 66:196–225.

————. 1784. Experiments on air. *Philosophical Transactions of the Royal Society* 74:119–153.

————. 1785. Experiments on air. *Philosophical Transactions of the Royal Society* 75:372–384.

Cavendish, Henry, William Watson, B. Franklin, and J. Robertson. 1773. A report of the committee appointed by the Royal Society, to consider of a method for securing the powder magazines at Purfleet. *Philosophical Transactions of the Royal Society* 63:42–47.

Chaldecott, J. A. 1951. *Handbook of the King George III Collection of Scientific Instruments.* Catalog of Exhibits with Descriptive Notes. London.

————. 1975. Josiah Wedgwood (1730–95)—scientist. *British Journal for the History of Science* 8:1–16.

Chambers, John. 1800. *A Pocket Herbal; containing the Medicinal Virtues and Uses of the Most Esteemed Native Plants; with Some Remarks on Bathing, Electricity, &c.* London.

Chaptal, Jean Antoine. 1807. *Chimie appliquée aux arts.* 4 vols. Paris.

Chapuis, Alfred, and Edmond Droz. 1958. *Automata: A Historical and Technological Study.* New York.

Clarke, John. 1793. *A Discourse Delivered before the Humane Society of the Commonwealth of Massachusetts.* Boston.

Clerk, Dugald. 1886. *The Gas Engine.* London.

Clifton, Gloria. 1991. The growth of the British scientific instrument trade, 1600–1850. In *Proceedings of the Eleventh International Scientific Instrument Symposium,* edited by G. Dragoni, A. McConnell, and G. L'E. Turner, 61–70. Bologna.

———. 1995. *Directory of British Scientific Instrument Makers, 1550–1851.* London.

Cohen, I. Bernard, editor. 1941. *Benjamin Franklin's Experiments: A New Edition of Franklin's Experiments and Observations on Electricity.* Cambridge, Massachusetts.

———. 1952. Prejudice against the introduction of lightning rods. *Journal of the Franklin Institute* 253(5):393–440.

———. 1954a. Neglected sources for the life of Stephen Gray (1666 or 1667–1736). *Isis* 45:41–50.

———. 1954b. Introduction to Commentary on the effects of electricity on muscular motion, by Luigi Galvani, 9–41. *Burndy Library,* publication no. 10. Norwalk, Connecticut.

———. 1956. *Franklin and Newton.* Philadelphia.

———. 1990. *Benjamin Franklin's Science.* Cambridge, Massachusetts.

———. 1995. *Science and the Founding Fathers: Science in the Political Thought of Jefferson, Franklin, Adams, and Madison.* New York.

Collier, Frank W. 1928. *John Wesley among the Scientists.* New York.

Conot, Robert. 1979. *Thomas A. Edison: A Streak of Luck.* New York.

Cook, Arthur B. 1951. *The Greeks and Their Gods.* Boston.

Cooper, Carolyn C. 1991. *Shaping Invention: Thomas Blanchard's Machinery and Patent Management in Nineteenth-Century America.* New York.

Corrigan, J. Frederick. 1924–1925. Stephen Gray (1696–1736): an early electrical experimenter. *Science Progress* 19:102–114.

Coulomb, Charles Augustin. 1785. *Construction et usage d'une balance électrique.* Vol. 1 of *Mémoires sur l'électricité et le magnétisme.* Paris.

Coulson, Thomas. 1943. Otto von Guericke: a neglected genius; part 2. *Journal of the Franklin Institute* 234(4):333–351.

Cullen, William. 1790. *First Lines of the Practice of Physic.* 3 vols. New ed. Worcester, Massachusetts.

A curious remark on electricity; from a gentleman in America. 1750. *The Gentleman's Magazine* 20:208.

Curry, James. 1792. *Popular Observations on Apparent Death from Drowning, Suffocation, &c. with an Account of the Means to be Employed for Recovery.* London.

Cuthbertson, John. 1786. *Abhandlung von der Elektrizität nebst einer genauen Beschreibung der dahingehörigen Werkzeuge und Versuche.* Leipzig.

———. 1807. *Practical Electricity, and Galvanism, containing a Series of Experiments Calculated for the Use of Those Who Are Desirous of Becoming Acquainted with That Branch of Science.* London.

———. 1821. *Practical Electricity, and Galvanism, containing a Series of Ex-*

periments Calculated for the Use of Those Who Are Desirous of Becoming Acquainted with That Branch of Science. 2d ed. London.

Dal Negro, Salvator D. 1799. *Nuovo metodo di costruire macchine elettriche di grandezza illimitata.* Venice.

Daumas, Maurice. 1989. *Scientific Instruments of the 17th & 18th Centuries and Their Makers.* London.

Davy, Humphry. 1807. The Bakerian Lecture, on some chemical agencies of electricity. *Philosophical Transactions of the Royal Society* 97:1–56.

de Clercq, Peter. 1997. *At the Sign of the Oriental Lamp: The Musschenbroek Workshop in Leiden, 1660–1750.* Rotterdam.

Delaborde, R. P. 1761. *Le clavessin électrique avec une nouvelle théorie du mécanisme et des phénomènes de l'électricité.* Paris.

de la Cepède, Bernard. 1781. *Essai sur l'électricité naturelle et artificielle.* 2 vols. Paris.

de Romas, Jacques. 1776. *Mémoire, sur les moyens de se garantir de la foudre dans les maisons.* Bordeaux.

Desaguliers, J. T. 1719. *Lectures of Experimental Philosophy.* 2d ed. London.
———. 1729. An attempt to solve the phaenomenon of the rise of vapours, formation of clouds and descent of rain. *Philosophical Transactions of the Royal Society* 36:6–22.
———. 1739. Some thoughts and experiments concerning electricity. *Philosophical Transactions of the Royal Society* 41:186–210.
———. 1741. Several electrical experiments, made at various times, before the Royal Society. *Philosophical Transactions of the Royal Society* 41:661–667.
———. 1742. *Dissertation concerning Electricity.* London.
———. 1763. *A Course of Experimental Philosophy.* 2 vols. 3d ed. London.

Description of the apparatus employed by Lavoisier to produce water from its component parts, oxygen and hydrogen. 1798a. *The Philosophical Magazine* 1:303–304; and plate 10.

Description of the apparatus employed by the Society for Philosophical Experiments and Conversations, for producing water by the combustion of hydrogen gas in oxygen gas; with an account of the process. 1798b. *The Philosophical Magazine* 2:148–155; and plate 4.

Description of the apparatus invented by Mr. John Cuthbertson for producing water by the combustion of hydrogen gas in oxygen gas. 1798c. *The Philosophical Magazine* 2:317–318; and plate 8.

De Solla Price, Derek J. 1984. Notes towards a philosophy of science/technology interaction. In *The Nature of Technological Knowledge,* edited by R. Laudan, 105–114. Dordrecht.

Dibner, Bern. 1957. Early electrical machines. *Burndy Library,* publication no. 14. Norwalk, Connecticut.
———. 1964. *Alessandro Volta and the Electric Battery.* New York.
———. 1967. Communications. In *Technology in Western Civilization,* edited by Melvin Kranzberg and Carroll Pursell, 1:452–468. New York.

————. 1970. Giovanni Aldini. In *Dictionary of Scientific Biography*, edited by Charles C. Gillispie, 1:107–108. New York.

————. 1971. Luigi Galvani. *Burndy Library*, publication no. 26. Norwalk, Connecticut.

————. 1976. Benjamin Franklin electrician. *Burndy Library*, publication no. 30. Norwalk, Connecticut.

Dollond, P., and G. Dollond. n.d. *A Catalogue of Optical, Mathematical, and Philosophical Instruments made by P. and G. Dollond, Opticians to His Majesty.* London. Two versions of this one-page flyer are on deposit at the Bakken Library, Minneapolis, Minnesota.

Donkin, Bryan. 1900. *A Text-book on Gas, Oil, and Air Engines.* 3d ed. London.

Donndorff, Johann A. 1784. *Die Lehre von der Elektricität theoretisch und praktisch aus einander gesezt.* 2 vols. Erfurt.

Doppelmayr, Johann Gabriel. 1744. *Neu-entdeckte Phaenomena von Bewunderns-würdigen Würckungen der Natur, welche bey der fast allen Körpern zukommenden Electrischen Krafft.* Nuremberg.

du Fay, Charles. 1737a. Premier mémoire sur l'électricité. *Mémoires de mathématique et de physique de l'Académie royale des sciences* (1733), 31–49. Amsterdam.

————. 1737b. Second mémoire sur l'électricité. *Mémoires de mathématique et de physique de l'Académie royale des sciences* (1733), 100–117. Amsterdam.

————. 1737c. Troisième mémoire sur l'électricité. *Mémoires de mathématique et de physique de l'Académie royale des sciences* (1733), 327–357. Amsterdam.

————. 1737d. Quatrième mémoire sur l'électricité. *Mémoires de mathématique et de physique de l'Académie royale des sciences* (1733), 617–643. Amsterdam.

————. 1738a. Cinquième mémoire sur l'électricité. *Mémoires de mathématique et de physique de l'Académie royale des sciences* (1734), 470–499. Amsterdam.

————. 1738b. Sixième mémoire sur l'électricité. *Mémoires de mathématique et de physique de l'Académie royale des sciences* (1734), 691–724. Amsterdam.

————. 1741. Septième mémoire sur l'électricité. *Mémoires de mathématique et de physique de l'Académie royale des sciences* (1737), 124–143. Amsterdam.

————. 1745. *Versuche und Abhandlungen von der Electricität derer Körper.* Erfurt.

Dulieu, Louis. 1970. Pierre Bertholon. In *Dictionary of Scientific Biography*, edited by Charles C. Gillispie, 2:82–83. New York.

Eisenberg, Mickey S. 1997. *Life in the Balance: Emergency Medicine and the Quest to Reverse Sudden Death.* Oxford.

Eklund, Jon. 1975. *The Incompleat Chymist: Being an Essay on the Eighteenth-Century Chemist in His Laboratory, with a Dictionary of Obsolete Chemical Terms of the Period.* Washington, D.C.

Eshet, Dan. 2001. Rereading Priestley: science at the intersection of theology and politics. *History of Science* 39:127–159.

Evans, Arthur F. 1932. *The History of the Oil Engine: A Review in Detail of the Development of the Oil Engine from the Year 1680 to the Beginning of the Year 1930.* London.

Fahie, John J. 1884. *A History of Electric Telegraphy, to the Year 1837.* London.

Falconer, William. 1790. Observations on the knowledge of the ancients respecting electricity. *Memoirs of the Literary and Philosophical Society of Manchester* 3:278–292.

Fara, Patricia. 1996. *Sympathetic Attractions: Magnetic Practices, Beliefs, and Symbolism in Eighteenth-Century England.* Princeton, New Jersey.

———. 2002. *An Entertainment for Angels: Electricity in the Enlightenment.* Cambridge.

Farrar, Kathleen R. 1963. A note on a eudiometer supposed to have belonged to Henry Cavendish. *British Journal for the History of Science* 1:375–380.

Faulwetter, Carl. 1791. *Kurze Gründfaze der Elektricitätslehre.* Vol. 2. Nuremberg.

Ferguson, James. 1775. *An Introduction to Electricity.* London.

———. 1778. *An Introduction to Electricity in Six Sections.* 3d ed. London.

Ferguson, James, and C. F. Partington. 1825. *Lectures on Electricity.* New ed. London.

Fields, Howard L., and Donald D. Price. 1997. Toward a neurobiology of placebo analgesia. In *The Placebo Effect: An Interdisciplinary Exploration,* edited by Anne Harrington, 93–116. Cambridge, Massachusetts.

Figuier, Louis. 1867. *Les merveilles de la science ou description populaire des inventions modernes.* 4 vols. Paris.

Finn, Bernard S. 1969. An appraisal of the origins of Franklin's electrical theory. *Isis* 60:362–369.

———. 1971. Output of eighteenth-century electrostatic machines. *British Journal for the History of Science* 5:289–291.

Flexner, James T. 1972. *George Washington: Anguish and Farewell (1793–1799).* Vol. 4. Boston.

Follini, Giorgio. 1791. *Teoria elettrica brevement esposta.* Ivrea.

Fortescue, John. [1927–1928] 1967. *The Correspondence of King George the Third from 1760 to December 1783.* 6 vols. London.

Fothergill, Anthony. 1796. *A New Inquiry into the Suspension of Vital Action, in Cases of Drowning and Suffocation.* 3d ed. Bath.

Fothergill, John. 1783. *The Works of John Fothergill, M.D.* Edited by Coakley Lettson. 3 vols. London.

Fourcroy, Antoine-François de. 1800–1802. *Système des connaissances chimiques, et de leurs applications aux phénomènes de la nature et de l'art.* 11 vols. Paris.

Fowler, Richard. 1793. *Experiments and Observations Relative to the Influence lately Discovered by M. Galvani and Commonly Called Animal Electricity.* Edinburgh.

Franklin, Benjamin. 1986. *Benjamin Franklin: The Autobiography and Other Writings*. New York.

———. [1769] 1996. *Experiments and Observations on Electricity, Made at Philadelphia in America*. New York.

Freke, John. 1752. *An Essay to Shew the Cause of Electricity and Why Some Things Are Non-electricable*. London.

Friedel, R., P. Israel, and B. Finn. 1986. *Edison's Electric Light: Biography of an Invention*. New Brunswick, New Jersey.

Frisinger, H. Howard. 1977. *The History of Meteorology: To 1800*. New York.

Fulton, John F., and Harvey Cushing. 1936. A bibliographical study of the Galvani and the Aldini writings on animal electricity. *Annals of Science* 1: 239–268.

Gale, T. 1802. *Theory and Practice of Medical Electricity; and Demonstrated to be an Infallible Cure of Fever, Inflammation, and Many Other Diseases*. Troy, New York.

Galison, Peter. 1987. *How Experiments End*. Chicago.

Gälle, Meingofus. 1813. *Beyträge zur Erweiterung und Vervollkommnung der Elektricitätslehre in theoretischer und practischer Hinsicht*. Salzburg.

Galvani, Luigi. [1791] 1953. Commentary on the effects of electricity on muscular motion. Translated by Margaret Glover Foley. *Burndy Library*, publication no. 10. Norwalk, Connecticut.

Gardane, J. J. 1768. *Conjectures sur l'électricité médicale*. Paris.

Gardini, Josephi F. 1780. *De effectis electricitatus in homine*. Genoa.

Garratt, G. R. M. 1958. Telegraphy. In *The Industrial Revolution, ca. 1750 to ca. 1850*. Vol. 4 of *A History of Technology*, edited by C. Singer, E. J. Holmyard, A. R. Hall, and T. I. Williams, 644–662. Oxford.

Geison, Gerald L. 1981. Scientific change, emerging specialties, and research schools. *History of Science* 19:20–40.

Gilbert, William. [1600] 1958. *De magnete*. Translated by P. Fleury Mottelay (1893). New York.

Gillispie, Charles C. 1983. *The Montgolfier Brothers and the Invention of Aviation, 1783–1784*. Princeton, New Jersey.

Gillmor, C. Stewart. 1971. Charles Augustin Coulomb. In *Dictionary of Scientific Biography*, edited by Charles C. Gillispie, 9:439–447. New York.

Giovanni, duca di Naja Cardafa. 1759. *Lettre sur la tourmaline, à Monsieur de Buffon*. Paris.

Glassie, Henry. 1999. *Material Culture*. Bloomington, Indiana.

Gluckman, Albert Gerard. 1996. *The Invention and Evolution of the Electro-technology to Transmit Electrical Signals without Wires*. 2d ed. Washington, D.C.

Golinski, Jan. 1992. *Science as Public Culture: Chemistry and Enlightenment in Britain, 1760–1820*. Cambridge.

Gooding, David. 1990. *Experiment and the Making of Meaning: Human Agency in Scientific Observation and Experiment*. Dordrecht.

Gooding, David, Trevor Pinch, and Simon Schaffer, editors. 1989. *The Uses of Experiment: Studies in the Natural Sciences.* Cambridge.

Gordon, Andreas. 1745. *Versuch einer Erklärung der Electricität.* 2d ed. Erfurt.

Graham, James. 1779. *The General State of Medical and Chirurgical Practice, Exhibited; Showing Them to Be Inadequate, Ineffectual, Absurd, and Ridiculous.* London.

Gray, Stephen. 1720. An account of some new electrical experiments. *Philosophical Transactions of the Royal Society* 31:104–107.

―――. 1731. A letter to Cromwell Mortimer, containing several experiments concerning electricity. *Philosophical Transactions of the Royal Society* 37: 18–44.

―――. 1732a. A letter concerning the electricity of water. *Philosophical Transactions of the Royal Society* 37:227–230.

―――. 1732b. A letter from Mr. Stephen Gray to Dr. Mortimer, containing a farther account of his experiments concerning electricity. *Philosophical Transactions of the Royal Society* 37:285–291.

―――. 1732c. Two letters from Mr. Stephen Gray to C. Mortimer, containing farther accounts of his experiments concerning electricity. *Philosophical Transactions of the Royal Society* 37:397–407.

―――. 1735a. Experiments and observations upon the light that is produced by communicating electrical attraction to animal or inanimate bodies, together with some of its most surprising effects. *Philosophical Transactions of the Royal Society* 39:16–24.

―――. 1735b. A letter from Stephen Gray, to Dr. Mortimer, containing some experiments relating to electricity. *Philosophical Transactions of the Royal Society* 39:166–170.

―――. 1736. Mr. Stephen Gray, his last letter to Granville Wheler, concerning the revolutions which small pendulous bodies will, by electricity, make round larger ones from west to east as the planets do round the sun. *Philosophical Transactions of the Royal Society* 39:220.

Gray, Stephen, and Cromwell Mortimer. 1736. An account of some electrical experiments intended to be communicated to the Royal Society by Mr. Stephen Gray. *Philosophical Transactions of the Royal Society* 39:400–403.

Grayson, Donald. 1983. *The Establishment of Human Antiquity.* New York.

Grylls, R. G. 1938. *Mary Shelley: A Biography.* London.

Guericke, Otto von. [1672] 1994. *The New (so-called) Magdeburg Experiments of Otto von Guericke.* Translated with preface by Margaret Glover Foley Ames. Dordrecht.

Guerlac, Henry. 1972a. Francis Hauksbee [the elder]. In *Dictionary of Scientific Biography,* edited by Charles C. Gillispie, 6:169–175. New York.

―――. 1972b. Francis Hauksbee [the younger]. In *Dictionary of Scientific Biography,* edited by Charles C. Gillispie, 6:175–176. New York.

Gütle, Johann Conrad. 1790. *Beschreibung eines mathematisch-physikalischen Maschinen und Instrumenten-Kabinets, mit zugehörigen Bersuchen zum Gebrauch für Schulen.* Leipzig.

———. 1791. *Bersuche Unterhaltungen und Belustigungen aus der natür-lichen Magie.* Leipzig.

Guyot, M. 1770. *Nouvelles récréations physiques et mathématiques.* Vol. 2, pt. 4. New ed. Paris.

Hackmann, W. D. 1971. Electrical researches. In *Martinus van Marum: Life and Work,* edited by R. J. Forbes, 3:329–378. Haarlem.

———. 1973. John and Jonathan Cuthbertson. The invention and development of the eighteenth-century plate electrical machine. *National Museum of the History of Science,* Communication no. 142. Leyden.

———. 1978a. *Electricity from Glass: The History of the Frictional Electrical Machine 1600–1850.* Alphen aan den Rijn.

———. 1978b. Eighteenth-century electrostatic measuring devices. *Istituto e museo di storia della scienza di Firenze, Annali* 3(2).

Hales, Stephen. 1733. *Statical Essays: Containing Haemastaticks.* London.

Hall, Richard W. 1806. *An Inaugural Essay on the Use of Electricity in Medicine.* Philadelphia.

Hankins, Thomas L. 1985. *Science and the Enlightenment.* Cambridge.

Hankins, Thomas L., and Robert J. Silverman. 1995. *Instruments and the Imagination.* Princeton, New Jersey.

Hardenberg, Horst O. 1992. Samuel Morey and his atmospheric engine. *Society of Automotive Engineers,* SP-922. Warrendale, Pennsylvania.

Harris, W. Snow. 1831. On the protection of ships from lightning. In *Papers on Naval Architecture, and Other Subjects connected with Naval Science,* edited by William Morgan and Augustin Creuze, 425–446. London.

Hart, Cheney. 1754. Part of a letter, giving some account of the effects of electricity in the county hospital at Shrewsbury. *Philosophical Transactions of the Royal Society* 48:786–788.

Hartmann, Johann F. 1766. *Electrische Experimente im lufleeren Raume.* Hannover.

———. 1770. *Die angewandte Electricität bey Krankheiten des menschlischen Körpers.* Hannover.

Hauksbee, Francis [the elder]. 1709. *Physico-mechanical Experiments on Various Subjects.* London.

———. 1716. *Esperienze fisico-meccaniche sopra varj soggetti.* Florence.

———. 1719. *Physico-mechanical Experiments on Various Subjects.* Rev. ed. London.

Hauksbee, Francis [the younger], and William Whiston. 1714. *A Course of Mechanical, Optical, Hydrostatical, and Pneumatical Experiments.* London.

Haüy, René J. 1787. *Exposition raisonnée de la théorie de l'électricité et du magnétisme.* Paris.

Hawes, W., editor. 1794. *Transactions of the Royal Humane Society from 1774–1784; with an Appendix of Miscellaneous Observations on Suspended Animation, to the Year 1794.* Vol. 1. London.

Hayden, Brian. 1998. Practical and prestige technologies: the evolution of material systems. *Journal of Archaeological Method and Theory* 5:1–55.

Haygarth, John. 1800. *Of the Imagination, as a Cause and as a Cure of Disorders of the Body; exemplified by Fictitious Tractors, and Epidemical Convulsions.* Bath.

Heidmann, Johann A. 1799. *Volständige auf Versuche und Vernunftschlüsse gegründete Theorie der Elektricität für Aerzte, Chymiker und Freunde der Naturkunde.* 2 vols. Vienna.

Heilbron, John L. 1970. Giambatista Beccaria. In *Dictionary of Scientific Biography*, edited by Charles C. Gillispie, 1:546–548. New York.

———. 1971a. Charles-François de Cisternai du Fay. In *Dictionary of Scientific Biography*, edited by Charles C. Gillispie, 4:214–217. New York.

———. 1971b. Tiberius Cavallo. In *Dictionary of Scientific Biography*, edited by Charles C. Gillispie, 3:153–154. New York.

———. 1972. Stephen Gray. In *Dictionary of Scientific Biography*, edited by Charles C. Gillispie, 5:515–517. New York.

———. 1974. Jean-Antoine Nollet. In *Dictionary of Scientific Biography*, edited by Charles C. Gillispie, 10:145–148. New York.

———. 1975. Georg Wilhelm Richmann. In *Dictionary of Scientific Biography*, edited by Charles C. Gillispie, 11:432–434. New York.

———. 1976a. William Watson. In *Dictionary of Scientific Biography*, edited by Charles C. Gillispie, 14:193–196. New York.

———. 1976b. Alessandro Guiseppe Antonio Anastasio Volta. In *Dictionary of Scientific Biography*, edited by Charles C. Gillispie, 14:69–82. New York.

———. 1979. *Electricity in the Seventeenth and Eighteenth Centuries: A Study in Early Modern Physics.* Berkeley.

———. 1981. The electrical field before Faraday. In *Conceptions of Ether*, edited by G. N. Cantor and M. J. S. Hodge, 187–213. Cambridge.

———. 1982. *Elements of Early Modern Physics.* Berkeley.

Helsham, Richard. 1755. *A Course of Lectures in Natural Philosophy.* 3d ed. London.

Henly (or Henley), William. 1772. An account of the death of a person destroyed by lightning in the chapel in Tottenham-Court-Road, and its effects on the building; as observed by Mr. William Henly, Mr. Edward Nairne, and Mr. William Jones. *Philosophical Transactions of the Royal Society* 62:131–136.

———. 1774. Experiments concerning the different efficacy of pointed and blunted rods, in securing buildings against the stroke of lightning. *Philosophical Transactions of the Royal Society* 64:133–151.

———. 1777. Experiments and observations in electricity. *Philosophical Transactions of the Royal Society* 67:85–143.

———. 1778. Observations and experiments tending to confirm Dr. Ingenhousz's theory of the electrophorus; and to shew the impermeability of glass to electric fluid. *Philosophical Transactions of the Royal Society* 68:1049–1055.

Henly, William, T. Lane, E. Nairne, and J. Planta. 1778. The report of the committee appointed by the Royal Society, for examining the effect of lightning, May 15, 1777, on the parapet-wall of the house of the Board of Ordnance, at Purfleet in Essex. *Philosophical Transactions of the Royal Society* 68:236–238.

Henry, William. 1800. Account of a series of experiments, undertaken with the view of decomposing the muriatic acid. *The Philosophical Magazine* 7: 211–218.

Hervieu, Jean. 1835. *Essai sur l'électricité atmospherique, et son influence dans les phénomènes météorologiques.* Paris.

Hibbert, Christopher. 1987. *The English: A Social History, 1066–1945.* Grafton, London.

Higgins, Thomas J. 1961. A biographical bibliography of electrical engineers and electrophysicists (pt. 1). *Technology and Culture* 2:28–32.

Himsel, Nicolaus von. 1759. The case of a paralytic patient cured by an electrical application. *Philosophical Transactions of the Royal Society* 51:179–185.

Histoire générale et particulière de l'électricité. 1752. 2 vols. Paris.

An historical account of the wonderful discoveries, made in Germany, &c. concerning electricity. 1745. *The Gentleman's Magazine* 15:193–197.

Hoadly, Benjamin, and Benjamin Wilson. 1759. *Observations on a Series of Electrical Experiments.* 2d ed. London.

Hobsbawm, Eric. [1962] 1996. *The Age of Revolution, 1789–1848.* New York.

Hoff, Hebbel H. 1936. Galvani and the pre-Galvanian electrophysiologists. *Annals of Science* 1:157–172.

Hoffmann, Johann C. 1798. *Praktische und gründliche Anweisung auf eine leichte und wohlfeile Art gute Elektrisirmaschinen zu bauen.* Leipzig.

Holmes, Frederic L. 1989. *Eighteenth-Century Chemistry as an Investigative Enterprise.* Berkeley.

———. 1998. *Antoine Lavoisier—the Next Crucial Year.* Princeton, New Jersey.

Home, R. W. 1970. Electricity and the nervous fluid. *Journal of the History of Biology* 3:235–251.

———. 1979. *Aepinus's Essay on the Theory of Electricity and Magnetism.* Princeton, New Jersey.

———. 1981. *The Effluvial Theory of Electricity.* New York.

———. 1992. *Electricity and Experimental Physics in Eighteenth-Century Europe.* Aldershot.

Hooper, W. 1774. *Rational Recreations, in which the Principles of Numbers and Natural Philosophy Are Clearly and Copiously Elucidated, by a Series of Easy, Entertaining, Interesting Experiments.* 4 vols. London.

Howe, M. A. DeWolfe. 1918. *The Humane Society of the Commonwealth of Massachusetts: An Historical Review, 1785–1916.* Cambridge, Massachusetts.

Huisman, Frank. 1999. Shaping the medical market: on the construction of quackery and folk medicine in Dutch historiography. *Medical History* 43:359–375.

Huizenga, John R. 1992. *Cold Fusion: The Scientific Fiasco of the Century.* Rochester, New York.

Hull, A. R. 1956. *The Scientific Revolution, 1500–1800: The Formation of the Modern Scientific Attitude.* Boston.

Imison, John. 1796. *The School of Arts; or, An Introduction to Useful Knowledge.* 4th ed. London.

———. 1803. *Elements of Science and Art: Being a Familiar Introduction to Natural Philosophy and Chemistry together with Their Application to a Variety of Elegant and Useful Arts.* New ed. 2 vols. London.

Ingen-Housz, Jan (John). 1778a. A ready way of lighting a candle, by a very moderate electrical spark. *Philosophical Transactions of the Royal Society* 68:1022–1026.

———. 1778b. Electrical experiments, to explain how far the phenomena of the electrophorus may be accounted for by Dr. Franklin's theory of positive and negative electricity. *Philosophical Transactions of the Royal Society* 68:1027–1048.

———. 1779a. Improvements in electricity. *Philosophical Transactions of the Royal Society* 69:659–673.

———. 1779b. Account of a new kind of inflammable air or gass, which can be made in a moment without apparatus, and is as fit for explosion as other inflammable gasses in use for that purpose; together with a new theory of gunpowder. *Philosophical Transactions of the Royal Society* 69:376–419.

———. 1784. *Vermischte Schriften physisch-medicinischen Inhalts.* Vienna.

———. 1785. *Nouvelles expériences et observations sur divers objets de physique.* Paris.

Irving, Washington. 1800. *The Life of Washington.* Vol. 2. New York.

Israel, Paul. 1998. *Edison: A Life of Invention.* New York.

Jackson, Mark. 1996. *New-born Child Murder: Women, Illegitimacy and the Courts in Eighteenth-century England.* Manchester.

Jacob, Margaret. 1997. *Scientific Culture and the Making of the Industrial West.* Oxford.

Jacquet, Louis S. 1775. *Précis de l'électricité ou extrait expérimental et théorétique des phénomènes électriques.* Vienna.

Jacyna, L. S. 1999. Galvanic influences: themes in the early history of British animal electricity. In *Luigi Galvani International Workshop, Proceedings,* edited by Marco Bresadola and Guiliano Pancaldi, 167–185. *Bologna Studies in History of Science,* no. 7.

Jaffe, Bernard. 1976. *Crucibles: The Story of Chemistry from Ancient Alchemy to Nuclear Fusion.* 4th ed. New York.

Jallabert, Jean. 1748a. *Expériences sur l'électricité, avec quelques conjectures sur la cause de ses effets.* Geneva.

———. 1748b. A palsy cured by electrising. *The Gentleman's Magazine* 18:487–488.

Jameson, Eric. 1961. *The Natural History of Quackery.* London.

Jardine, Nick. 2000. Uses and abuses of anachronism in the history of the sciences. *History of Science* 38:251–270.

Jarvis, C. Mackechnie. 1956. The origin and development of the electric telegraph. *Journal of the Institution of Electrical Engineers*, n.s., 2:130–137.

Joerges, B., and T. Shinn. 2001. *Instrumentation: Between Science, State and Industry*. Dordrecht.

Johnson, Ann. 2000. Engineering culture and the production of knowledge: an intellectual history of anti-lock braking systems. Ph.D. dissertation, Department of History, Princeton University.

———. 2001. From dynamometers to simulations: transforming brake testing technology into antilock braking systems. In *Instrumentation: Between Science, State and Industry*, edited by B. Joerges and T. Shinn, 199–218. Dordrecht.

Jones, H. Bence. 1852. *On Animal Electricity: Being an Abstract of the Discoveries of Emil du Bois-Reymond*. London.

Jones, Thomas P. 1828. Some remarks upon explosion and vapour engines. *Journal of the Franklin Institute and American Mechanics' Magazine* 5:18–29.

Josephson, Matthew. 1959. *Edison: A Biography*. New York.

Jungnickel, Christa, and Russell McCormmach. 1996. Cavendish. *The American Philosophical Society, Memoirs* 220.

Karrass, T. 1909. *Geschichte der Telegraphie*. Vol. 1. Braunschweig.

Keithley, Joseph F. 1998. *The Story of Electrical and Magnetic Measurements: From 500 B.C. to the 1940s*. New York.

King-Hele, D. 1977. *Doctor of Revolution: The Life and Genius of Erasmus Darwin*. London.

———, editor. 1981. *The Letters of Erasmus Darwin*. Cambridge.

———. 1984. *Shelley: His Thought and Work*. 3d ed. London.

Kipnis, Naum. 1987. Luigi Galvani and the debate on animal electricity. *Annals of Science* 44:107–142.

Kite, Charles. 1788. *An Essay on the Recovery of the Apparently Dead*. London.

Knight, Gowin. 1759. Some remarks on the preceding letter. *Philosophical Transactions of the Royal Society* 51:294–299.

Krüger, Johann G. 1744. *Zuschrifft an seine Zuhörer worinnen er ihnen seine Gedancken von der Electricität mittheilet und Ihnen zugleich seine künstige Lectionen bekant macht*. Halle.

Kryzhanovsky, Leonid N. 1990. The lightning rod in 18th-century St. Petersburg: a note on the occasion of the bicentennial of the death of Benjamin Franklin. *Technology and Culture* 31:813–817.

Kühn, Karl Gottlob. 1783. *Geschichte der medizinischen und physikalischen Elektricität*. 2 vols. Leipzig.

Kuhn, Thomas. 1970. *The Structure of Scientific Revolutions*. 2d ed. Chicago.

Labaree, Leonard, William B. Wilcox, Barbara B. Oberg, and Ellen R. Cohn, editors. 1959–2001. *The Papers of Benjamin Franklin*. New Haven, Connecticut.

La Beaume, Michael. 1820. *Remarks on the History and Philosophy, but Par-*

ticularly on the Medical Efficacy of Electricity, in the Cure of Nervous and Chronic Disorders. London.

———. 1821. *An Account of the New and Successful Treatment of Indigestion, Bilious and Nervous Complaints, Deafness, Blindness, &c*. London.

LaMotta, Vincent M., and Michael B. Schiffer. 2001. Behavioral Archaeology: Towards a New Synthesis. In *Archaeological Theory Today*, edited by Ian Hodder, 14–64. Cambridge.

Lampadius, W. A. 1793. *Bersuche und Beobachtungen über die Elektrizität und Wärme der Atmosphäre*. Berlin.

Landriani, Marsilio. 1784. *Dell'utilità dei conduttori elettrici*. Milan.

Lane, T. 1767. Description of an electrometer invented by Mr. Lane; with an account of some experiments made by him with it. *Philosophical Transactions of the Royal Society* 57:451–460.

Langenbucher, Jacob. 1780. *Beschreibung einer beträchtlish verbesserten Electrisiermaschine*. Augsburg.

———. 1783. *Richtige Begriffe vom Blitz und von Blitzableitern*. Augsburg.

———. 1788. *Praktische Elektricitätslehre*. Augsburg.

La Perrière du Roiffé, Jacques C. 1761. *Mécanismes de l'électricité et de l'univers*. 2 vols. Paris.

Lave, J., and E. Wenger. 1991. *Situated Learning: Legitimate Peripheral Participation*. Cambridge.

Lavoisier, Antoine. 1783. *Essays, on the Effects Produced by Various Processes on Atmospheric Air*. London.

———. 1793. *Elements of Chemistry in a New Systematic Order, containing All the Modern Discoveries*. Translated by Robert Kerr. 2d ed. Edinburgh.

———. [1789] 1952. Elements of chemistry. In *Great Books of the Western World*, edited by Robert Maynard Hutchins, 45:ix–133. Chicago.

Lemay, J. A. L. 1964. *Ebenezer Kinnersley, Franklin's Friend*. Philadelphia.

Leone, Mark P., and Paul A. Shackel. 1987. Forks, clocks, and power. In *Mirror and Metaphor: Material and Social Constructions of Reality*, edited by Daniel W. Ingersoll, Jr., and Gordon Bronitsky, 46–61. Lanham, Maryland.

Licht, Sidney. 1967. History of electrotherapy. In *Therapeutic Electricity and Ultraviolet Radiation*, edited by S. Licht, 1–70. New Haven, Connecticut.

Lichtenberg, Georg Christoph. 1778. *De nova methodo naturam ac motum fluidi electrici investigandi*. Göttingen.

———. 1779. *De nova methodo naturam ac motum fluidi electrici investigandi*. Göttingen.

Lienhard, John H. 1979. The rate of technological improvement before and after the 1830s. *Technology and Culture* 20:515–530.

Life of Joseph Toaldo. 1801. *The Philosophical Magazine* 11:257–261.

Lining, John. 1754. Extract of a letter from John Lining; with his answers to several queries sent to him concerning his experiment of electricity with a kite. *Philosophical Transactions of the Royal Society* 48:757–764.

Loesser, Arthur. 1954. *Men, Women and Pianos: A Social History*. New York.

Louis, Antoine. 1747. *Observations sur l'électricité*. Paris.

Lovett, R. 1756. *The Subtil Medium Prov'd; or, That Wonderful Power of Nature, So Long Ago Conjectur'd by the Most Ancient and Remarkable Philosophers, which They Call'd Sometimes Aether but Oftener Elementary Fire, Verify'd*. London.

———. 1760. *The Reviewers Review'd; or, The Bush-fighters Exploded: Being a Reply to the Animadversions, Made by the Authors of the Monthly Review, on a Late Pamphlet, entitled Sir Isaac Newton's Aether Realiz'd*. London.

Lowndes, Francis. 1787. *Observations on Medical Electricity*. London.

Lyons, Henry. 1944. *The Royal Society, 1660–1940: A History of Its Administration under Its Charters*. Cambridge.

McClellan, John E., III. 1985. *Science Reorganized: Scientific Societies in the Eighteenth Century*. New York.

McCormmach, Russell. 1971. Henry Cavendish. In *Dictionary of Scientific Biography*, edited by Charles C. Gillispie, 3:155–159. New York.

McEvoy, John G. 1979. Electricity, knowledge, and the nature of progress in Priestley's thought. *British Journal for the History of Science* 12:1–30.

MacGregor, William. 1885. *Gas Engines*. London.

McKelvey, James L. 1973. *George III and Lord Bute: The Leicester House Years*. Durham, North Carolina.

McKendrick, Neil. 1973. The role of science in the industrial revolution: a study of Josiah Wedgwood as a scientist and industrial chemist. In *Changing Perspectives in the History of Science: Essays in Honour of Joseph Needham*, edited by Mikulas Teich and Robert Young, 274–319. Dordrecht.

McMahon, A. Michal. 1984. *The Making of a Profession: A Century of Electrical Engineering in America*. New York.

Maggiotto, Francesco. 1781. *Sopra una nuova construzione di macchina elettrica*. Venice.

Mahon, Charles [earl of Stanhope]. 1779. *Principles of Electricity, containing Divers New Theorems and Experiments, Together with an Analysis of the Superior Advantages of High and Pointed Conductors*. London.

Mainstone, Rowland J. 1981. The Eddystone lighthouse. In *John Smeaton, FRS*, edited by A. W. Skempton, 83–102. London.

Maluf, Ramez Bahige. 1985. *Jean-Antoine Nollet and Experimental Natural Philosophy in Eighteenth-Century France* [Ph.D. dissertation, University of Oklahoma]. Ann Arbor.

Mangin, Msr. 1752. *Histoire générale et particulière de l'électricité*. Paris.

Marat, Jean Paul. 1782. *Recherches physiques sur l'électricité*. Paris.

———. 1784. *Mémoire sur l'électricité médicale*. Académie royale des sciences, belles-lettres et arts de Rouen. Paris.

Marland, E. A. 1964. *Early Electrical Communication*. London.

Martin, Benjamin. 1743. *A Course of Lectures in Natural and Experimental Philosophy, Geography and Astronomy*. Reading.

———. 1746. *An Essay on Electricity*. London.

———. 1754. *A Plain and Familiar Introduction to the Newtonian Philosophy*. London.

————. 1759a. *A Supplement to the Philosophia Britannica*. Appendix 1, Containing *New Experiments in Electricity, and the Method of Making Artificial Magnets*. London.

————. 1759b. *The Young Gentleman and Lady's Philosophy in a Continued Survey of the Works of Nature and Art; by Way of a Dialogue*. Vol. 1. London.

Marvin, Carolyn. 1988. *When Old Technologies Were New: Thinking about Electrical Communication in the Late 19th Century*. Oxford.

Masars de Cazeles, François. 1780–85. *Mémoire sur l'électricité médicale, et histoire du traitement de vingt malades traités, et la plupart guéris par l'électricité*. Paris.

Mascart, M. E. 1876. *Traité d'électricité statique*. 2 vols. Paris.

Mauduyt, Pierre. 1784. *Mémoire sur les différentes manières d'administrer l'électricité, et observations sur les effets qu'elles ont produits*. Paris.

Maxwell, J. Clerk, editor. [1879] 1967. *The Electrical Researches of the Honourable Henry Cavendish*. London.

Mazeas, G. 1753. Observations upon the electricity of the air, made at the château de Maintenon, during the months of June, July, and October, 1753. *Philosophical Transactions of the Royal Society* 48:377–384.

Mazzolari, G. M. 1772. *Josephi Mariani parthenii S. J. commentarii*. Rome.

Meredith, Nicholas. 1789. *Considerations on the Utility of Conductors for Lightning*. London.

Middleton, W. E. Knowles, editor. 1965. *A History of the Theories of Rain and Other Forms of Precipitation*. London.

Millburn, John R. 1976. *Benjamin Martin: Author, Instrument-maker, and 'Country Showman.'* Leyden.

————. 2000. *Adams of Fleet Street, Instrument Makers to King George III*. Aldershot.

Millburn, John R., and Henry C. King. 1988. *Wheelright of the Heavens: The Life and Work of James Ferguson, FRS*. London.

Miller, David P. 1988. 'My favourite studdys': Lord Bute as naturalist. In *Lord Bute: Essays in Re-interpretation*, edited by K. W. Schweizer, 213–239. Leicester.

Milner, Thomas. 1783. *Experiments and Observations on Electricity*. London.

Molenier, Jacob. 1768. *Essai sur le mécanisme de l'électricité, et l'utilité que l'on peut en tirer pour la guérison de quelques maladies*. Bordeaux.

Moore, Arthur D. 1973, editor. *Electrostatics and Its Applications*. New York.

————. 1997. *Electrostatics: Exploring, Controlling, and Using Static Electricity*. 2d ed. Morgan Hill, California.

Moore, C. B., G. D. Aulich, and W. Rison. 2000. Measurements of lightning rod responses to nearby strikes. *Geophysical Research Letters* 27:1487–1490.

Moratelli, D. Giambatista. 1805. *Memorie fisico-chimiche*. Venice.

Morin, Jean. 1748. *Nouvelle dissertation sur l'électricité des corps*. Chartres.

Morse, William Northrop. 1934. Lectures on electricity in colonial times. *The New England Quarterly* 8:364–374.

Morton, Alan Q. 1990. Lectures on natural philosophy in London, 1750–1765: S. C. T. Demainbray (1710–1782) and the 'inattention' of his countrymen. *British Journal for the History of Science* 23:411–434.

Morton, Alan Q., and Jane A. Wess. 1993. *Public and Private Science: The King George III Collection.* Oxford.

Morus, Iwan R. 1998. *Frankenstein's Children: Electricity, Exhibition, and Experiment in Early-Nineteenth-Century London.* Princeton, New Jersey.

Mottelay, Paul Fleury. 1922. *Bibliographical History of Electricity and Magnetism.* London.

Multhauf, Robert P., and David Davies. 1961. *A Catalogue of Instruments and Models in the Possession of the American Philosophical Society.* Philadelphia.

Muntendam, Alida M. 1969. Dr. Martinus van Marum. In *Martinus van Marum: Life and Work,* edited by R. J. Forbes, 1:1–72. Haarlem.

Musschenbroek, Petrus van. 1739a. *Beginsels der Natuurkunde, beschreeven ten dienste der Landgenooten, door Petrus van Musschenbroek.* 3 vols. Leyden.

———. 1739b. *Essai de physique.* 2 vols. Leyden.

———. 1744. *The Elements of Natural Philosophy: Chiefly Intended for the Use of Students in Universities.* 2 vols. London.

———. 1751. *Essai de physique.* 2 vols. Leyden.

———. 1769. *Cours de physique expérimentale et mathématique.* 2 vols. Paris.

Nagel, Ernest. 1961. *The Structure of Science.* New York.

Nairne, Edward. 1773. *Directions for Using the Electrical Machine, as Made and Sold by Edward Nairne, Optical, Philosophical, and Mathematical Instrument-maker.* London.

———. 1774. Electrical experiments made with a machine of his own workmanship, a description of which is prefixed. *Philosophical Transactions of the Royal Society* 64:79–89.

———. 1778. Experiments on electricity, being an attempt to shew the advantage of elevated pointed conductors. *Philosophical Transactions of the Royal Society* 69:823–860.

———. 1783. *The Description and Use of Nairne's Patent Electrical Machine.* London.

———. 1784. *Description de la machine électrique negative et positive de M. Nairne.* Paris.

———. 1793. *The Description and Use of Nairne's Patent Electrical Machine.* 4th ed. London.

Neale, John. 1747. *Directions for Gentlemen, Who Have Electrical Machines, How to Proceed in Making Their Experiments.* London.

Needham, John Turberville. 1746. A letter from Paris, concerning some new electrical experiments made there. *Philosophical Transactions of the Royal Society* 44:247–263.

Nicholson, William. 1782. *An Introduction to Natural Philosophy.* 2 vols. London.

————. 1788. A description of an instrument which, by the turning of a winch, produces the two states of electricity without friction or communication with the earth. *Philosophical Transactions of the Royal Society* 78:403–407.

————. 1790. *The First Principles of Chemistry.* London.

————. 1795. *A Dictionary of Chemistry.* 2 vols. London.

Nickson, Edward. 1778. Letters to Charles Frederick and Lord Amherst relative to Purfleet incident. *Philosophical Transactions of the Royal Society* 68: 233–235.

Nitchie, E. 1953. *Mary Shelley: Author of "Frankenstein."* New Brunswick, New Jersey.

Noad, Henry M. 1857. *Manual of Electricity.* Part 2, *Magnetism and the Electric Telegraph.* London.

Nollet, Jean-Antoine. 1738. *Programme ou idée générale d'un cours de physique expérimentale, avec un catalogue raisonné des instrumens qui servent aux expériences.* Paris.

————. 1745–65. *Leçons de physique expérimentale.* 8 vols. Amsterdam.

————. 1747. *Ensayo sobre la electricidad de los cuerpos.* Translated by Joseph Vazquez y Morales. Madrid.

————. 1748. *Lectures in Experimental Philosophy.* Translated by John Colson. London.

————. 1749a. An examination of certain phaenomena in electricity. *Philosophical Transactions of the Royal Society* 46:370–397.

————. 1749b. *Recherches sur les causes particulières des phénomènes électriques, et sur les effets nuisibles ou avantageux qu'on peut en attendre.* Paris.

————. 1751–1752. Extracts of two letters, relating to the extracting of electricity from the clouds. *Philosophical Transactions of the Royal Society* 47: 553–558.

————. 1754. *Recherches sur les causes particulières des phénomènes électriques, et sur les effets nuisibles ou avantageux qu'on peut en attendre.* New ed. Paris.

————. 1764. *Essai sur l'électricité des corps.* 4th ed. Paris.

————. 1770. *L'art des expériences.* 3 vols. Paris.

Northmore, Thomas. 1799. *A Quadruplet of Inventions.* Exeter.

Ostwald, Wilhelm. [1896] 1980. *Electrochemistry: History and Theory.* New Delhi.

Ozanam, Jacques. 1790. *Récréations mathématiques et physiques.* 4 vols. Paris.

Pain, Nesta. 1975. *George III at Home.* London.

Pancaldi, Guiliano. 1990. Electricity and life: Volta's path to the battery. *Historical Studies in the Physical and Biological Sciences* 21:123–160.

Park, Robert. 2000. *Voodoo Science: The Road from Foolishness to Fraud.* Oxford.

Partington, J. R. 1961. *A History of Chemistry.* 4 vols. London.

Paulian, Aimé-Henri. 1781. *Dictionnaire de physique.* 4 vols. 8th ed. Nîmes.

Pearson, George. 1797. Experiments and observations, made with the view of ascertaining the nature of the gas produced by passing electric discharges through water. *Philosophical Transactions of the Royal Society* 87: 142–158.

Pera, Marcello. 1992. *The Ambiguous Frog: The Galvani-Volta Controversy on Animal Electricity*. Princeton, New Jersey.

Perrin, C. E. 1973. Lavoisier, Monge, and the synthesis of water, a case of pure coincidence? *British Journal for the History of Science* 6:424–428.

Pivati, Giovanni F. 1749. *Riflessioni fisiche sopra la medicina elettrica*. Venice.

Poggendorff, J. C. 1863. *Biographisch-literärisches Handwörterbuch*. 2 vols. Leipzig.

Poirer, Jean-Pierre. 1998. *Lavoisier: Chemist, Biologist, Economist*. Philadelphia.

Polinière, Pierre. 1741. *Expériences de physique*. 2 vols. 5th ed. Paris.

Poncelet, Polycarpe. 1766. *La nature dans la formation du tonnerre, et la reproduction des êtres vivans, pour servir d'introduction aux vrais principes de l'agriculture*. 2 vols. Paris.

Post, Robert C. 1976. *Physics, Patents, and Politics: A Biography of Charles Grafton Page*. New York.

Priestley, Joseph. 1769. Experiments on the lateral force of electrical explosions. *Philosophical Transactions of the Royal Society* 59:57–62.

———. 1779. *Experiments and Observations relating to the Various Branches of Natural Philosophy; with a Continuation of the Observations on Air*. London.

———. 1781. *Experiments and Observations on Different Kinds of Air*. 3 vols. 3d ed. London.

———. 1786. *A Familiar Introduction to the Study of Electricity*. 4th ed. London.

———. 1802. Observations and experiments relating to the pile of Volta. *Journal of Natural Philosophy, Chemistry, and the Arts* 1:198–203.

———. [1775] 1966. *The History and Present State of Electricity, with Original Experiments*. 2 vols. 4th ed. New York.

Pringle, J., W. Watson, H. Cavendish, W. Henly, S. Horsley, T. Lane, Mahon, E. Nairne, Joseph Priestley. 1778. A report of the committee, appointed by the Royal Society, to consider of the most effectual method of securing the powder magazines at Purfleet against the effects of lightning. *Philosophical Transactions of the Royal Society* 68:313–317.

Purrington, Robert D. 1997. *Physics in the Nineteenth Century*. New Brunswick, New Jersey.

Rabiqueau, Charles. 1785. *Le spectacle du feu élémentaire, ou cours d'électricité expérimentale*. Paris.

Rackstrow, B. 1748. *Miscellaneous Observations, Together with a Collection of Experiments on Electricity*. London.

Raj, Kapil. 2000. 18th-century-Pacific voyages of discovery, "big science," and the shaping of an European scientific and technological culture. *History and Technology: An International Journal* 17:79–98.

Ramsden, Jesse. 1766. *Directions for Using the New Invented Electrical Machine*. London.

Rathje, William L., and Cullen Murphy. 1992. *Rubbish!* New York.

Rathje, William L., and Michael B. Schiffer. 1982. *Archaeology*. New York.

Rattansi, P. M. 1972. The social interpretation of science in the seventeenth century. In *Science and Society, 1600–1900*, edited by Peter Mathias, 1–32. Cambridge.

Read, John. 1793. *A Summary View of the Spontaneous Electricity of the Earth and Atmosphere*. London.

———. 1794. Experiments and observations made with the doubler of electricity. *Philosophical Transactions of the Royal Society* 84:266–274.

Reimarus, Johann A. 1794. *Neuere Bemerkungen vom Blitze*. Hamburg.

Remarkable effects of the late violent storms of lightning. 1787. *The Gentlemen's Magazine* 57:820–824.

Ribright, George. 1779. *A Curious Collection of Experiments, to be Performed on the Electrical Machines, made by Geo. Ribright and Son*. London.

Riddle, John. 1997. *Eve's Herbs: A History of Contraception and Abortion in the West*. Cambridge, Massachusetts.

Ritter, J. W. 1805. *Das electrische System der Körper*. Leipzig.

Robbins, Chandler. 1796. *A Discourse Delivered before the Humane Society of the Commonwealth of Massachusetts*. Boston.

Roberts, Lissa. 1998. Eudiometer. In *Instruments of Science*, edited by Robert Bud and D. J. Warner, 232–234. New York.

———. 1999. Science becomes electric: Dutch interaction with the electrical machine during the eighteenth century. *Isis* 90:680–714.

Ronalds, Francis. 1823. *Descriptions of an Electrical Telegraph, and of Some Other Electrical Apparatus*. London.

Ronayne, Thomas. 1772. A letter inclosing an account of some observations on atmospherical electricity; in regard of fogs, mists, &c. with some remarks. *Philosophical Transactions of the Royal Society* 62:137–146.

Rouland, Msr. 1785. *Description des machines électriques à taffetas*. Amsterdam.

Rowbottom, Margaret, and Charles Susskind. 1984. *Electricity and Medicine: History of Their Interaction*. San Francisco.

Royal Humane Society. 1809. *Annual Report of the Royal Humane Society*. London.

Royal Society for Improving Natural Knowledge, London. 1776. Meteorological journal kept at the house of the Royal Society. *Philosophical Transactions of the Royal Society* 66:319–352.

Rueger, Alexander. 1997. Experiments, nature and aesthetic experience in the eighteenth century. *The British Journal of Aesthetics* 37:305–323.

Ruestow, Edward. 1973. *Physics at Seventeenth and Eighteenth-Century Leiden: Philosophy and the New Science in the University*. The Hague.

Sabine, Robert. 1869. *The History and Progress of the Electric Telegraph with Descriptions of Some of the Apparatus*. 2d ed. New York.

Sans, Msr. 1772. *Guérison de la paralysie, par l'électricité.* Paris.

Schäffer, Jacob C. 1778. *Abbildung und Beschreibung der electrischen Pistole und eines kleinen zu bersuchen sehr bequemen Electricitätträgers.* Regensburg.

———. 1780. *Bersuche mit dem beständigen Electricitätträger.* Regensburg.

Schäffer, Johann Gottlieb. 1766. *Die electrische Medicin.* Regensburg.

Schaffer, Simon. 1983. Natural philosophy and public spectacle in the eighteenth century. *History of Science* 21:1–43.

Schechter, David C. 1983. *Exploring the Origins of Electrical Cardiac Stimulation.* Minneapolis, Minnesota.

Schiebinger, Londa. 1989. *The Mind Has No Sex? Women in the Origins of Modern Science.* Cambridge, Massachusetts.

Schiffer, Michael Brian. 1991. *The Portable Radio in American Life.* Tucson.

———. 1992. *Technological Perspectives on Behavioral Change.* Tucson.

———. 1993. Cultural imperatives and product development: the case of the shirt-pocket radio. *Technology and Culture* 34:98–113.

———. 1995. Social theory and history in behavioral archaeology. In *Expanding Archaeology,* edited by J. M. Skibo, W. H. Walker, and A. E. Nielsen, 22–35. Salt Lake City.

———. 1996a. *Formation Processes of the Archaeological Record.* Salt Lake City.

———. 1996b. Formation processes of the archaeological and historical records. In *Learning from Things: Method and Theory in Material Culture Studies,* edited by W. D. Kingery, 73–80. Washington, D.C.

———. 1996c. Pathways to the present: in search of shirt-pocket radios with subminiature tubes. In *Learning from Things: Method and Theory in Material Culture Studies,* edited by W. D. Kingery, 81–88. Washington, D.C.

———. 2000. Indigenous theories, scientific theories and product histories. In *Matter, Materiality and Modern Culture,* edited by Paul Graves-Brown, 72–96. London.

———. 2001a. The explanation of long-term technological change. In *Anthropological Perspectives on Technology,* edited by M. B. Schiffer, 215–235. Albuquerque.

———, editor. 2001b. *Anthropological Perspectives on Technology.* Albuquerque.

———. 2002. Studying technological differentiation: the case of 18th-century electrical technology. *American Anthropologist* 104:1148–1161.

Schiffer, Michael Brian, Tamara C. Butts, and Kimberly K. Grimm. 1994. *Taking Charge: The Electric Automobile in America.* Washington, D.C.

Schiffer, Michael Brian, Theodore Downing, and Michael McCarthy. 1981. Waste not, want not: an ethnoarchaeological study of reuse in Tucson, Arizona. In *Modern Material Culture: The Archaeology of Us,* edited by R. A. Gould and M. B. Schiffer, 67–86. New York.

Schiffer, Michael Brian, and Andrea R. Miller. 1999. *The Material Life of Human Beings: Artifacts, Behavior, and Communication.* London.

Schiffer, Michael Brian, and James M. Skibo. 1987. Theory and experiment in the study of technological change. *Current Anthropology* 28:595–622.

———. 1997. The explanation of artifact variability. *American Antiquity* 62: 27–50.

Schiffer, Michael Brian, James M. Skibo, Janet Griffitts, Kacy Hollenback, and William A. Longacre. 2001. Behavioral archaeology and the study of technology. *American Antiquity* 66:729–737.

Schmidt, Georg C. 1784. *Beschreibung gemeinnütziger Maschinen.* Jena.

Schofield, Robert E. 1963. *The Lunar Society of Birmingham: A Social History of Provincial Science and Industry in Eighteenth-century England.* Oxford.

———. 1975. Joseph Priestley. In *Dictionary of Scientific Biography,* edited by Charles C. Gillispie, 11:139–147. New York.

Schonland, B. F. J. 1952. The work of Benjamin Franklin on thunderstorms and the development of the lightning rod. *Journal of the Franklin Institute* 253(5):375–392.

Schottler, R. 1902. *Die Gasmaschine.* Braunschweig.

Schreiben eines Nuturforschers an den K. K. Hernn Hofrath von Gr. von der Beschaffenbeit der immerwährenden Elektrophores. 1776. Vienna.

Schweizer, Karl W., editor. 1988. *Lord Bute: Essays in Re-interpretation.* Leicester.

Scudery, G. 1775. *Das Fernglas der Arzeneywissenschaft.* Münster.

Secord, James A. 1985. Newton in the nursery: Tom Telescope and the philosophy of toys and balls, 1761–1838. *History of Science* 23:127–151.

Seiferheld, G. H. 1791a. *Sammlung electrischer Spielwerke für junge Electriker.* No. 1. Nuremberg.

———. 1791b. *Sammlung electrischer Spielwerke für junge Electriker.* No. 2. Nuremberg.

———. 1791c. *Sammlung electrischer Spielwerke für junge Electriker.* No. 3. Nuremberg.

———. 1795. *Sammlung electrischer Spielwerke für junge Electriker.* No. 6. Nuremberg.

———. 1796. *Sammlung electrischer Spielwerke für junge Electriker.* No. 7. Nuremberg.

———. 1799. *Sammlung electrischer Spielwerke für junge Electriker.* No. 8. Nuremberg.

———. 1802a. *Zauber-Bersuche für Freunde der Electricität.* Nuremberg.

———. 1802b. *Sammlung electrischer Spielwerke für junge Electriker.* No. 4. 2d ed. Nuremberg.

Seligardi, Raffaella. 1999. Luigi Galvani between chemistry and physiology. In Luigi Galvani International Workshop, Proceedings, edited by Marco Bresadola and Guiliano Pancaldi, 83–98. *Bologna Studies in History of Science,* no. 7.

Seymour, Miranda. 2000. *Mary Shelley.* London.

's Gravesande, Willem Jacob. 1731. *Mathematical Elements of Natural Philos-*

ophy, Confirm'd by Experiments; or, An Introduction to Sir Isaac Newton's Philosophy. Translated by J. T. Desaguliers. 2 vols. 4th ed. London.

―――. 1735. *An Explanation of the Newtonian Philosophy in Lectures Read to the Youth of Leyden*. Translated by J. T. Desaguliers. London.

Sguario, Eusebio. 1746. *Dell'elettricismo*. Venice.

Shaffner, Taliaferro P. 1859. *The Telegraphy Manual*. New York.

Shapin, Steven. 1996. *The Scientific Revolution*. Chicago.

Shapin, Steven, and Simon Schaffer. 1985. *Leviathan and the Air-Pump: Hobbes, Boyle, and the Experimental Life*. Princeton, New Jersey.

Shapiro, Arthur K., and Elaine Shapiro. 1997. The placebo: is it much ado about nothing? In *The Placebo Effect: An Interdisciplinary Exploration*, edited by Anne Harrington, 12–36. Cambridge, Massachusetts.

Shelley, Mary. [1831] 1995. *Frankenstein or the Modern Prometheus*. Cologne.

Sidaway, G. H. 1975. Some early experiments in electro-culture. *Journal of Electrostatics* 1 : 389–393.

Sigaud de La Fond, Joseph. 1776. *Traité de l'électricité*. Paris.

―――. 1802. *De l'électricité médicale*. Paris.

Silva, José Alvares da. 1756. *Investigaçaõ das causas proximas do Terremoto, succedido em Lisboa*. Lisbon.

Simili, Raffaella. 1999. Luigi Galvani. In Luigi Galvani International Workshop, Proceedings, edited by Marco Bresadola and Guiliano Pancaldi, 33–63. *Bologna Studies in History of Science*, no. 7.

Simon, John S. 1927. *John Wesley the Master-builder*. London.

Singer, George John. 1814. *Elements of Electricity and Electro-chemistry*. London.

Sketch of a treatise entitled "A dissertation on the uncertainty of the signs of death, and the folly of precipitant burials, and embalments," mentioned by Dr. Mead. 1745. *The Gentleman's Magazine* 15 : 311–313.

Skibo, James M., and Michael B. Schiffer. 2001. Understanding artifact variability and change: a behavioral framework. In *Anthropological Perspectives on Technology*, edited by M. B. Schiffer, 139–162. Albuquerque.

Small, Christopher. 1973. *Mary Shelley's Frankenstein: Tracing the Myth*. Pittsburgh.

Smith, E. 1746. *The Compleat Housewife; or, Accomplish'd Gentlewoman's Companion*. London.

Smyth, Albert H., editor. 1970. *The Writings of Benjamin Franklin*. 10 vols. New York.

Snorrason, E. 1974. *C. G. Kratzenstein and His Studies on Electricity during the Eighteenth Century*. Odense.

Socin, Abel. 1777. *Anfangsgründe der Elektricität*. Hanau.

Sowerby, E. Millicent, compiler. 1952. *Catalogue of the Library of Thomas Jefferson*. Vol. 1. Washington, D.C.

Spengler, Lorenz. 1754. *Briefe welche einige Erfahrungen der electrischen Wirkungen in Krankheiten enthalten*. Copenhagen.

Spiro, Howard. 1997. Clinical reflections on the placebo phenomenon. In *The*

Placebo Effect: An Interdisciplinary Exploration, edited by Anne Harrington, 37–55. Cambridge, Massachusetts.

Stevenson, D. Alan. 1959. *The World's Lighthouses before 1820.* London.

Stopes, Marie Carmichael. 1928. *Contraception (Birth Control): Its Theory, History and Practice; a Manual for the Medical and Legal Professions.* London.

Struik, D. J. 1974. Petrus van Musschenbroek. In *Dictionary of Scientific Biography,* edited by Charles C. Gillispie, 9:594–597. New York.

Suchman, Lucy A. 1987. *Plans and Situated Actions.* Cambridge.

Sutton, Geoffrey. 1981. Electric medicine and mesmerism. *Isis* 72:375–392.

———. 1995. *Science for a Polite Society: Gender, Culture, and the Demonstration of Enlightenment.* Boulder, Colorado.

Swinden, J. H. van. 1784. *Recueil de mémoires sur l'analogie de l'électricité et du magnétisme.* Vol 1. The Hague.

Symes, Richard. 1771. *Fire Analysed.* Bristol.

Takahashi, Yuzo. 1979. Two hundred years of Lichtenberg figures. *Journal of Electrostatics* 6:1–13.

Tegg, William. 1878. *Posts and Telegraphs, Past and Present: with an Account of the Telephone and Phonograph.* London.

Telescope, Tom (pseudonym). 1798. *The Newtonian System of Philosophy; explained by Familiar Objects, in an Entertaining Manner, for the Use of Young Ladies and Gentlemen.* Improved ed. London.

Thompson, C. J. S. 1929. *The Quacks of Old London.* London.

Thomson, Thomas. 1840. *An Outline of the Sciences of Heat and Electricity.* 2d ed. London.

Toaldo, D. Giuseppe. 1779. *Mémoires sur les conducteurs pour préserver les édifices de la foudre.* Strasbourg.

———. 1783. Sur une nouvelle machine électrique. *Nouveaux Mémoires de l'Académie des sciences et belles-lettres* (1781), 30. Berlin.

Toderini, Del P. Giambattista. 1771. *Filosofia Frankliniana, della punte preservatrici dal fulmine, particolarmente applicate alle polviere, alle Navi, e a Santa Barbara in mare.* Modena.

Torlais, J. 1954. *L'abbé Nollet, 1700–1770: un physicien au siècle des lumières.* Paris.

———. 1956. Une grande controverse scientifique au xviiie siècle: l'abbé Nollet et Benjamin Franklin. *Revue de l'histoire des sciences* 9:339–349.

Tressan, le comte de. 1786. *Essai sur la fluide électrique considerée comme agent universel.* Paris. 2 vols.

Troostwyk, A. Paets van, and C. R. T. Krayenhoff. 1788. *De l'application de l'électricité à la physique et à la médecine.* Amsterdam.

Trumpler, Maria. 1999. From tabletops to triangles: increasing abstraction in the depiction of experiments in animal electricity from Galvani to Ritter. In *Luigi Galvani International Workshop, Proceedings,* edited by Marco Bresadola and Guiliano Pancaldi, 115–145. *Bologna Studies in History of Science,* no. 7.

Turner, Anthony. 1987. *Early Scientific Instruments, Europe 1400–1800*. London.

Turner, Gerard L'E. 1973. Van Marum's scientific instruments in Teyler's Museum. In *Martinus van Marum: Life and Work*, edited by E. Lefebvre and J. G. de Bruijn, 4:127–396. Leyden.

———. 1976. Benjamin Wilson. In *Dictionary of Scientific Biography*, edited by Charles C. Gillispie, 14:418–420. New York.

———. 1991. *Scientific Instruments and Experimental Philosophy, 1550–1850*. Aldershot.

Turner, R. 1746. *Electricology; or, A Discourse upon Electricity*. 2d ed. Worcester, Massachusetts.

Turney, Jon. 1998. *Frankenstein's Footsteps: Science, Genetics, and Popular Culture*. New Haven.

Tyerman, L. 1870. *The Life and Times of the Rev. John Wesley, M.A., Founder of the Methodists*. 2 vols. London.

Valli, Eusebius. 1793. *Experiments on Animal Electricity, with Their Application to Physiology, and Some Pathological and Medical Observations*. London.

van der Pas, P. W. 1973. Jan Ingen-Housz. In *Dictionary of Scientific Biography*, edited by Charles C. Gillispie, 7:11–16. New York.

Van Doren, Carl. 1938. *Benjamin Franklin*. New York.

van Marum, Martinus. 1798. *Description de quelques appareils chimiques nouveaux ou perfectionnés de la fondation Teylerienne et des expériences faites avec ces appareils*. Haarlem.

———. 1802. Letter from Dr. Van Marum to Mr. Volta, professor at Pavia, containing experiments on the electric pile, made by him and Professor Pfaff, in the Teylerian Laboratory at Haarlem, in November, 1801. *Journal of Natural Philosophy, Chemistry, and the Arts* 1:173–180.

———. [1785] 1974a. Description of a very large electrical machine installed in Teyler's Museum at Haarlem and of the experiments performed with it. In *Martinus van Marum: Life and Work*, edited by E. Lefebvre and J. G. de Bruijn, 5:1–56. Leyden.

———. [1787] 1974b. First sequel to the experiments performed with Teyler's electrical machine. In *Martinus van Marum: Life and Work*, edited by E. Lefebvre and J. G. de Bruijn, 5:57–142. Leyden.

———. [1791] 1974c. Description of an electrical machine. In *Martinus van Marum: Life and Work*, edited by E. Lefebvre and J. G. de Bruijn, 5:215–237. Leyden.

———. [1798] 1974d. Description of some new or perfected chemical instruments belonging to Teyler's Foundation and of experiments carried out with these instruments. In *Martinus van Marum: Life and Work*, edited by E. Lefebvre and J. G. de Bruijn, 5:239–298. Leyden.

———. [1795] 1974e. Second sequel to the experiments performed with Teyler's electrical machine. In *Martinus van Marum: Life and Work*, edited by E. Lefebvre and J. G. de Bruijn, 5:143–237. Leyden.

Veau Delaunay, Claude. 1809. *Manuel de l'électricité, comprenant les principes élémentaires, l'exposition des systèmes, la description et l'usage des différens appareils électriques, avec l'exposé des méthodes employées dans l'électricité médicale.* Paris.

Veratti, Giuseppe. 1748. *Osservazioni fisico-mediche intorno all'elettricita.* Bologna.

———. 1750. *Observations physico-médicales sur l'électricité.* Geneva.

Volta, Alessandro. 1782. Of the method of rendering very sensible the weakest natural or artificial electricity. *Philosophical Transactions of the Royal Society* 72 (appendix): vii–xxxiii.

———. 1800. On the electricity excited by the mere contact of conducting substances of different kinds. *Philosophical Transactions of the Royal Society* 90:403–431.

———. 1802. Letter of Professor Volta to J. C. Delamethrie, on the galvanic phenomena. *Journal of Natural Philosophy, Chemistry, and the Arts* 1: 135–142.

———. [1775] 1926. Articolo di una lettera del signor Alessandro Volta al signor Dottore Guiseppe Priestley. In *Le opere di Alessandro Volta*, 3:95–108. Milan.

———. [1787] 1929. Descrizione ed uso di un accendilume ad aria inflammabile. In *Le opere di Alessandro Volta*, 7:153–157. Milan.

Vulliamy, C. E. 1931. *John Wesley.* London.

Walker, Adam. 1799. *A System of Familiar Philosophy: in Twelve Lectures.* London.

Walker, W. C. 1936. The detection and estimation of electric charges in the eighteenth century. *Annals of Science* 1:66–100.

Walker, William H. 1995. Ceremonial trash? In *Expanding Archaeology*, edited by J. M. Skibo, W. H. Walker, and A. E. Nielsen, 67–79. Salt Lake City.

———. 2001. Ritual technology in an extranatural world. In *Anthropological Perspectives on Technology*, edited by M. B. Schiffer, 87–106. Albuquerque.

Walker, William H., James M. Skibo, and Axel Nielsen. 1995. Introduction. In *Expanding Archaeology*, edited by J. M. Skibo, W. H. Walker, and A. E. Nielsen, 1–12. Salt Lake City.

Walsh, John. 1773. Of the electric property of the torpedo. *Philosophical Transactions of the Royal Society* 63:461–477.

Warner, Deborah J. 1990. What is a scientific instrument, when did it become one, and why? *British Journal for the History of Science* 23:83–93.

———. 1998. Edward Nairne: scientist and instrument maker. *Rittenhouse* 12:65–93.

Watkins, Francis. 1747. *A Particular Account of the Electrical Experiments Hitherto Made Publick, with Variety of New Ones, and Full Instructions for Performing Them.* London.

Watson, William. 1745. Experiments and observations, tending to illustrate the nature and properties of electricity. *Philosophical Transactions of the Royal Society* 43:481–501.

———. 1746a. *Experiments and Observations tending to Illustrate the Nature and Properties of Electricity.* 3d ed. London.

———. 1746b. *A Sequel to the Experiments and Observations tending to Illustrate the Nature and Properties of Electricity: wherein it is Presumed, by a Series of Experiments, that the Source of the Electrical Power, and Its Manner of Acting Are Demonstrated.* Addressed to the Royal Society. London.

———. 1747. A sequel to the experiments and observations tending to illustrate the nature and properties of electricity. *Philosophical Transactions of the Royal Society* 44:704–749.

———. 1748. A collection of electrical experiments communicated to the Royal Society, read at several meetings between October 29, 1747 and Jan. 21, following. *Philosophical Transactions of the Royal Society* 45: 49–120.

———. 1749. A letter from Mr. William Watson, declaring that he as well as many others have not been able to make odours pass thro' glass by means of electricity; and giving a particular account of Professor Bose at Wittenberg, his experiment of beatification, or causing a glory to appear round a man's head by electricity. *Philosophical Transactions of the Royal Society* 46:348–356.

———. 1751–1752a. An account of Mr. Benjamin Franklin's treatise. *Philosophical Transactions of the Royal Society* 47:202–211.

———. 1751–1752b. An account of Professor Winkler's experiments relating to odours passing through electrified globes and tubes. *Philosophical Transactions of the Royal Society* 47:231–241.

———. 1753. An account of a treatise, presented to the Royal Society, intituled [sic] "Letters concerning electricity; in which the latest discoveries upon this subject, and the consequences which may be deduced from them, are examined; by the Abbé Nollet." *Philosophical Transactions of the Royal Society* 48:201–216.

———. 1754. An answer to Dr. Lining's query relating to the death of Professor Richman [sic]. *Philosophical Transactions of the Royal Society* 48: 765–773.

Watson, W., B. Franklin, B. Wilson, John Canton, and Edward Delaval. 1769. Report from the committee appointed to consider of the properest means to secure the Cathedral of St. Paul's from the effects of lightning. *Philosophical Transactions of the Royal Society* 59:160–169.

Watt, W. 1751. Cure by electricity. *The Gentleman's Magazine* 21:152.

Webers, Joseph. 1781. *Neue Erfahrungen idiolektrische Körper: ohne einiges Reiben zu elektrisiren.* Augsburg.

Webster, Roderick S. 1974. Edward Nairne. In *Dictionary of Scientific Biography,* edited by Charles C. Gillispie, 9:607–608. New York.

Weld, Charles R. 1848. *History of the Royal Society, with Memoirs of the Presidents.* 2 vols. London.

Wells, Walter A. 1927. Last illness and death of George Washington. *Virginia Medical Monthly* 53(10):629–642.

Wesley, John. 1771. *The desideratum; or, Electricity Made Plain and Useful.* London.

———. 1792. *Primitive Physic; or an Easy and Natural Method of Curing Most Diseases.* 24th ed. London.

Wheler, Granville. 1739. A letter, containing some remarks on the late Stephen Gray, his electrical circular experiment. *Philosophical Transactions of the Royal Society* 41:118–125.

Wheler, Granville, and C. Mortimer. 1739. An account of some of the electrical experiments, made at the Royal Society's house, on May 11, 1737. *Philosophical Transactions of the Royal Society* 41:112–117.

Whittaker, Edmund T. [1910] 1951. *A History of the Theories of Aether and Electricity.* Vol. 1. New York.

Whytt, Robert. 1768. *The Works of Robert Whytt, M.D.* Edinburgh.

Wiesner, Julius. 1905. *Jan Ingen-Housz: sein Leben un sein Wirken als Naturforscher und Arzt.* Vienna.

Wilkinson, C. H. 1799. *An Analysis of a Course of Lectures on the Principles of Natural Philosophy.* London.

Willcox, William B. 1966. *The Age of Aristocracy, 1688–1830.* Edited by Lacey Baldwin Smith. Vol. 3 of *A History of England.* Boston.

Wilson, Benjamin. 1752. *A Treatise on Electricity.* 2d ed. London.

———. 1773. Observations upon lightning, and the method of securing buildings from its effects. *Philosophical Transactions of the Royal Society* 63:49–65.

———. 1778a. Mr. Wilson's dissent from the above report [i.e., from Henly et al. 1778]. *Philosophical Transactions of the Royal Society* 68:239–242.

———. 1778b. New experiments and observations on the nature and use of conductors. *Philosophical Transactions of the Royal Society* 68:245–313.

———. 1778c. *An Account of Experiments Made at the Pantheon, on the Nature and Use of Conductors; to which are Added, some New Experiments with the Leyden Phial.* London.

Wilson, Geoffrey. 1976. *The Old Telegraphs.* London.

Winchester, C. T. 1906. *The Life of John Wesley.* New York.

Winkler, Johann. H. 1745. Regiae Societati Anglicanae Scientarium guaedam electricitatis recens observata exhibet. *Philosophical Transactions of the Royal Society* 43:307–314.

———. 1746. Extract of a letter, concerning the effects of electricity upon himself and his wife. *Philosophical Transactions of the Royal Society* 44:211–212.

———. 1748. *Essai sur la nature, les effets et les causes de l'électricité, avec une description de deux nouvelles machines à électricité.* Paris.

Winn, J. L. 1770. A letter giving an account of the appearance of lightning on a conductor fixed from the summit of the mainmast of a ship, down to the water. *Philosophical Transactions of the Royal Society* 60:188–191.

Wollaston, William H. 1801. Experiments on the chemical production and agency of electricity. *Philosophical Transactions of the Royal Society* 91: 427–434.

Wood, Clive, and Beryl Suitters. 1970. *The Fight for Acceptance: A History of Contraception.* Aylesbury.

Zilsel, Edgar. 2000. The origins of William Gilbert's scientific method. In *The Social Origins of Modern Science,* edited by D. Raven, W. Krohn, and R. S. Cohen, 71–95. Dordrecht.

Index

Page numbers in *italic* refer to figures.

Compositor:	G & S Typesetters, Inc.
Indexer:	Andrew Joron
Text:	10/13 Aldus
Display:	Aldus
Printer and binder:	Edwards Brothers, Inc.